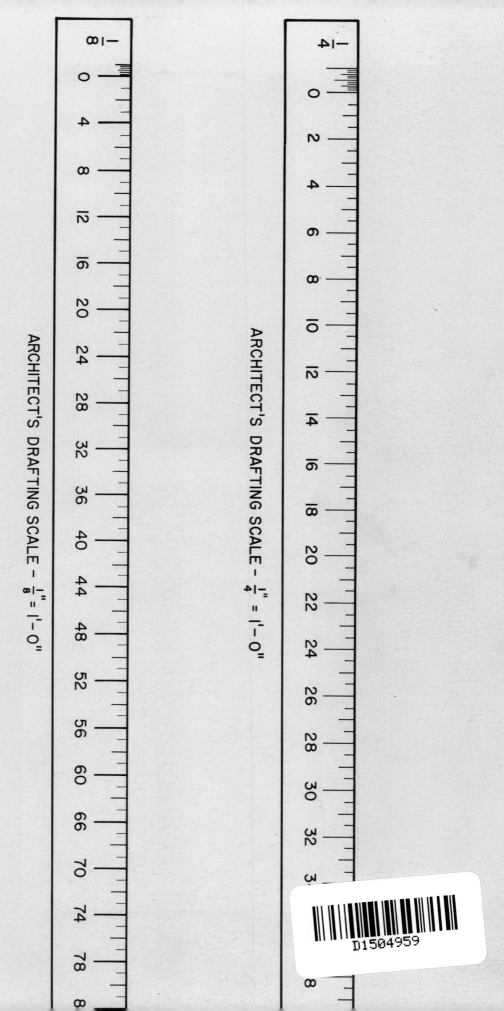

ARCHITECT'S DRAFTING SCALE – $\frac{1}{4}"$ = $1'$–$0"$

ARCHITECT'S DRAFTING SCALE – $\frac{1}{8}"$ = $1'$–$0"$

Reading Construction Drawings

Paul I. Wallach
Instructor
San Carlos High School
San Carlos, California

Donald E. Hepler
Executive Editor
Trade and Technical Occupations
Gregg Division
McGraw-Hill Book Company

McGraw-Hill Book Company

New York   Mexico
St. Louis   Montreal
San Francisco   New Delhi
Auckland   Panama
Bogotá   Paris
Hamburg   São Paulo
Johannesburg   Singapore
London   Sydney
Madrid   Tokyo
  Toronto

About The Authors

Paul I. Wallach received his undergraduate education at the University of California at Santa Barbara and did his graduate work at Los Angeles State College. He has acquired extensive experience in the drafting, designing, and construction phases of architecture. He has studied and taught for several years in Europe. He taught architecture and drafting for 27 years in California at both the secondary school and community college levels.

Donald E. Hepler completed his undergraduate work at California State College, California, Pennsylvania, and his graduate work at the University of Pittsburgh. He has worked on the architectural staffs of Admiral Homes, Inc.; Rust Engineering; Patterson, Emerson, and Comstock, Engineers; and Union Switch and Signal Company. He also served as an officer with the United States Army Corps of Engineers; headed the Industrial Arts Department, Avonworth High School, Pittsburgh, Pennsylvania; and spent four years on the faculty of California State College, California, Pennsylvania. He is presently Executive Editor for Trade and Technical Occupations publications in the Gregg Division, McGraw-Hill Book Company, New York.

Library of Congress Cataloging in Publication Data

Wallach, Paul I
    Reading construction drawings.

    Includes index.
    1. Structural drawing. 2. Blue-prints.
I. Hepler, Donald E., joint author. II. Title.
TH431.W34 1981 692'.1 80-19633
ISBN 0-07-067940-1

1 2 3 4 5 6 7 8 9 0 HDHD 8 9 8 7 6 5 4 3 2 1

CONTENTS

PREFACE

Construction workers and technicians need to know how to interpret architectural drawings, but occupants of commercial structures, factories, homes, and apartments also need to understand the basics of interpreting architectural drawings. Thus, the need for reading architectural plans permeates every aspect of private and public life. The person who cannot read and interpret the graphic language today is as illiterate as the one who could not read the written language a few decades ago.

Today the construction of a relatively simple residence requires a minimum of six different drawings, while the completion of a large building complex may require hundreds of thousands of drawings to ensure that the structure will be built as planned. Construction workers of all types, carpenters, masons, plasterers, electricians, plumbers, sheet-metal workers, and everyone involved in the building process all must interpret drawings consistently to ensure that buildings are constructed precisely as designed, and with the materials specified.

Reading Construction Drawings provides an introduction to reading and interpreting construction (architectural) plans. Since the ability to read drawings is vital to all phases of the construction industry, this book covers both basic principles and specific practices related to all areas of construction.

Reading Construction Drawings is divided into 16 chapters.

Chapters 1 through 4 cover the functions of both the graphic language and the written language in understanding construction plans. Architectural symbols, the shorthand of building and drafting, are presented as they appear on all types of drawings. Construction terms are clearly defined, and symbols are shown pictorially and are related to each architectural term, synonym, and abbreviation.

Chapters 5 and 6 include the basic principles and practices necessary for reading floor plans and elevation drawings. Methods and procedures for understanding and interpreting the use of materials and for the determination of sizes and dimensions on both general plans and working drawings are covered here.

Chapters 7 through 14 show how sectional drawings, foundation plans, framing drawings, location plans, electrical plans, plumbing diagrams, and climate control plans are used to communicate the designer's ideas specifically to the builder. This part also covers procedures for reading sets of plans with emphasis on understanding the relationship of plans and indexing systems used to facilitate the location of specialized drawings.

Chapters 15 and 16 illustrate how and why documents such as schedules, specifications, building codes, contracts, and bids are related to construction drawings and the construction process.

The illustrations are an integral part of the presentation. Where appropriate, a pictorial drawing accompanies each plan, elevation, and sectional view. This approach helps the reader better visualize and understand the meaning and relationship of views. It also helps maximize the retention of basic graphic concepts. All illustrations have been prepared to reinforce and amplify the principles described. Wherever possible, each principle has been reduced to its most elementary form for easy comprehension. Progression within each unit is from the simple to the complex and from the familiar to the unfamiliar.

Since communication in all areas of building depends largely on the understanding of the vocabulary of architecture and construction, new terms, abbreviations, and symbols are defined when they first appear, and their definitions are reinforced throughout the remainder of the book.

The metric system of measurement is used selectively in this text even though most architectural work is not currently designed in metric units. The introduction of the metric system as a secondary system of measure is offered so that the reader may begin to understand the system, which will be used extensively in the next decade.

Because metric standards for building materials and modular units have not yet been officially established, the metric dimensions used are not necessarily those which will become standard. Usually the dimensions are conversions of present standards. The metric equivalents are approximate, not exact. For example, in a plan which includes a reference to 100 square feet (sq ft), the exact metric equivalent is 9.290 square meters (m²), but, as a practical approach, 10 m² is used. Similarly, the equivalent for 48 inches (in.) is given as 1220 millimeters (mm) rather than 1219 mm.

As a guide to accuracy, meters are carried to one decimal place for approximate size and to three decimal places for more accurate dimensions.

The metric practice of using a space rather than a comma to separate numbers of five digits or more at three-digit intervals is followed throughout.

The authors wish to acknowledge the contribu-tions and the drawings of Home Planners, Inc.; Karren and Seals, Architects, Inc.; and landscape architect Dana Hepler. Special thanks to Howard Hull for his review of the technical accuracy of the material.

Paul Wallach
Don Hepler

CHAPTER 1

Understanding the Graphic Language

An old Chinese proverb says, "One picture is worth ten thousand words." This point is well illustrated in Fig. 1-1. Could this structure be adequately described in ten thousand words? Certainly not. And it absolutely could not be built precisely the way it was designed using any number of words alone to describe its structure and design. Although words are used extensively to communicate design ideas, drawings are the main vehicles used in construction communication. There are two types of drawings used to describe the construction of architectural structures: pictorial drawings and multiview drawings. Pictorial drawings are picturelike drawings. They show several sides of an object in one drawing. Multiview drawings show different sides, or views, of an object in separate drawings. For this reason multiview drawings are used almost exclusively as working drawings. A working drawing is any drawing used as a basis for construction or manufacturing and includes all necessary information concerning the size, shape, and materials used in a structure.

Unit 1
Pictorial Drawings
Two types of pictorial drawings are used extensively in architectural presentations: the perspective drawing and the isometric drawing (Fig. 1-2). The perspective drawing, the more popular of the two, has lines that recede to vanishing points, thus giving the drawing a more realistic, though a technically inaccurate, appearance. Isometric drawings show true dimensions and create an optical illusion of distortion, since the human eye is accustomed to seeing long objects recede. Isometric drawings are used primarily for small construction details.

Exterior Pictorial Drawings
As you look down a railroad track, the track appears to come together and vanish at a point on the distant horizon. Similarly, the horizontal lines in a perspective drawing, such as those of the building shown in Fig. 1-2, appear to come together. More than any other kind of drawing, a perspective drawing of a structure resembles a photograph of the exterior of the structure. The receding lines of the building are purposely drawn closer together on one or several sides of the building to create the illusion of depth. The point at which these receding lines intersect on a perspective drawing is known as the vanishing point. Just as railroad tracks appear to come together at the horizon, the vanishing points on a perspective drawing are always found on a horizon line. In reading perspective drawings, always

Fig. 1-1. "One picture is worth ten thousand words."
(Home Planners, Inc.)

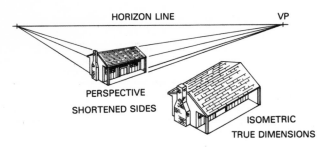

Fig. 1-2. In perspective drawings, true distances are shortened to make the drawing look more realistic.

locate the position of the horizon line. If the horizon line is placed through the building, the walls of the building will appear at eye level. If the horizon line is placed below the building, the structure will appear to be above eye level. If the horizon line is placed above the building, the structure will appear to be below the line of sight. Figure 1-3 shows how illustrators position a structure in reference to the horizon to show and emphasize either the top portion, the lower portion, the left side, or the right side of the building. Notice that the horizon line and accompanying vanishing points are placed above the structure in Fig. 1-2; thus the building appears below the line of sight. In Fig. 1-4, however, the illustrator wanted to show the appearance of the building from a lower angle; therefore, the horizon is

Fig. 1-4. Use of a pictorial drawing to show terrain and its relationship with other buildings. (Home Planners, Inc.)

placed below the structure, and the viewer sees the building from that angle. The people in the foreground in Fig. 1-4 are placed on the horizon line.

Because perspective drawings show several angles of a building and can be distorted to emphasize or deemphasize selected features of the architectural design, they are used extensively

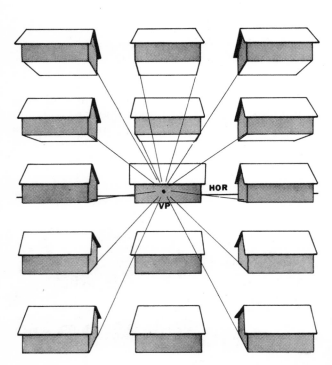

Fig. 1-3. One-point perspective drawings are prepared from many vantage points.

Fig. 1-5. Adding interesting surroundings helps adjust a plan for a specific purpose. (Home Planners, Inc.)

for selling purposes. For this same reason perspective drawings are also used to communicate basic design concepts to those who cannot read and interpret the working drawings of the structure. As shown in Fig. 1-4, surrounding terrain and its relationship to buildings can be effectively shown by perspective drawings. Also, adding unique or interesting surroundings to the drawing often helps depict how the plan can be used or adjusted for specific purposes or locations (see Fig. 1-5).

One of the most popular uses of pictorial drawings is the illustration of design alternatives (Fig. 1-6). The three perspective exterior drawings in this plan utilize the same floor plan. Only the exterior design is changed to make the structure appear different. This is a popular technique utilized in selling tract houses, allowing the homeowner to select alternative exterior designs to eliminate the appearance of duplication.

Pictorial drawings are also used to show design variations and level differences between different

Fig. 1-6. Pictorial drawings are often used to show design alternatives. (Home Planners, Inc.)

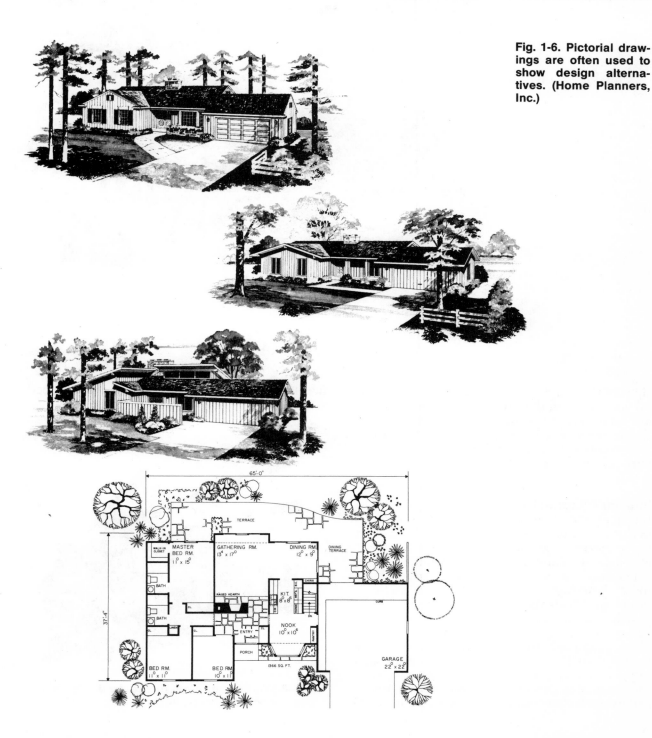

Fig. 1-7. Level differences are often shown with pictorial drawings. (Home Planners, Inc.)

sides of the same structure. For example, in Fig. 1-7 two pictorial drawings of the same structure are used. One shows the front and right sides of the design at one level, and the other shows the rear and left sides, revealing the level difference between the front and rear of the structure.

Interior Pictorial Drawings

Interior pictorial drawings such as the one shown in Fig. 1-8 are used almost exclusively for marketing purposes, since many prospective occupants of a structure cannot effectively visualize the finished appearance of its interior from architectural working drawings. Interior pictorial drawings can also be used to show the design and positioning of furniture and equipment, which are not normally included as part of the basic plan for the building. Here the illustrator also uses the position of the horizon line and vanishing points to emphasize or deemphasize parts of the interior design. For example, in Fig. 1-8 the extension of the beams and the position of the

fireplace on the horizon emphasize those aspects of that design.

Fig. 1-8. Furnished interiors are best shown pictorially. (Home Planners, Inc.)

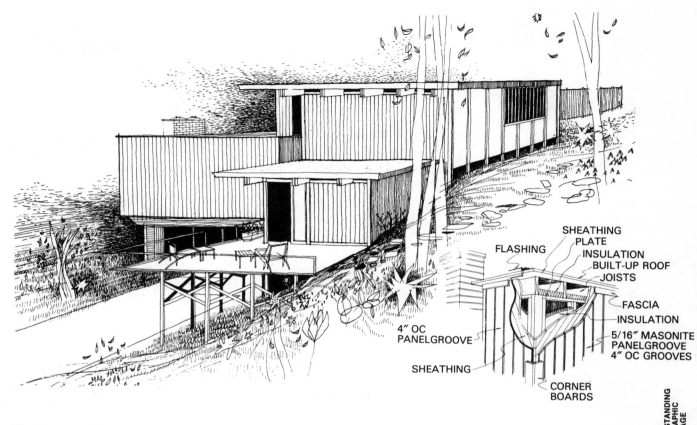

FLASHING
SHEATHING
PLATE
INSULATION
BUILT-UP ROOF
JOISTS
FASCIA
INSULATION
5/16" MASONITE
PANELGROOVE
4" OC GROOVES
4" OC
PANELGROOVE
SHEATHING
CORNER
BOARDS

Fig. 1-9. Construction details are sometimes shown in pictorial form. (Masonite Corporation.)

Construction Drawings

Because pictorial drawings are made with angular lines of imprecise length, they are rarely used for construction purposes. However, construction details are occasionally shown in pictorial form to help builders visualize construction techniques (Fig. 1-9). When this practice is used, dimensions are usually omitted. Occasionally, framing methods may be shown to a layperson with a pictorial drawing (Fig. 1-10). When new methods of construction are developed, often exploded pictorial views (Fig. 1-11) are used to show the relationship of components to each other.

Since they are time-consuming to produce, pictorial drawings are used for construction purposes only when the designer thinks multiview drawings may not be interpreted correctly.

Space Relationships

Pictorial drawings are sometimes used to establish space relationships between buildings, as in the arrangement of buildings in Fig. 1-12. The size relationships among buildings, as shown in Fig. 1-13, is also effectively shown to the layperson through the use of pictorial drawings.

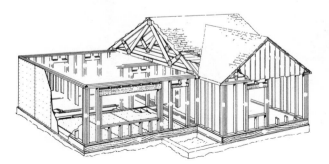

Fig. 1-10. Framing methods may be shown with pictorial drawings.

Unit 2
Multiview Interpretation
Although pictorial drawings are effective in showing at a glance the general design of a structure, they are not sufficiently precise to accurately describe all elements of the design. For this reason, multiview (several view) drawings are provided for use in the actual construction of buildings. Multiview drawings are sometimes called orthographic drawings. Orthographic drawings represent the exact form and size of each side of an object in two or more views (or planes) usually at right angles

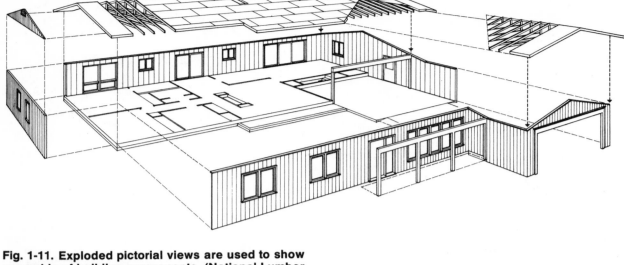

Fig. 1-11. Exploded pictorial views are used to show assembly of building components. (National Lumber Manufacturers Assoc.)

to each other. Figure 2-1 shows the comparison of a three-view orthographic drawing of a building with several pictorial drawings of the same building. To visualize and understand multiview (orthographic) projection, imagine a structure surrounded by imaginary transparent planes, as shown in Fig. 2-2. If you draw the outline of the structure on the imaginary transparent planes,

Fig. 1-12. Pictorial drawings are used to show space relationships between buildings. (Karren and Seals Architects, Inc.)

Fig. 1-13. Size comparisons shown with pictorial drawings. (Port Authority of New York.)

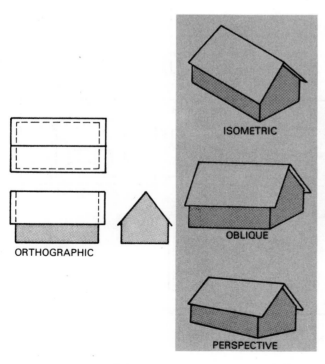

ISOMETRIC

OBLIQUE

PERSPECTIVE

ORTHOGRAPHIC

Fig. 2-1. Orthographic drawings are the basis of architectural working drawings.

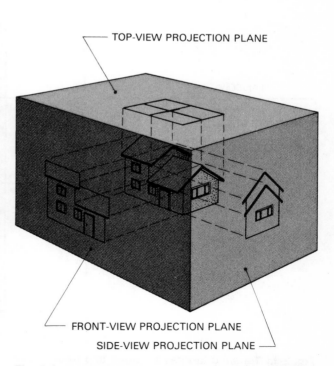

TOP-VIEW PROJECTION PLANE

FRONT-VIEW PROJECTION PLANE

SIDE-VIEW PROJECTION PLANE

Fig. 2-2. A projection box shows three planes (surfaces) of a building.

you create the various orthographic views: the front view on the front plane, the side view on the side plane, and the top view on the top plane. When the planes of the top, bottom, and sides are hinged (swung) out from the front plane, as shown in Fig. 2-3, the six views of an object are shown exactly as they are positioned on an orthographic drawing. Study the position of each view as it relates to the front view: the right side is to the right of the front view, the left side is to the left, the top (roof) view is on the top, the bottom view is on the bottom, and the rear view is to the left of the left-side view, since, when this view hinges around to the back, it would fall into this position.

Notice that the length of the front view, top (roof) view, and bottom view are exactly the same as the length of the rear view. Notice also that the heights and alignments of the front view, right side, left side, and rear view are the same. Memorize the position of these views and remember that the lengths of the front, bottom, and top views are *always* the same. Similarly, the heights of the rear, left, front, and right side are *always* the same.

All six views are rarely used to depict architectural structures. Instead, only four elevations (sides) are usually shown, and the top view is usually replaced with a section through the structure called a floor plan. The roof plan is also developed from the top view. The bottom view is never developed in a construction drawing.

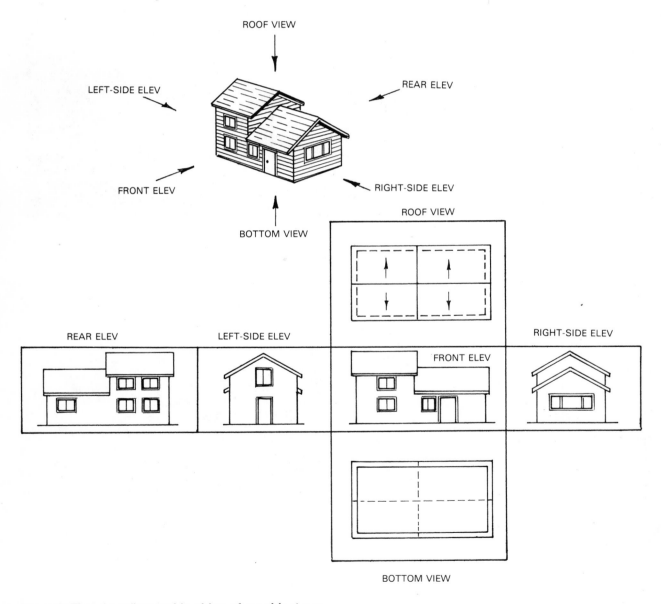

Fig. 2-3. The six orthographic sides of an object are shown when the projection box is opened and laid flat.

CHAPTER 2

Interpreting Architectural Symbols

Architectural symbols are the alphabet and shorthand of the building industry. A thorough understanding of each symbol and the material, component, or feature it represents is absolutely vital to personnel involved in all phases and types of construction. It is not necessary to develop total recall for each symbol, but it is important to recognize what material or component each symbol represents when reading construction drawings. The symbols in this chapter represent materials or features which appear on drawings viewed from the top (plan view) and as they appear from the side (elevation view). Plan views are often sectional drawings. Therefore, the plan symbol represents materials in their sectioned (cut through) form.

Symbols are designed to approximate the appearance of the material, fixture, or component. Some symbols, especially of elevation materials, represent very accurately the material they depict. Others, usually because of size and space

restrictions, are assigned a design symbol which must be memorized if drawings are to be meaningful and consistently interpreted by each reader.

This chapter is divided into specialized units, each containing symbols which represent similar components found on construction drawings: doors and windows, appliances and fixtures, sanitation facilities, plumbing, electrical, and climate control components, building materials, topographic features, and reference callouts. The practical application of key symbols is presented in following units to further reinforce their use on various construction drawings.

Since construction drawings include written terms and abbreviations, each symbol is shown in both plan and elevation form and is included with the term, abbreviation, or synonym, as well as a pictorial drawing of what the term and symbol represents.

Selected architectural materials symbols.

Unit 3
Door and Window Symbols

NAME	ABR	SYMBOL	ELEVATION	PICTORIAL
INTERIOR HINGED DOOR HOLLOW CORE	DR		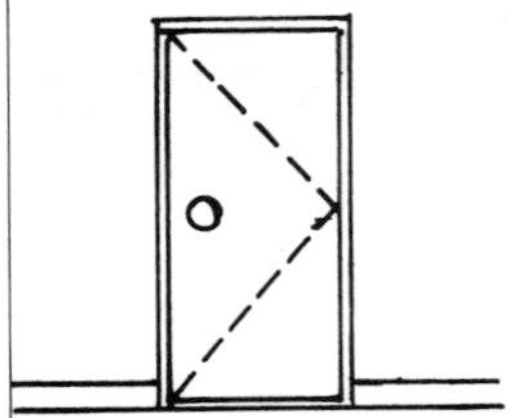	
EXTERIOR HINGED DOOR SOLID CORE	DR	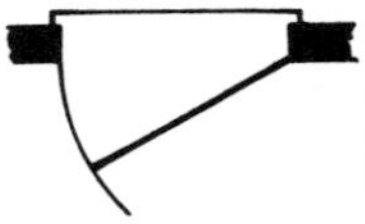	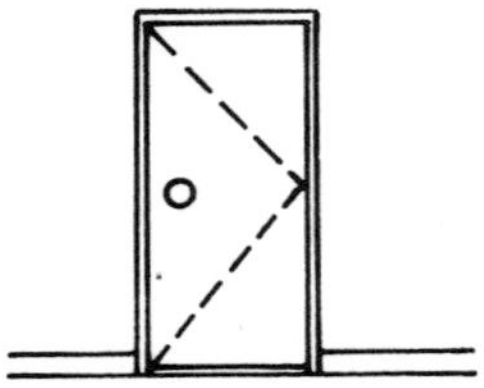	
DOUBLE ACTION DOOR	DBL AC DR	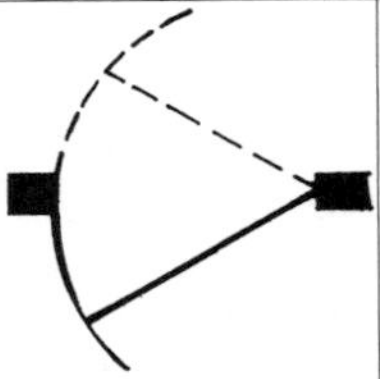	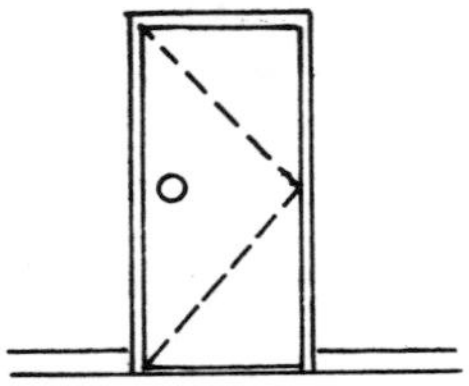	
BYPASSING SLIDING DOOR	BP SLDG DR		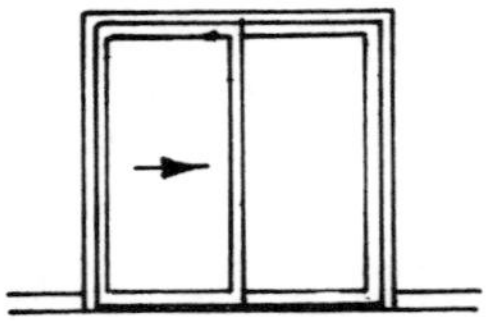	
DOUBLE FRENCH DOORS	DBL FR DR		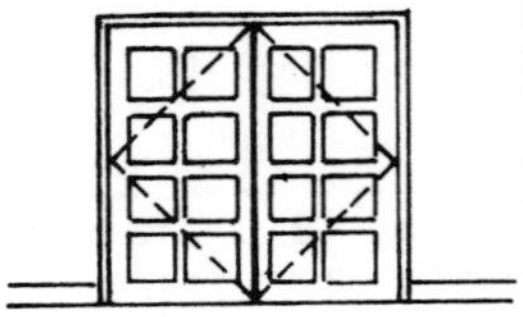	
SLIDING POCKET DOOR	SLDG PK DR		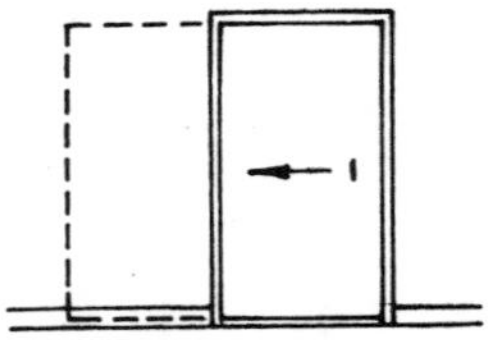	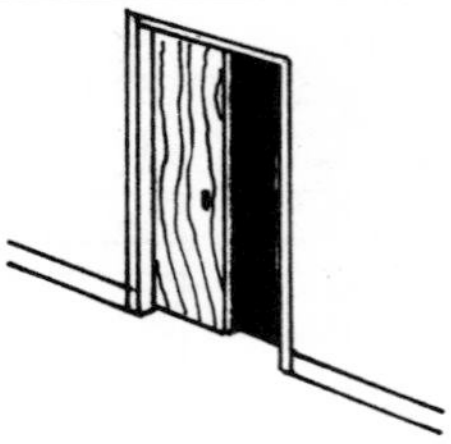
BIFOLDING DOORS	BI-FLD DR		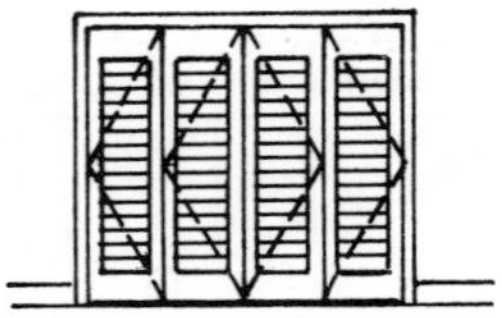	

Fig. 3-1. Door symbols.

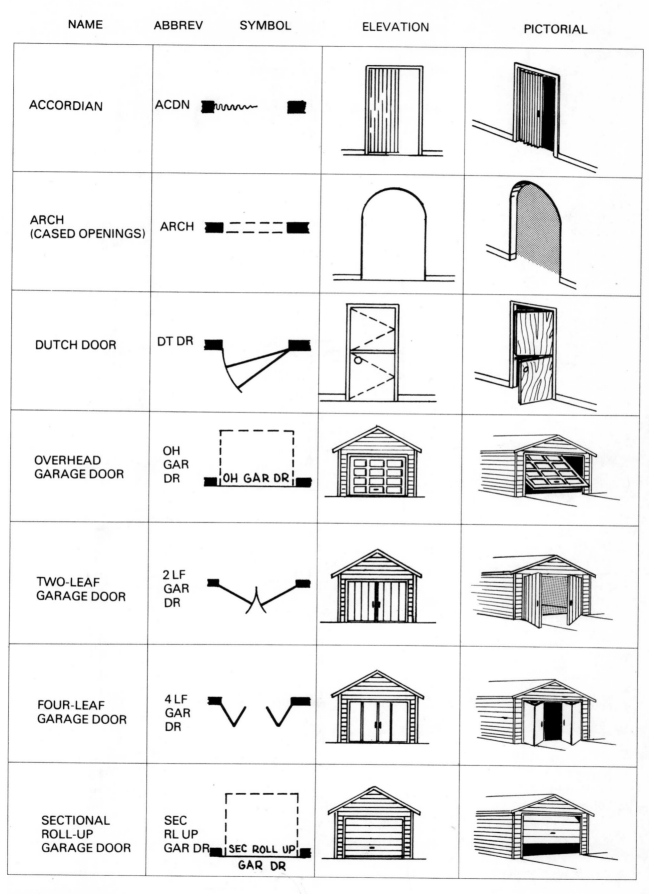

NAME	ABBREV	SYMBOL	ELEVATION	PICTORIAL
ACCORDIAN	ACDN			
ARCH (CASED OPENINGS)	ARCH			
DUTCH DOOR	DT DR			
OVERHEAD GARAGE DOOR	OH GAR DR	OH GAR DR		
TWO-LEAF GARAGE DOOR	2 LF GAR DR			
FOUR-LEAF GARAGE DOOR	4 LF GAR DR			
SECTIONAL ROLL-UP GARAGE DOOR	SEC RL UP GAR DR	SEC ROLL UP GAR DR		

Fig. 3-2. Door symbols.

NAME	ABR	SYMBOL	ELEVATION	PICTORIAL
DOUBLE-HUNG WINDOW	DHW			
HORIZONTAL SLIDING WINDOW	SLD WDW			
AWNING WINDOW	AWN WDW			
SWINGING CASEMENT WINDOW	CSMT WDW			
HOPPER WINDOW	HOP WDW			
JALOUSIE WINDOW	JAL WDW			
DOUBLE DOUBLE-HUNG WINDOW	DDHW			

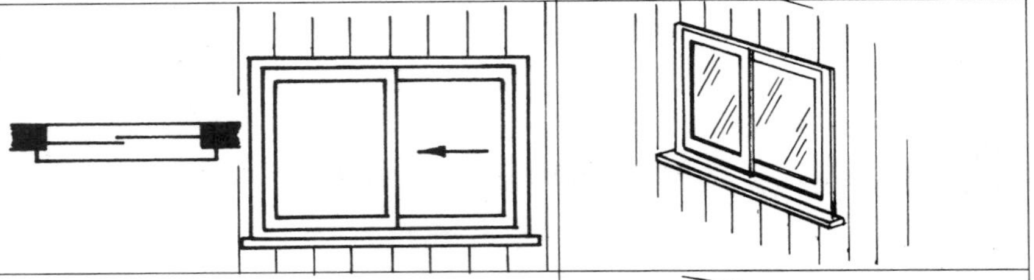

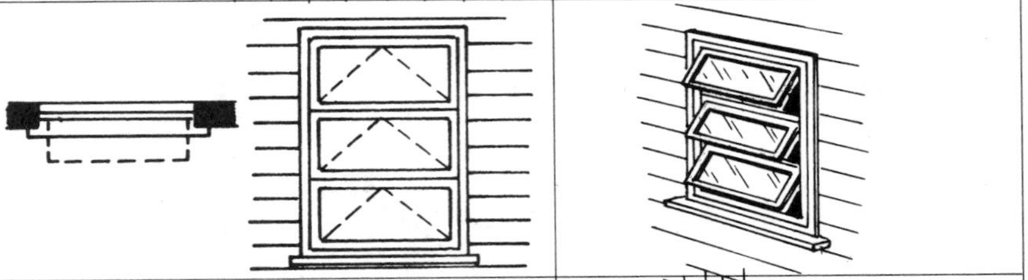

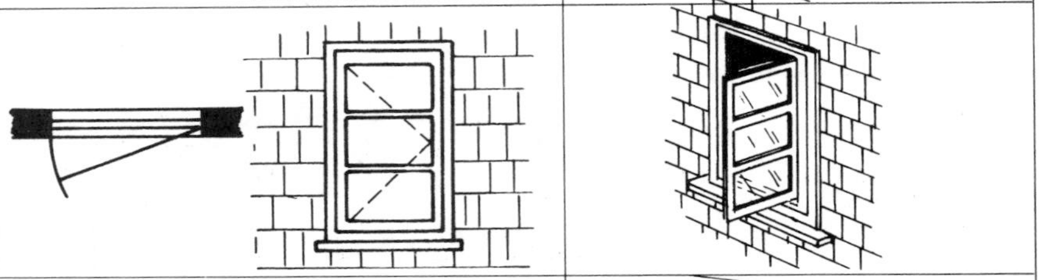

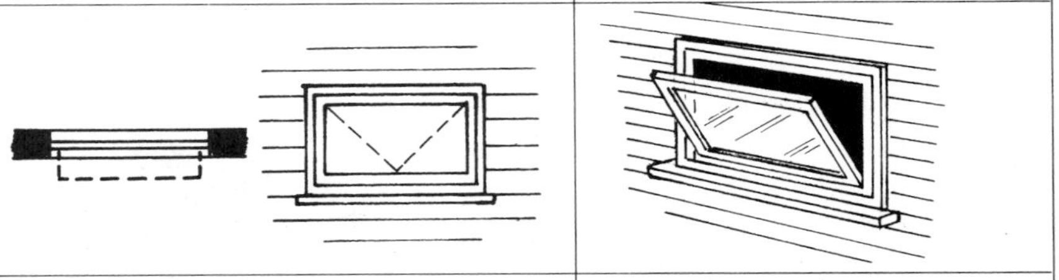

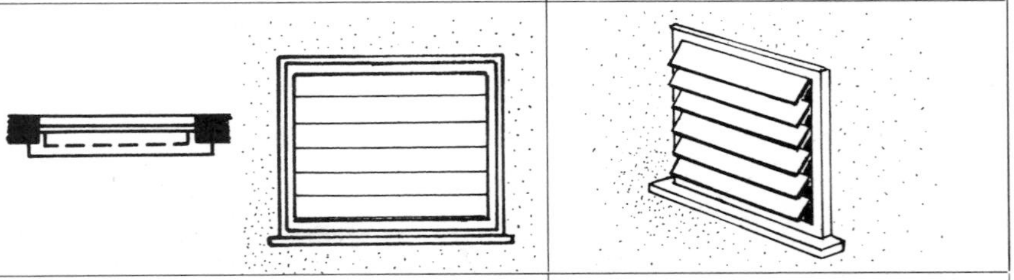

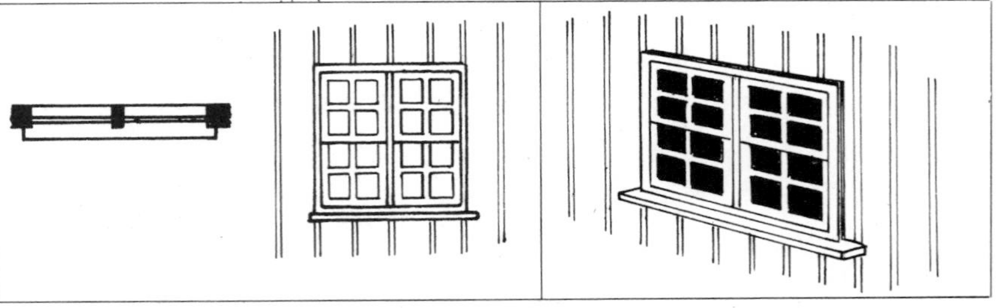

Fig. 3-3. Window symbols.

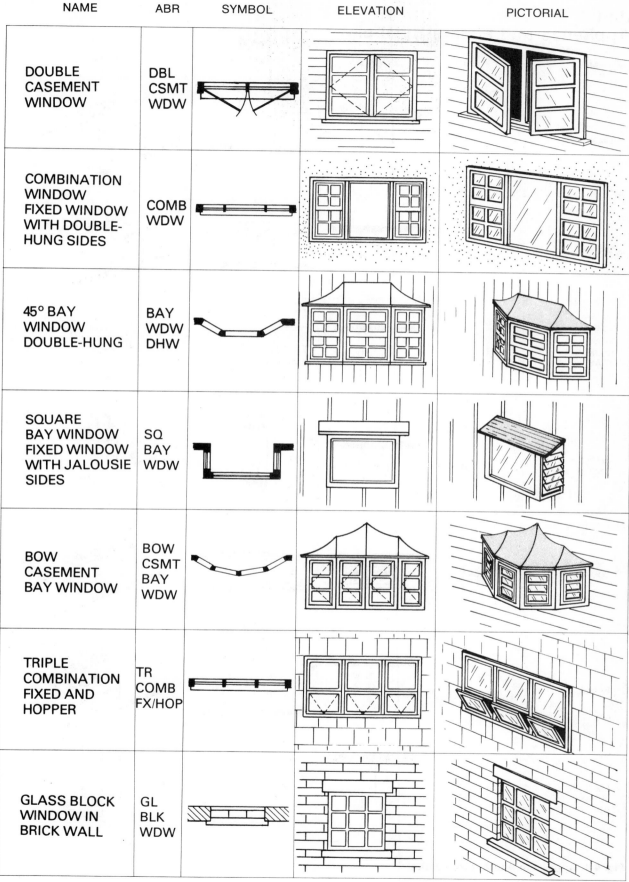

NAME	ABR	SYMBOL	ELEVATION	PICTORIAL
DOUBLE CASEMENT WINDOW	DBL CSMT WDW			
COMBINATION WINDOW FIXED WINDOW WITH DOUBLE-HUNG SIDES	COMB WDW			
45° BAY WINDOW DOUBLE-HUNG	BAY WDW DHW			
SQUARE BAY WINDOW FIXED WINDOW WITH JALOUSIE SIDES	SQ BAY WDW			
BOW CASEMENT BAY WINDOW	BOW CSMT BAY WDW			
TRIPLE COMBINATION FIXED AND HOPPER	TR COMB FX/HOP			
GLASS BLOCK WINDOW IN BRICK WALL	GL BLK WDW			

Fig. 3-4. Window symbols.

Unit 4
Appliance and Fixture Symbols

NAME	ABBREV	SYMBOL	ELEVATION	PICTORIAL
WASHER	W			
DRYER	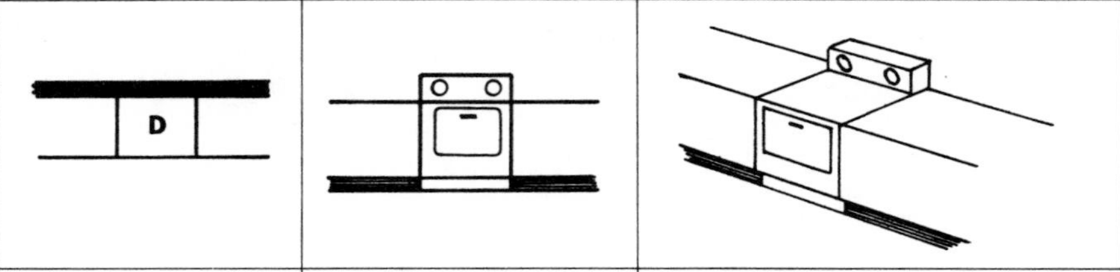 D			
LAUNDRY TRAY	LT			
WATER HEATER	WH			
COOK TOP	CK TP			
RANGE	R			
FOLD-UP IRONING BOARD	I BRD			

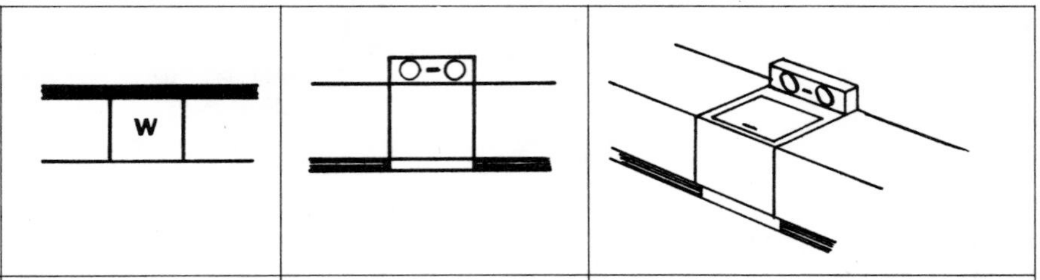

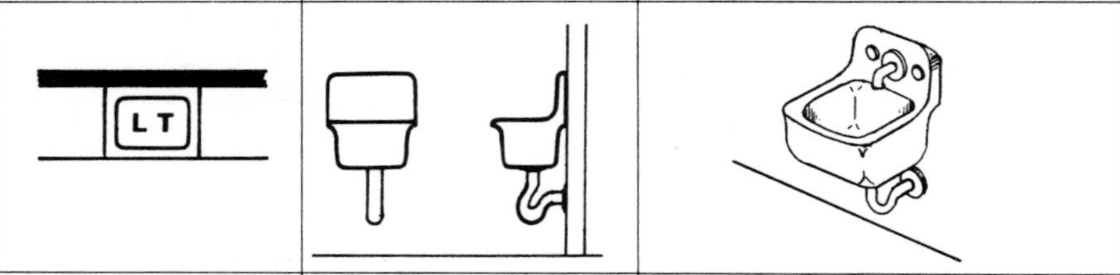

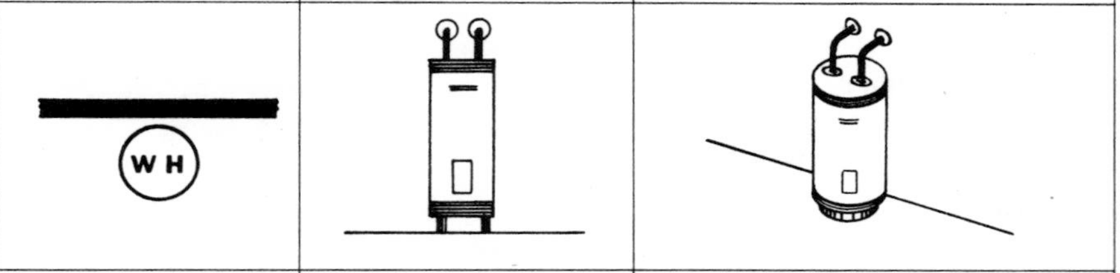

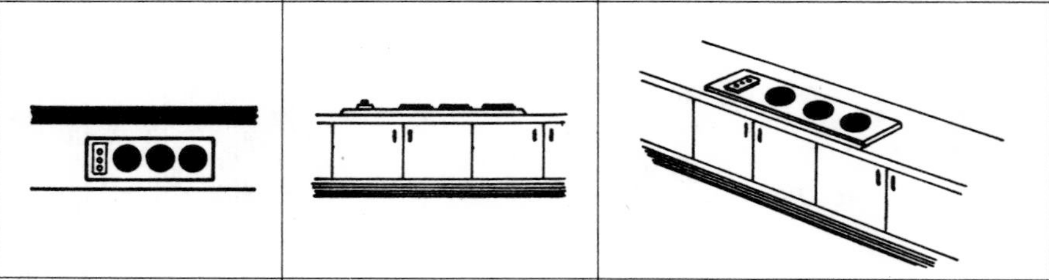

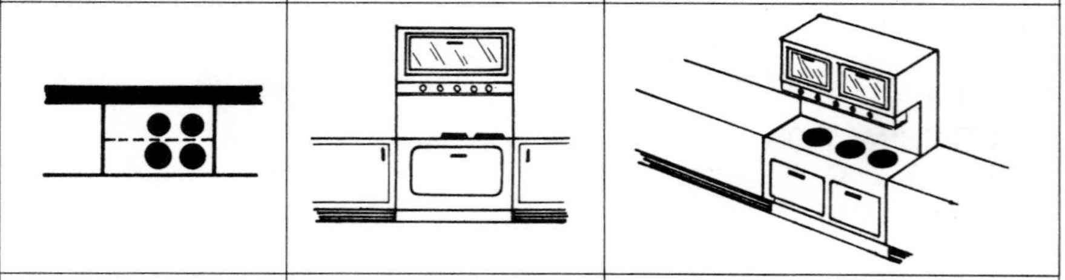

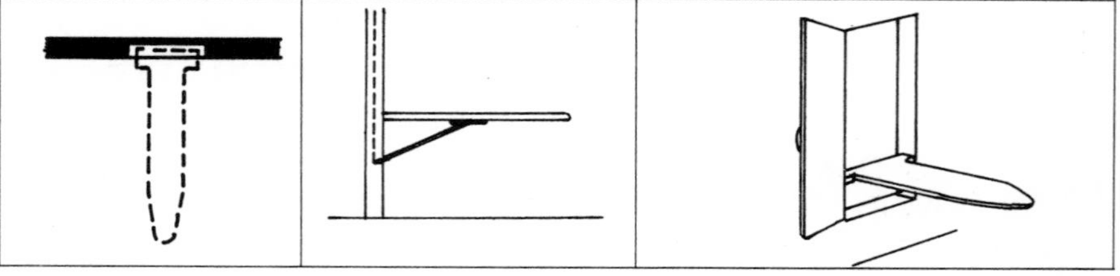

Fig. 4-1. Appliance and fixture symbols.

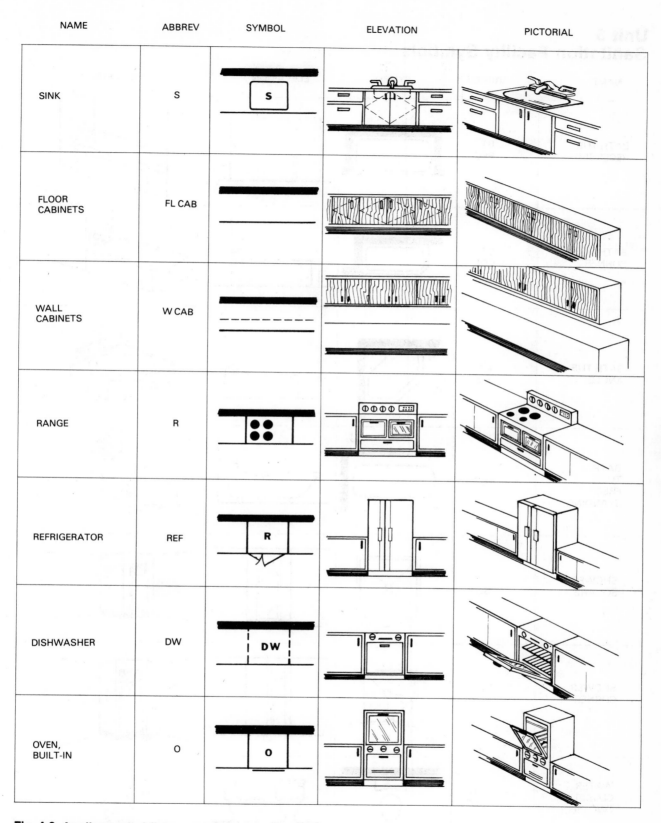

NAME	ABBREV	SYMBOL	ELEVATION	PICTORIAL
SINK	S			
FLOOR CABINETS	FL CAB			
WALL CABINETS	W CAB			
RANGE	R			
REFRIGERATOR	REF			
DISHWASHER	DW			
OVEN, BUILT-IN	O			

Fig. 4-2. Appliance and fixture symbols.

Unit 5
Sanitation Facility Symbols

NAME	ABBREV	SYMBOL	ELEVATION	PICTORIAL
BATH TUB RECESSED	BT REC		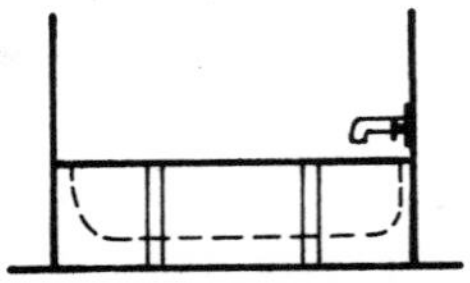	
BATH TUB CORNER	BT COR		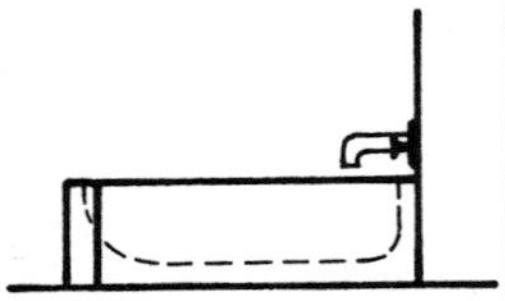	
BATH TUB ANGLE	BT ANG	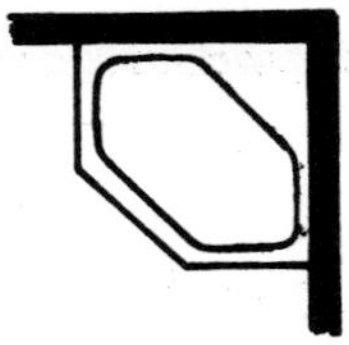	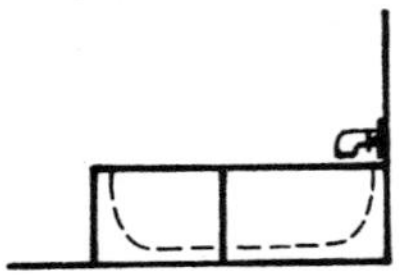	
BATH TUB FREE-STANDING	FST	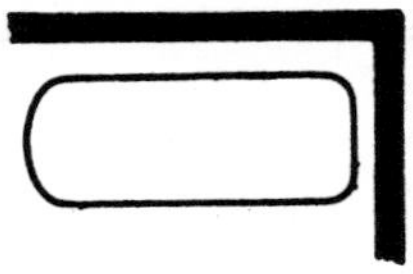	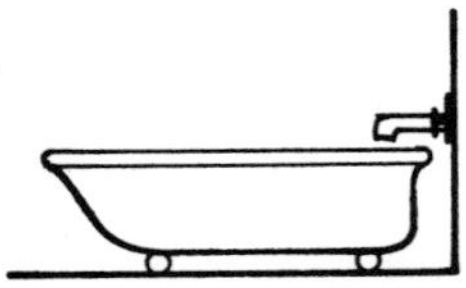	
SHOWER SQUARE	SH SQ	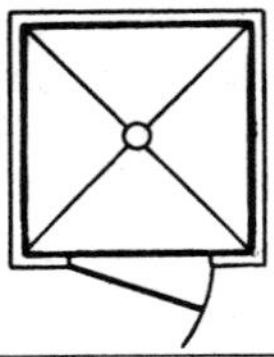		
SHOWER CORNER	SH COR			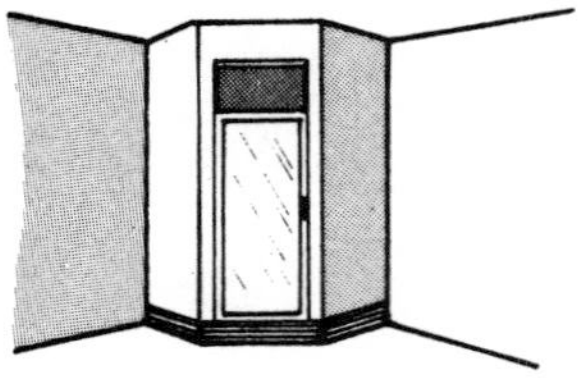
WATER CLOSET TWO PIECE	WC 2 PC		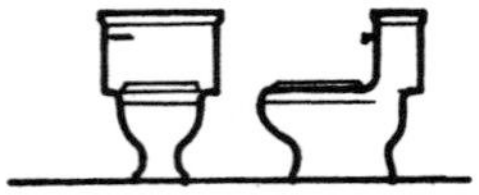	

Fig. 5-1. Sanitation facility symbols.

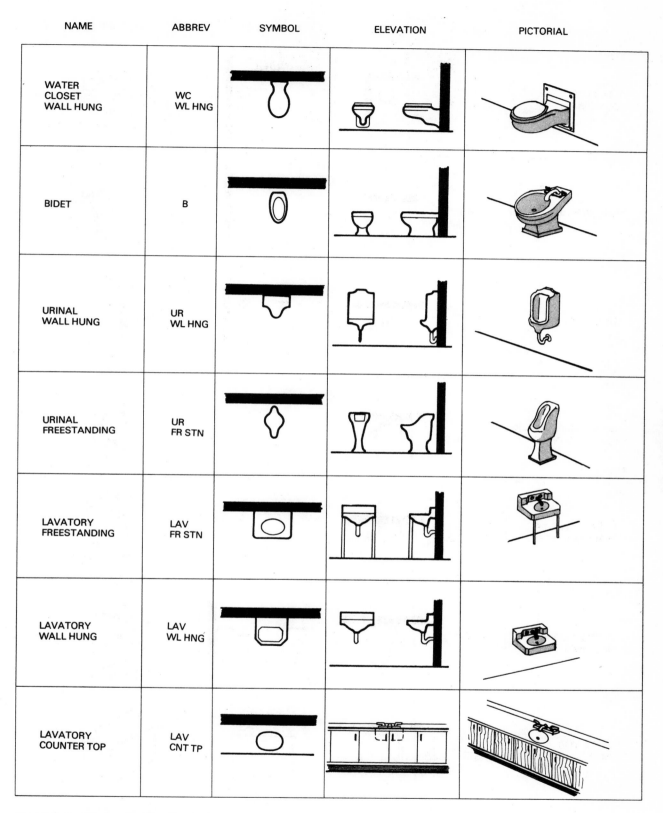

NAME	ABBREV	SYMBOL	ELEVATION	PICTORIAL
WATER CLOSET WALL HUNG	WC WL HNG			
BIDET	B			
URINAL WALL HUNG	UR WL HNG			
URINAL FREESTANDING	UR FR STN			
LAVATORY FREESTANDING	LAV FR STN			
LAVATORY WALL HUNG	LAV WL HNG			
LAVATORY COUNTER TOP	LAV CNT TP			

Fig. 5-2. Sanitation facility symbols.

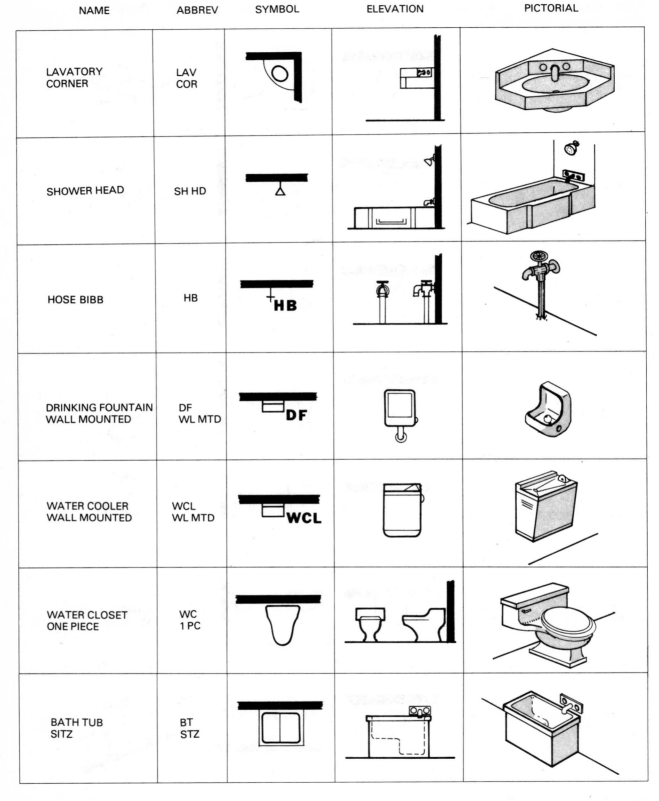

NAME	ABBREV	SYMBOL	ELEVATION	PICTORIAL
LAVATORY CORNER	LAV COR			
SHOWER HEAD	SH HD			
HOSE BIBB	HB	HB		
DRINKING FOUNTAIN WALL MOUNTED	DF WL MTD	DF		
WATER COOLER WALL MOUNTED	WCL WL MTD	WCL		
WATER CLOSET ONE PIECE	WC 1 PC			
BATH TUB SITZ	BT STZ			

Fig. 5-3. Sanitation facility symbols.

Unit 6
Plumbing Symbols

NAME	ABBREV	SYMBOL	ELEVATION	PICTORIAL
COUPLING	CPLG			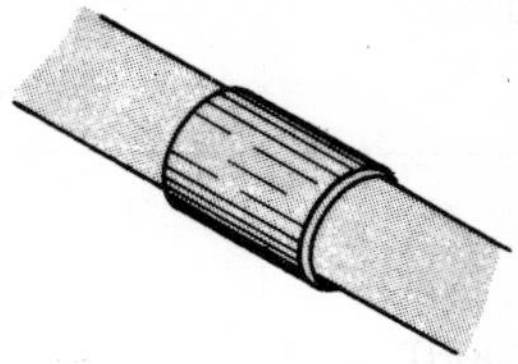
ELBOW 90°	EL	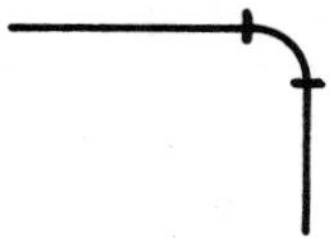	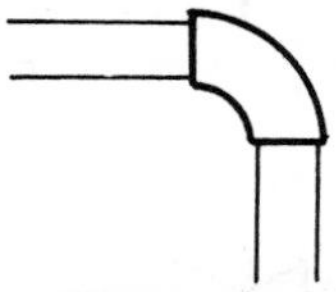	
ELBOW 45°	EL	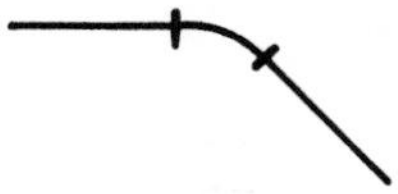	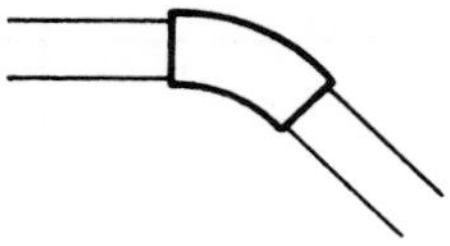	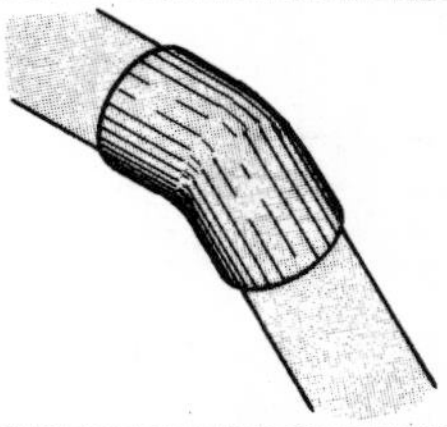
TEE 90°	T	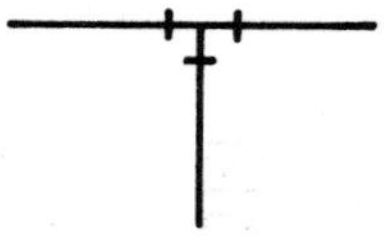	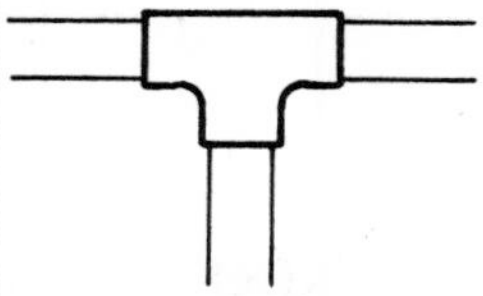	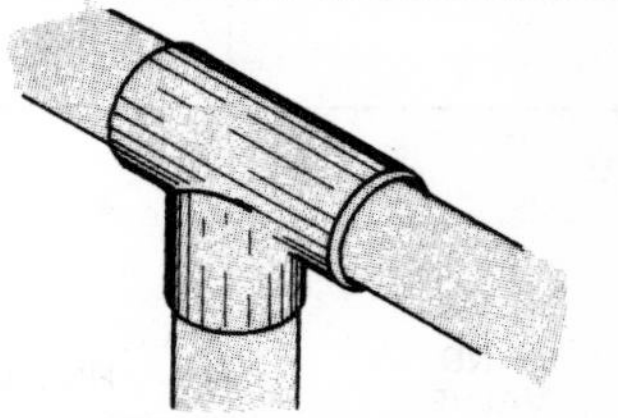
LATERAL 45°	LAT	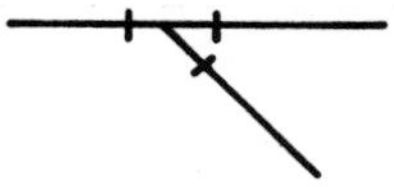	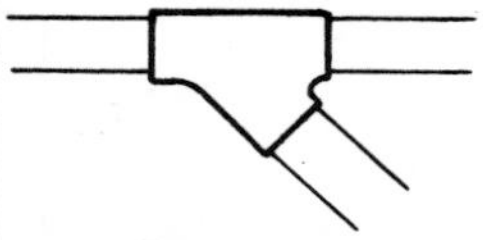	
CLEAN OUT	CO	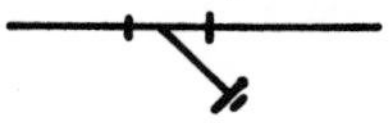	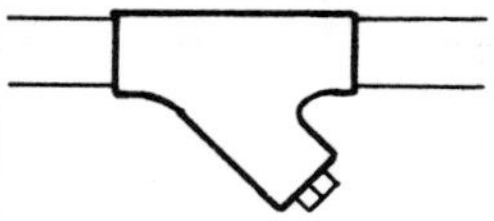	
REDUCER	RED		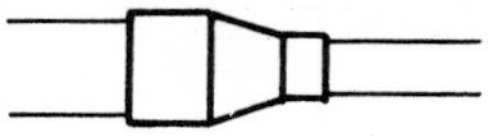	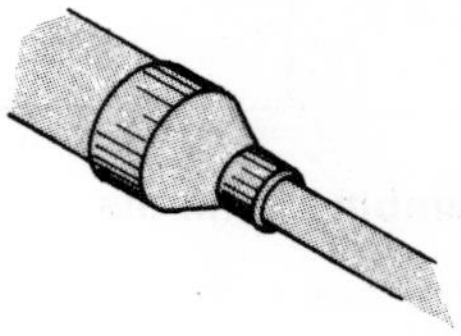

Fig. 6-1. Plumbing symbols.

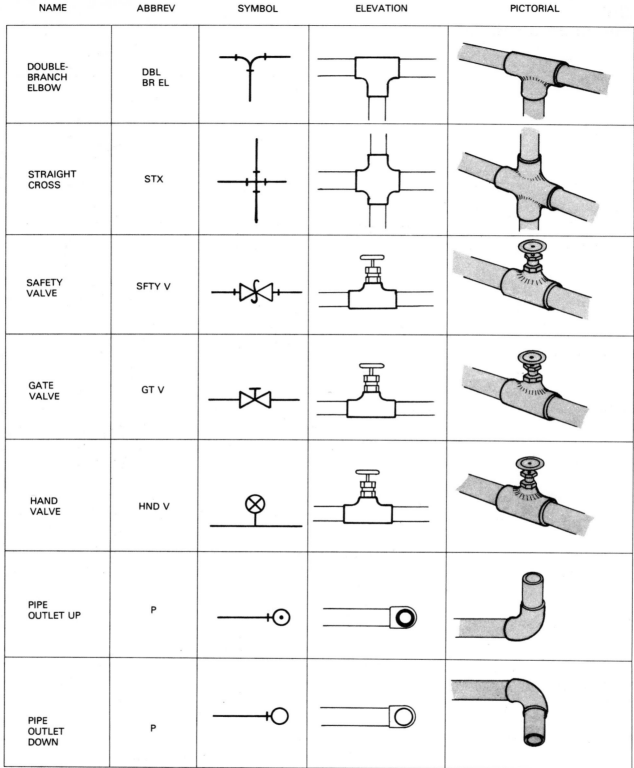

NAME	ABBREV	SYMBOL	ELEVATION	PICTORIAL
DOUBLE-BRANCH ELBOW	DBL BR EL			
STRAIGHT CROSS	STX			
SAFETY VALVE	SFTY V			
GATE VALVE	GT V			
HAND VALVE	HND V			
PIPE OUTLET UP	P			
PIPE OUTLET DOWN	P			

Fig. 6-2. Plumbing symbols.

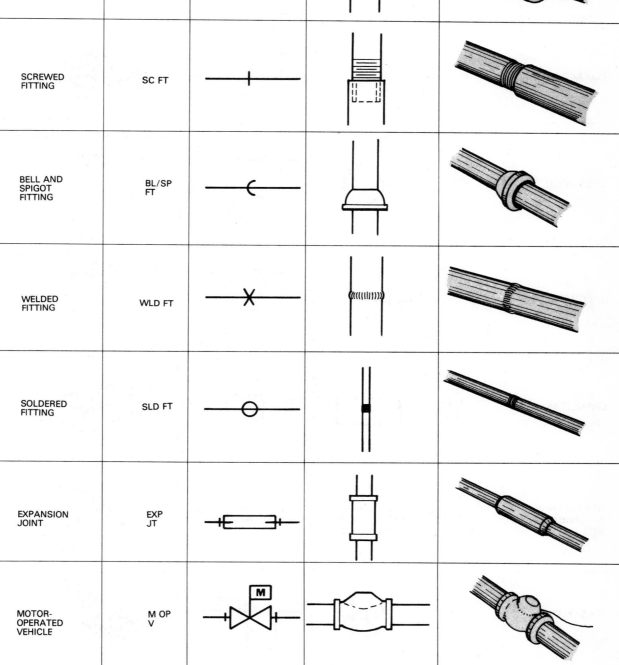

NAME	ABBREV	SYMBOL	ELEVATION	PICTORIAL
FLANGED FITTING	FL FT			
SCREWED FITTING	SC FT			
BELL AND SPIGOT FITTING	BL/SP FT			
WELDED FITTING	WLD FT			
SOLDERED FITTING	SLD FT			
EXPANSION JOINT	EXP JT			
MOTOR-OPERATED VEHICLE	M OP V			

Fig. 6-3. Plumbing symbols.

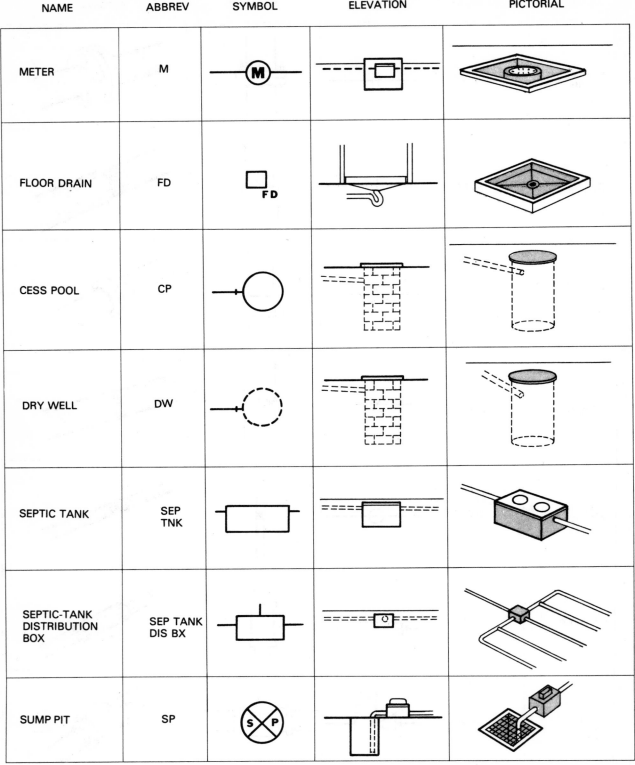

NAME	ABBREV	SYMBOL	ELEVATION	PICTORIAL
METER	M			
FLOOR DRAIN	FD	FD		
CESS POOL	CP			
DRY WELL	DW			
SEPTIC TANK	SEP TNK			
SEPTIC-TANK DISTRIBUTION BOX	SEP TANK DIS BX			
SUMP PIT	SP			

Fig. 6-4. Plumbing symbols.

NAME	ABBREV	SYMBOL	NAME	ABBRV	SYMBOL
COLD-WATER LINE	CW	——— – — – —	AIR-PRESSURE RETURN LINE	APR	— — — — — —
HOT-WATER LINE	HW	— – — – — – —	ICE-WATER LINE	IW	——— IW ———
GAS LINE	G	—G——G—	DRAIN LINE	D	—D———D—
VENT	V	– – – – – – –	FUEL-OIL RETURN LINE	FOF	——FOF——
SOIL STACK PLAN VIEW	SS	═══○═══	FUEL-OIL FLOW LINE	FOR	– – –FOR– – –
SOIL LINE ABOVE GRADE	SL	———————	REFRIGERANT LINE	R	—+—+—+—+—
SOIL LINE BELOW GRADE	SL	— — — — —	STEAM LINE MEDIUM PRESSURE	SL	/—/—/—/—/
CAST-IRON SEWER	S-CI	——S-CI——	STEAM RETURN LINE—MEDIUM PRESSURE	SRL	/—/—/—/—/
CLAY-TILE SEWER	S-CT	——S-CT——	PNEUMATIC TUBE	PT	═══════
LEACH LINE	LEA	—I—I—I—I—	INDUSTRIAL SEWAGE	IS	—I—I—I—I
SPRINKLER LINE	SPR	—S——S—	CHEMICAL WASTE LINE	CW	——\——\——\—
VACUUM LINE	VAC	—V—V——	FIRE LINE	F	—F—F—
COMPRESSED AIR LINE	COMP	—A——A—	ACID WASTE LINE	AC WST	——ACID——
AIR-PRESSURE LINE FLOW	APF	➤——➤——➤	HUMIDIFICATION LINE	HUM	– – —H—– – —

Fig. 6-5. Plumbing symbols.

Unit 7
Electrical Symbols

NAME	ABBREV	SYMBOL	ELEVATION	PICTORIAL
SWITCH SINGLE-POLE	S	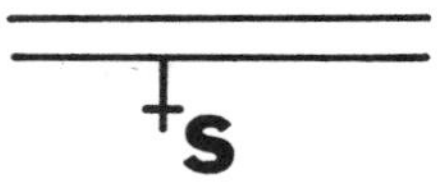 S		
SWITCH DOUBLE-POLE	S_2	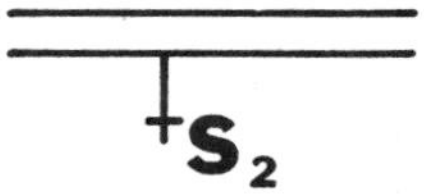 S_2		
SWITCH THREE-WAY	S_3	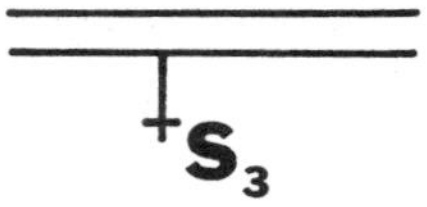 S_3		
SWITCH FOUR-WAY	S_4	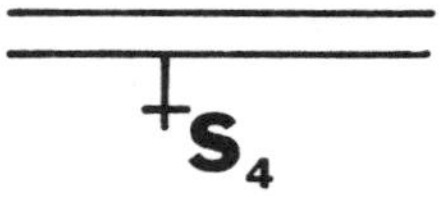 S_4		
SWITCH WEATHERPROOF	S_{wp}	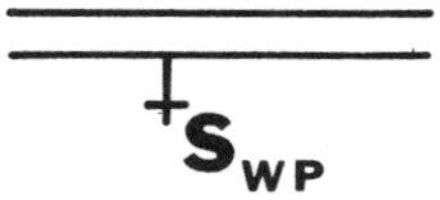 S_{WP}		
SWITCH AUTOMATIC DOOR	S_D	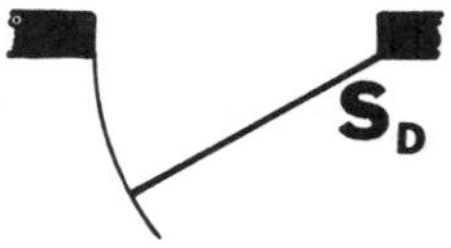 S_D	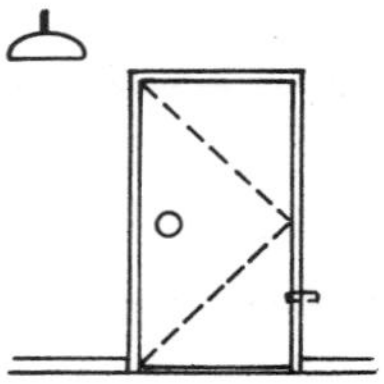	
SWITCH PILOT LIGHT	S_p	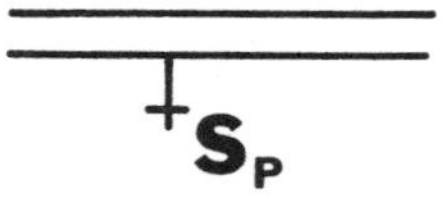 S_P	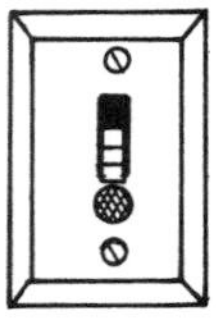	

Fig. 7-1. Electrical symbols.

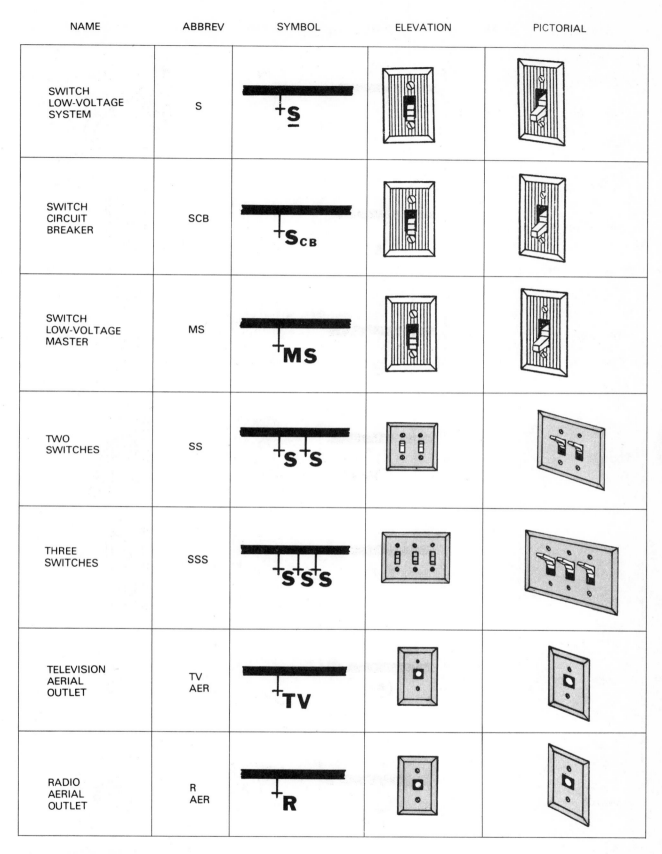

NAME	ABBREV	SYMBOL	ELEVATION	PICTORIAL
SWITCH LOW-VOLTAGE SYSTEM	S	$\underline{S}$		
SWITCH CIRCUIT BREAKER	SCB	S_{CB}		
SWITCH LOW-VOLTAGE MASTER	MS	MS		
TWO SWITCHES	SS	S S		
THREE SWITCHES	SSS	S S S		
TELEVISION AERIAL OUTLET	TV AER	$\underline{TV}$		
RADIO AERIAL OUTLET	R AER	R		

Fig. 7-2. Electrical symbols.

NAME	ABBREV	SYMBOL	ELEVATION	PICTORIAL
DUPLEX OUTLET	DUP OUT	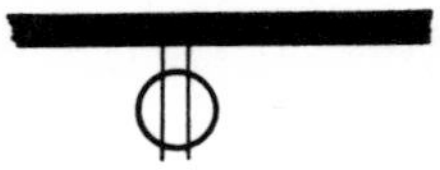		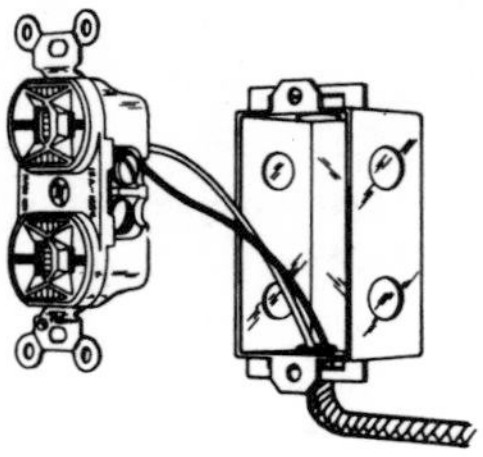
SINGLE OUTLET	S OUT	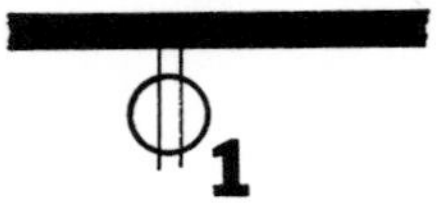 1		
TRIPLE OUTLET	TR OUT	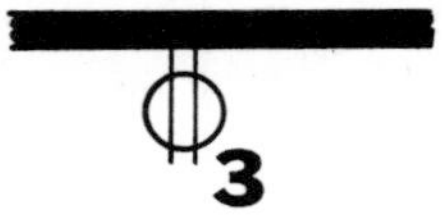 3	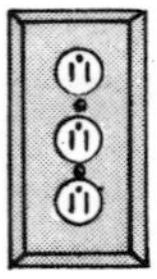	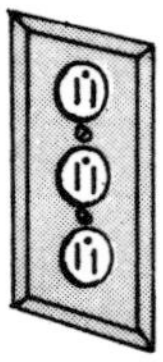
WEATHERPROOF OUTLET	WP OUT	 WP		
SPLIT WIRE OUTLET	SPT WR OUT	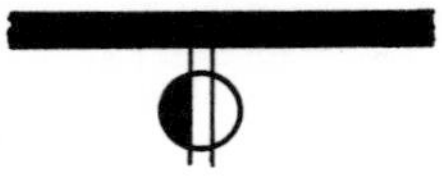	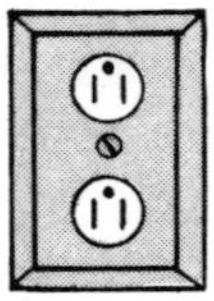	
FLOOR OUTLET	FL OUT		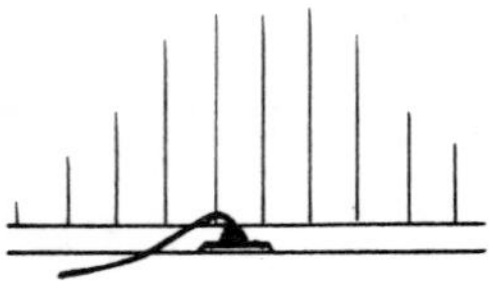	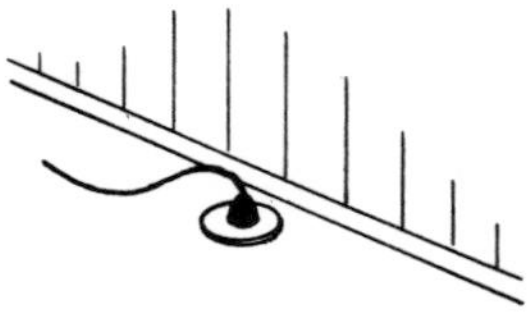
OUTLET WITH SWITCH	OUT/S	 S		

Fig. 7-3. Electrical symbols.

NAME	ABBREV	SYMBOL	ELEVATION	PICTORIAL
HEAVY-DUTY OUTLET 220 VOLTAGE	HVY DTY OUT			
SPECIAL-PURPOSE OUTLET 110 VOLTAGE	SP PUR OUT	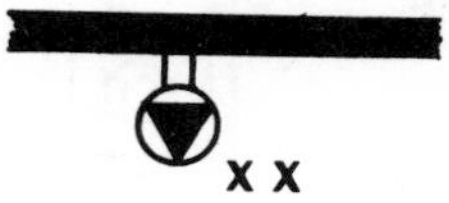 x x		
RANGE OUTLET	R OUT	 R		
REFRIGERATOR OUTLET	REF OUT	 R	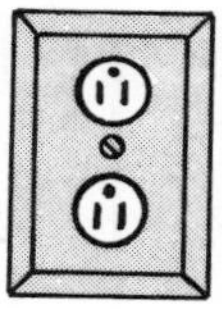	
WATERHEATER OUTLET	WH OUT	 WH		
GARBAGE-DISPOSAL OUTLET	GD OUT	 GD		
DISHWASHER OUTLET	DW OUT	 DW		

Fig. 7-4. Electrical symbols.

NAME	ABBREV	SYMBOL	ELEVATION	PICTORIAL
IRON OUTLET WITH PILOT LIGHT	I OUT/PL	 IP	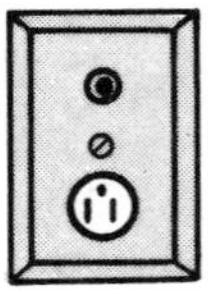	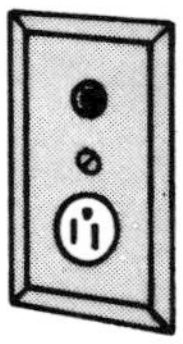
WASHER OUTLET	W OUT	 W		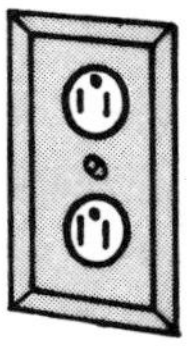
DRYER OUTLET	D OUT	 D		
MOTOR OUTLET	M OUT	 M		
STRIP OUTLET	STP OUT	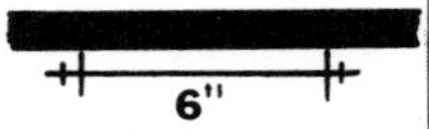 6"		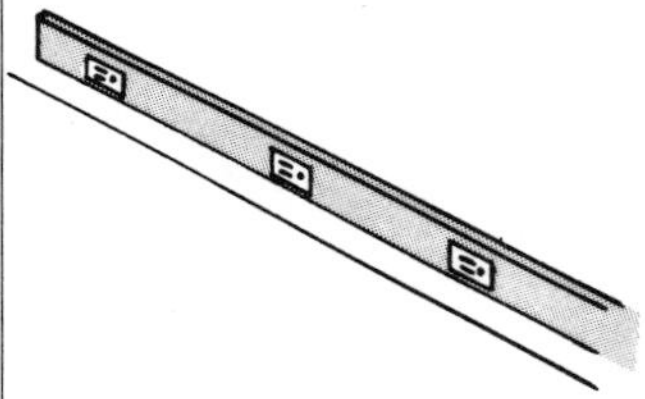
GROUNDED OUTLET	GRD OUTLET	 GR		
FREEZER OUTLET	FR OUT	 FR		

Fig. 7-5. Electrical symbols.

NAME	ABBREV	SYMBOL	ELEVATION	PICTORIAL
LIGHTING OUTLET — CEILING	LT OUT CLG		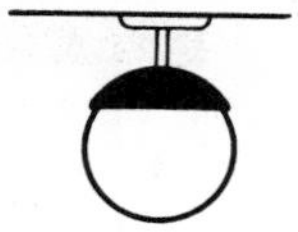	
LIGHTING OUTLET — RECESSED	LT OUT REC			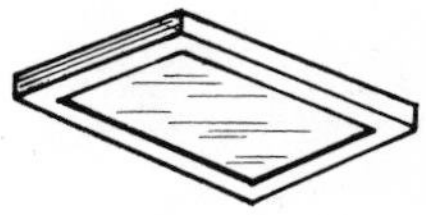
LIGHTING OUTLET — WALL	LT OUT WALL			
LIGHTING OUTLET — CEILING PULL SWITCH	PS	 PS	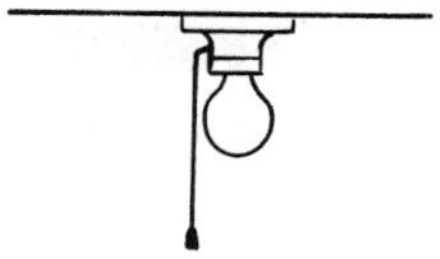	
FLOOD LIGHT	FL			
SPOT LIGHT	SL			
LIGHTING OUTLET — VAPOR PROOF	LT OUT	 VP	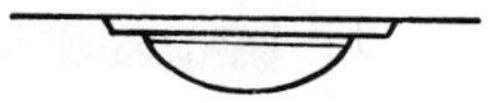	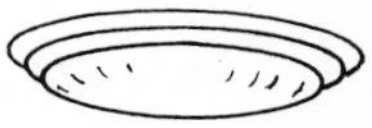

Fig. 7-6. Electrical service outlet and lighting symbols.

NAME	ABBREV	SYMBOL	ELEVATION	PICTORIAL

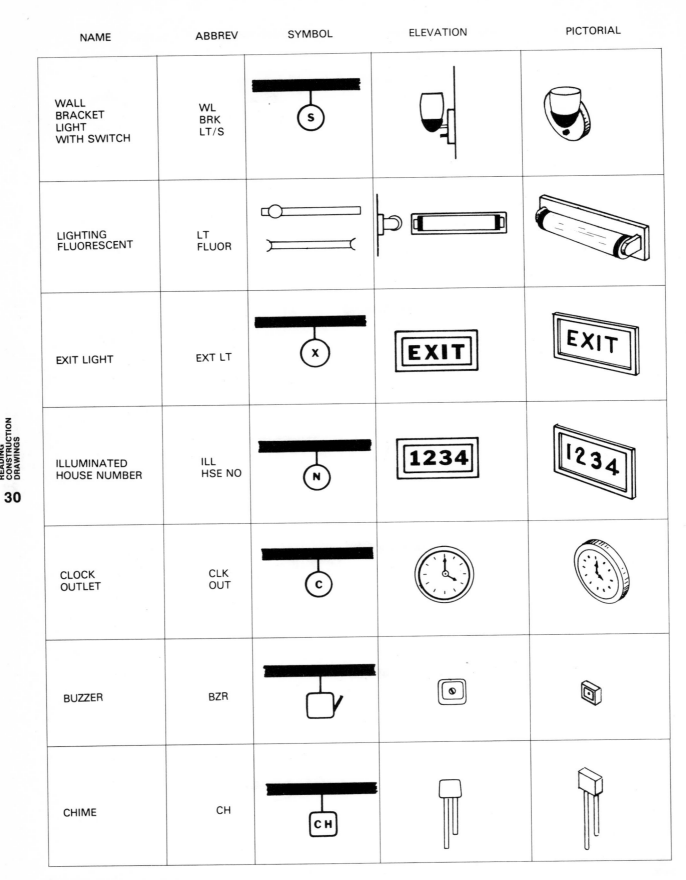

NAME	ABBREV	SYMBOL	ELEVATION	PICTORIAL
WALL BRACKET LIGHT WITH SWITCH	WL BRK LT/S	(S)		
LIGHTING FLUORESCENT	LT FLUOR			
EXIT LIGHT	EXT LT	(X)	EXIT	EXIT
ILLUMINATED HOUSE NUMBER	ILL HSE NO	(N)	1234	1234
CLOCK OUTLET	CLK OUT	(C)		
BUZZER	BZR			
CHIME	CH	CH		

Fig. 7-7. Electrical lighting and special outlet symbols.

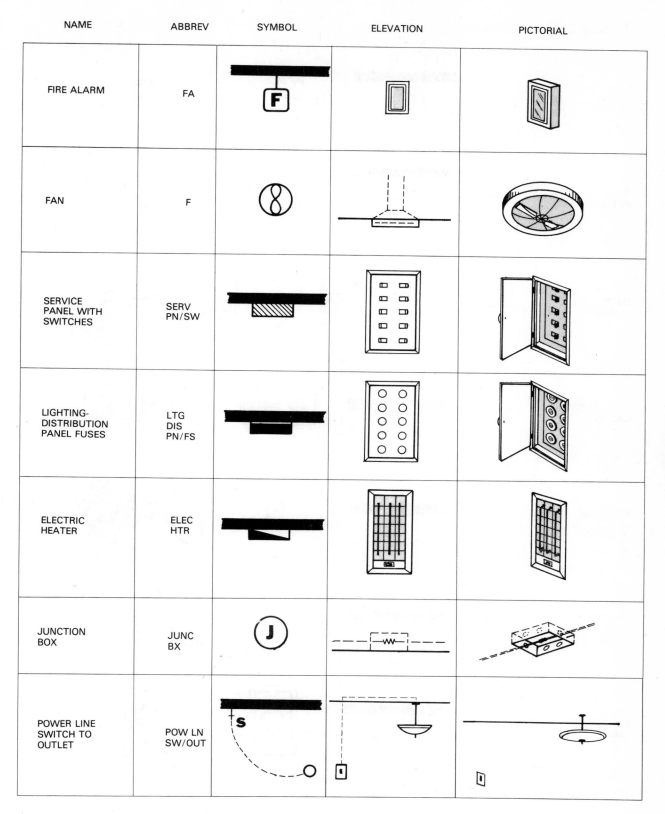

NAME	ABBREV	SYMBOL	ELEVATION	PICTORIAL
FIRE ALARM	FA			
FAN	F			
SERVICE PANEL WITH SWITCHES	SERV PN/SW			
LIGHTING-DISTRIBUTION PANEL FUSES	LTG DIS PN/FS			
ELECTRIC HEATER	ELEC HTR			
JUNCTION BOX	JUNC BX			
POWER LINE SWITCH TO OUTLET	POW LN SW/OUT			

Fig. 7-8. Electrical special outlet symbols.

NAME	ABBREV	SYMBOL	ELEVATION	PICTORIAL
BELL	BL	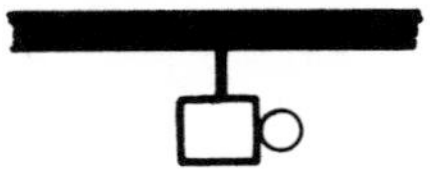		
PUSH BUTTON	PB			
ELECTRIC DOOR OPENER	ELEC DR OP	 D	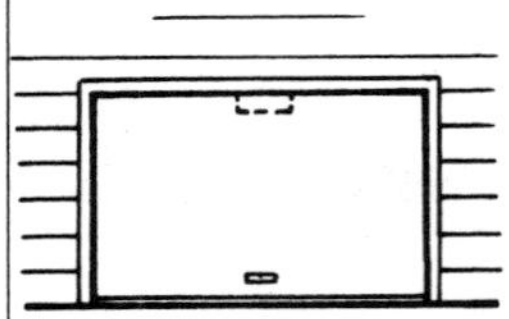	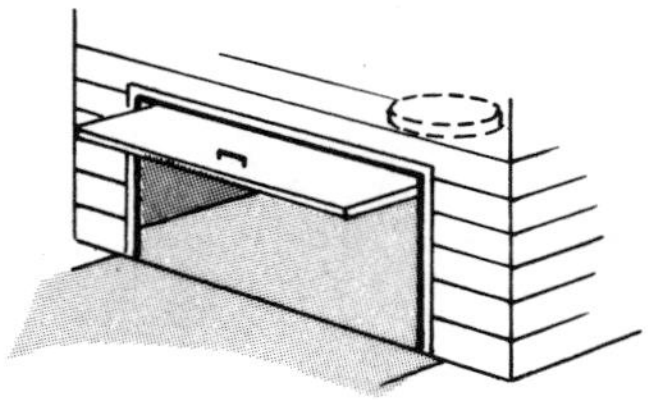
INTERCOMMUNICATION	INTERCOM		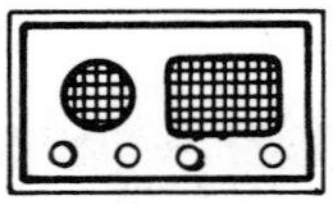	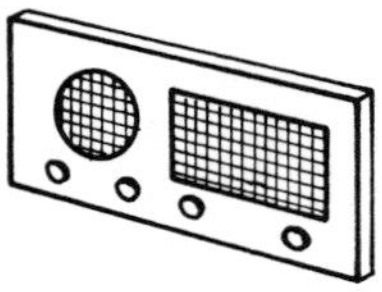
TELEPHONE OUTLET	TEL OUT			
TELEPHONE JACK	TEL JK			
DIMMER SWITCH	DM SW	 S_{DM}	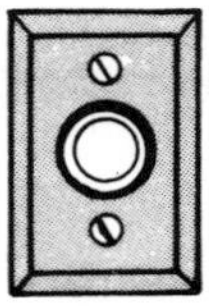	

Fig. 7-9. Electrical special outlet symbols.

Unit 8
Climate Control Symbols

NAME	ABBREV	SYMBOL	NAME	ABBRV	SYMBOL
DUCT SIZE & FLOW DIRECTION	DCT/FD	 ←	HEAT REGISTER	R	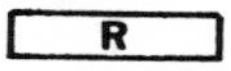 R
DUCT SIZE CHANGE	DCT/SC	 ←	THERMOSTAT	T	 T
DUCT LOWERING	DCT/LW	 →D →D	RADIATOR	RAD	 RAD
DUCT RISING	DCT/RS	 R← R←	CONVECTOR	CONV	 CONV
DUCT RETURN	DCT/RT	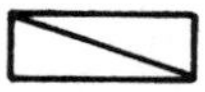	ROOM AIR CONDITIONER	RAC	 RAC
DUCT SUPPLY	DCT SUP	 S	HEATING PLANT FURNACE	HT PLT FUR	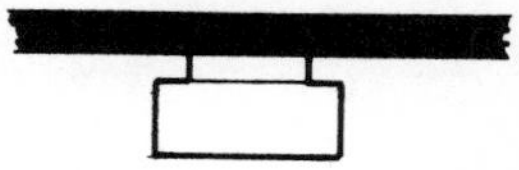
CEILING-DUCT OUTLET	CLG DCT OUT	○	FUEL-OIL TANK	FOT	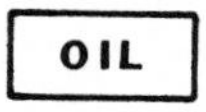 OIL
WARM-AIR SUPPLY	WA SUP	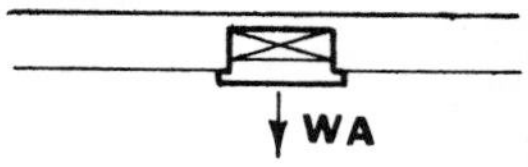 ↓WA	HUMIDSTAT	H	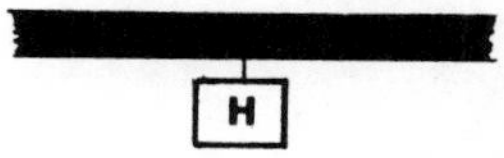 H
SECOND-FLOOR SUPPLY	2nd FL SUP	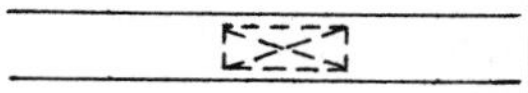	HEAT PUMP	HP	 HP
COLD-AIR RETURN	CA RET	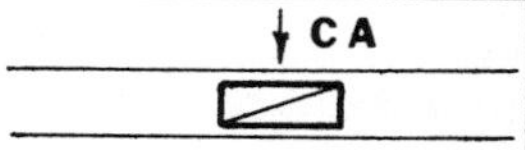 ↓CA	THERMOMETER	T	 T
SECOND-FLOOR RETURN	2 FL RET	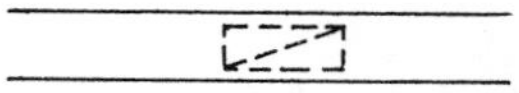	PUMP	P	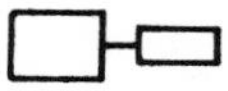
GAS OUTLET	G OUT	 ↑G	GAGE	GA	
HEAT OUTLET	HT OUT	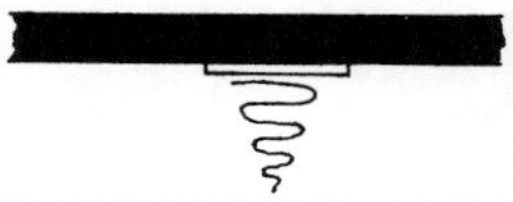	FORCED CONVECTION	FRC CONV	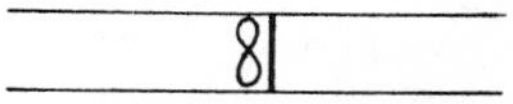

Fig. 8-1. Climate control symbols.

Unit 9
Building Materials Symbols

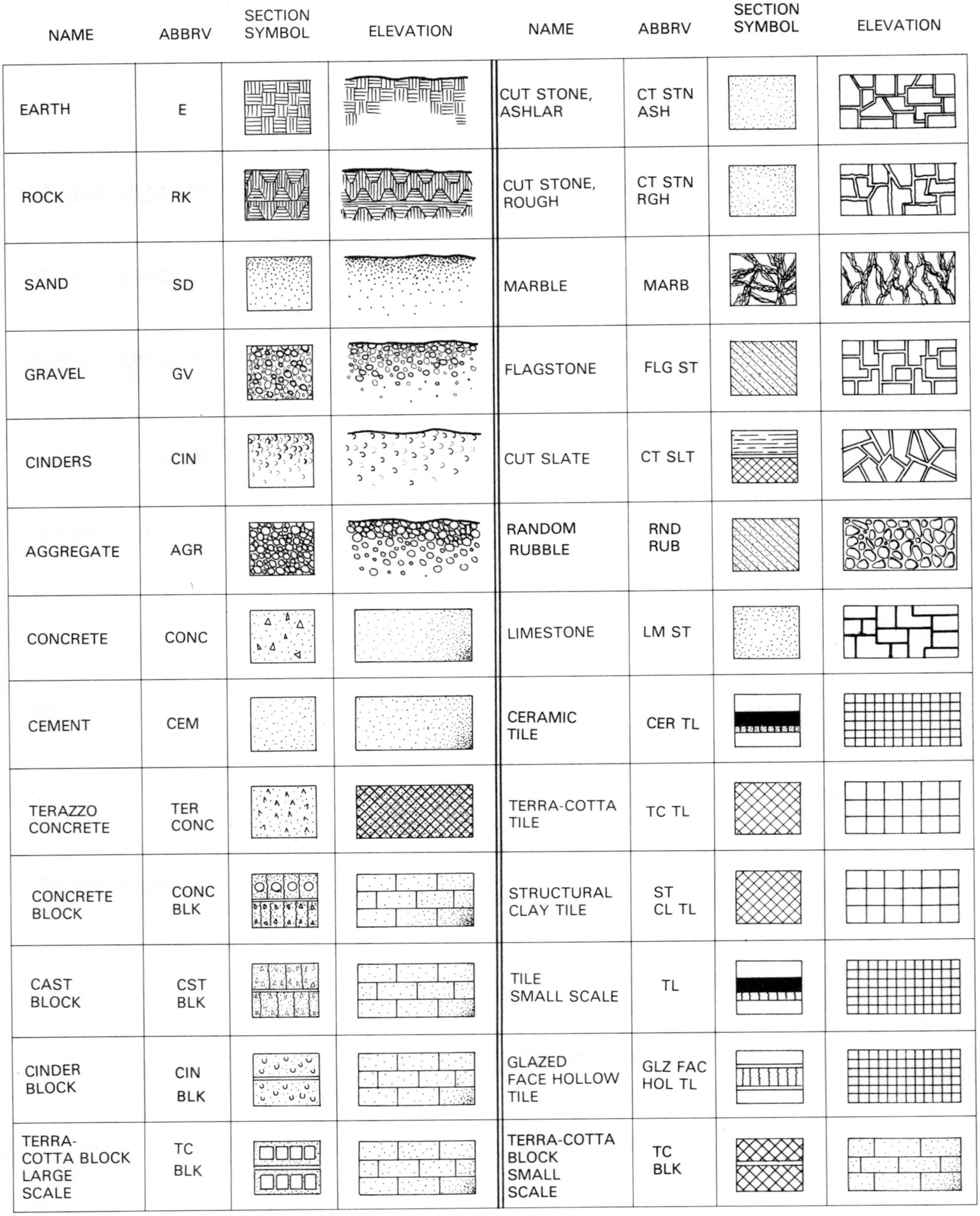

NAME	ABBRV	SECTION SYMBOL	ELEVATION	NAME	ABBRV	SECTION SYMBOL	ELEVATION
EARTH	E			CUT STONE, ASHLAR	CT STN ASH		
ROCK	RK			CUT STONE, ROUGH	CT STN RGH		
SAND	SD			MARBLE	MARB		
GRAVEL	GV			FLAGSTONE	FLG ST		
CINDERS	CIN			CUT SLATE	CT SLT		
AGGREGATE	AGR			RANDOM RUBBLE	RND RUB		
CONCRETE	CONC			LIMESTONE	LM ST		
CEMENT	CEM			CERAMIC TILE	CER TL		
TERAZZO CONCRETE	TER CONC			TERRA-COTTA TILE	TC TL		
CONCRETE BLOCK	CONC BLK			STRUCTURAL CLAY TILE	ST CL TL		
CAST BLOCK	CST BLK			TILE SMALL SCALE	TL		
CINDER BLOCK	CIN BLK			GLAZED FACE HOLLOW TILE	GLZ FAC HOL TL		
TERRA-COTTA BLOCK LARGE SCALE	TC BLK			TERRA-COTTA BLOCK SMALL SCALE	TC BLK		

Fig. 9-1. Building material symbols.

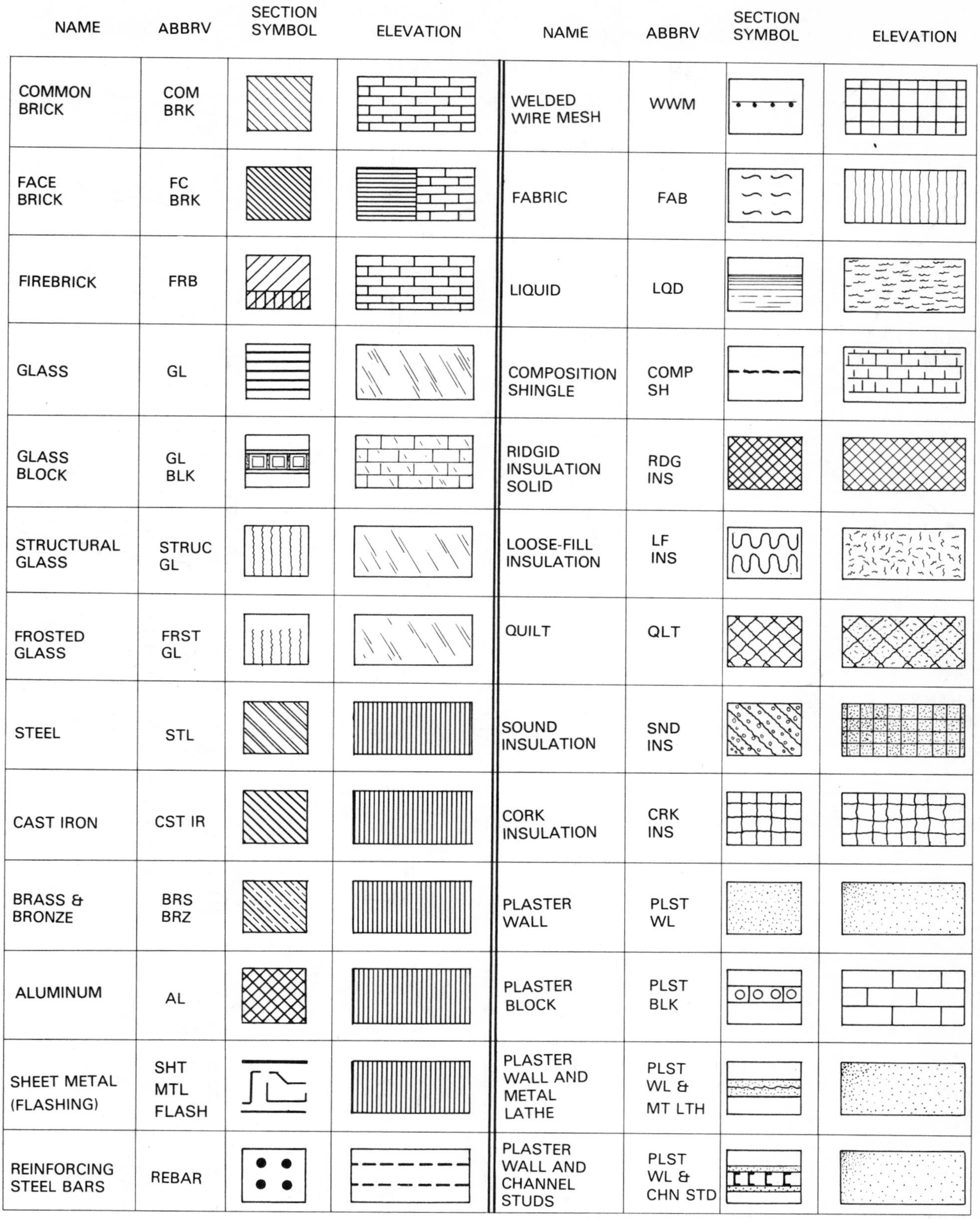

NAME	ABBRV	SECTION SYMBOL	ELEVATION	NAME	ABBRV	SECTION SYMBOL	ELEVATION
COMMON BRICK	COM BRK			WELDED WIRE MESH	WWM		
FACE BRICK	FC BRK			FABRIC	FAB		
FIREBRICK	FRB			LIQUID	LQD		
GLASS	GL			COMPOSITION SHINGLE	COMP SH		
GLASS BLOCK	GL BLK			RIDGID INSULATION SOLID	RDG INS		
STRUCTURAL GLASS	STRUC GL			LOOSE-FILL INSULATION	LF INS		
FROSTED GLASS	FRST GL			QUILT	QLT		
STEEL	STL			SOUND INSULATION	SND INS		
CAST IRON	CST IR			CORK INSULATION	CRK INS		
BRASS & BRONZE	BRS BRZ			PLASTER WALL	PLST WL		
ALUMINUM	AL			PLASTER BLOCK	PLST BLK		
SHEET METAL (FLASHING)	SHT MTL FLASH			PLASTER WALL AND METAL LATHE	PLST WL & MT LTH		
REINFORCING STEEL BARS	REBAR			PLASTER WALL AND CHANNEL STUDS	PLST WL & CHN STD		

Fig. 9-2. Building material symbols.

NAME	ABBREV	SYMBOL	ELEVATION SECTION	PICTORIAL
I BEAM	IBM	I		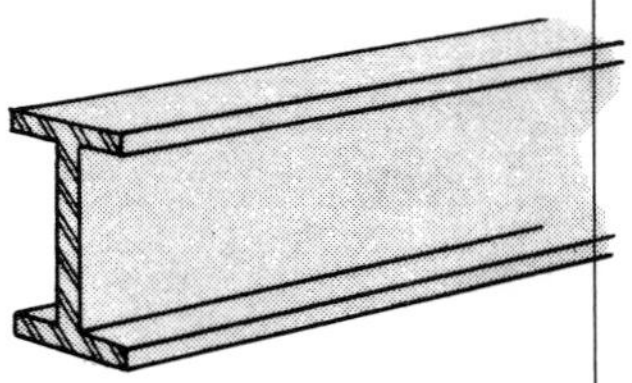
WIDE FLANGE BEAM	WD FL BM	WF		
ANGLE BEAM	ANG BM	⌐		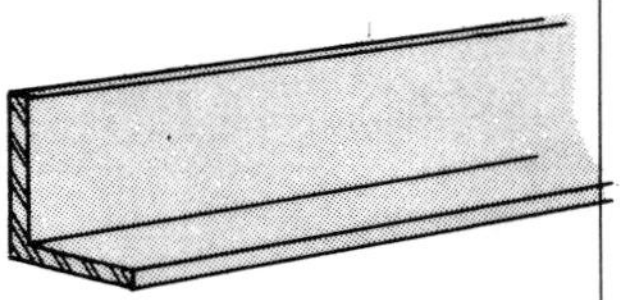
CHANNEL BEAM	CHN BM	C		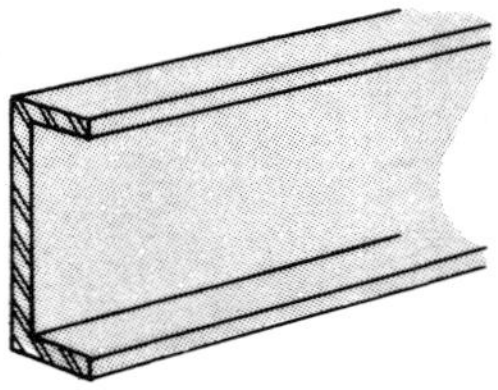
STEEL TUBE	STL TB	▢		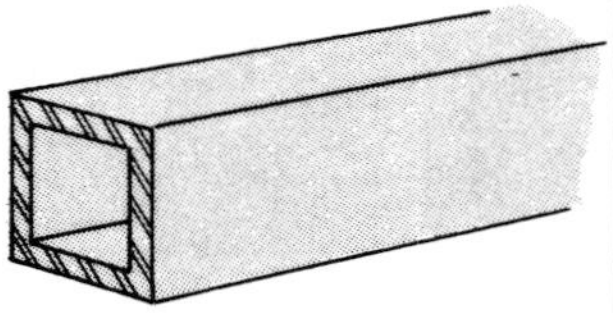
STRUCTURAL TEE BEAM	STRUC T BM	T		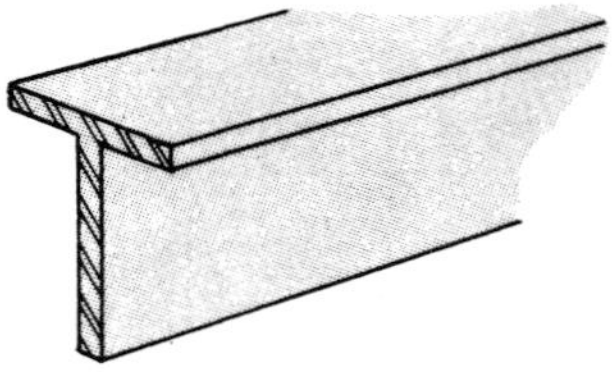
BULB TEE BEAM	BLB T BM	BT		

Fig. 9-3. Building material symbols.

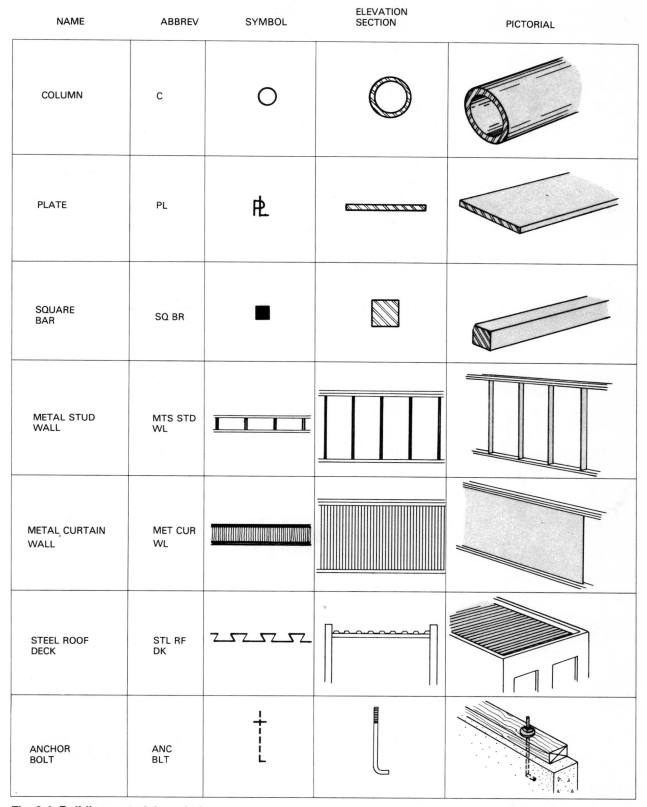

NAME	ABBREV	SYMBOL	ELEVATION SECTION	PICTORIAL
COLUMN	C			
PLATE	PL			
SQUARE BAR	SQ BR			
METAL STUD WALL	MTS STD WL			
METAL CURTAIN WALL	MET CUR WL			
STEEL ROOF DECK	STL RF DK			
ANCHOR BOLT	ANC BLT			

Fig. 9-4. Building material symbols.

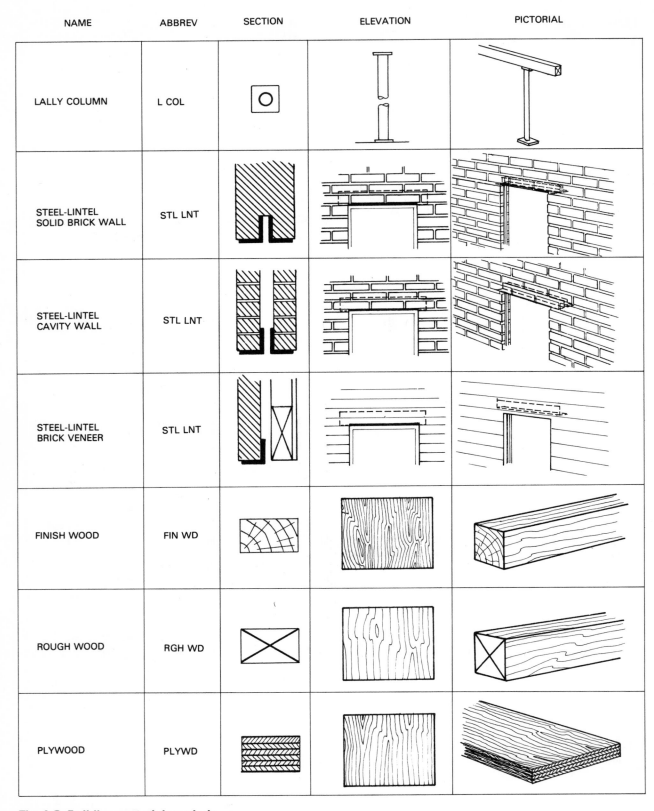

NAME	ABBREV	SECTION	ELEVATION	PICTORIAL
LALLY COLUMN	L COL			
STEEL-LINTEL SOLID BRICK WALL	STL LNT			
STEEL-LINTEL CAVITY WALL	STL LNT			
STEEL-LINTEL BRICK VENEER	STL LNT			
FINISH WOOD	FIN WD			
ROUGH WOOD	RGH WD			
PLYWOOD	PLYWD			

Fig. 9-5. Building material symbols.

Unit 10
Topographic Symbols

NAME	ABBREV	SYMBOL	NAME	ABBREV	SYMBOL
TREES	TR	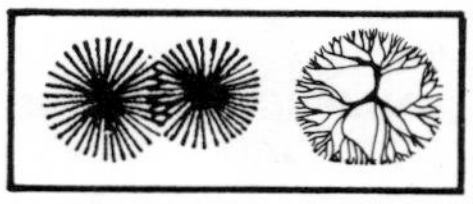	GRAVEL	GRV	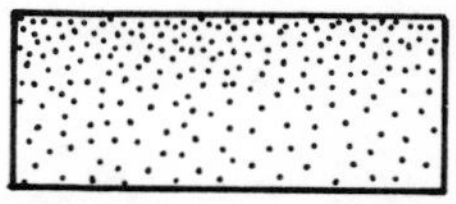
GROUND COVER	GRD CV	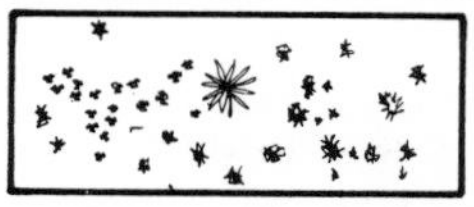	CULTIVATED AREA	CULT	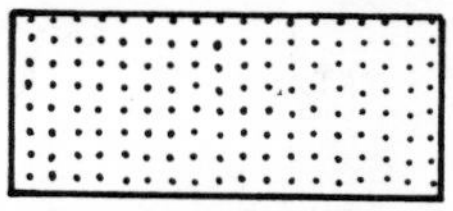
BUSHES SHRUBS	BSH SH		WATER	WT	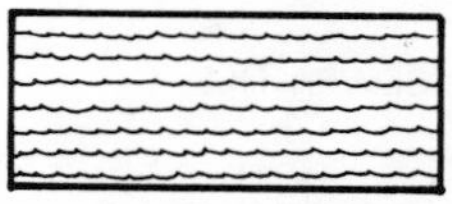
OPEN WOODLAND	OP WDL	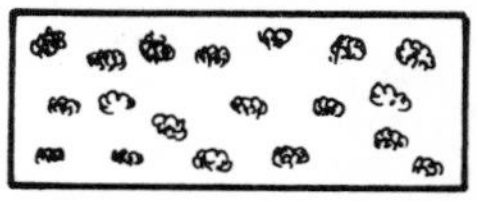	WELL	W	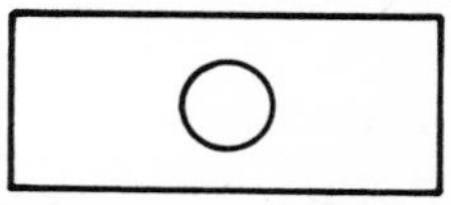
ORCHARD	OR	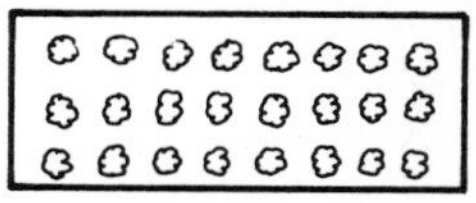	NORTH-MERIDIAN ARROWS	N MER ARR	
MARSH	MRS	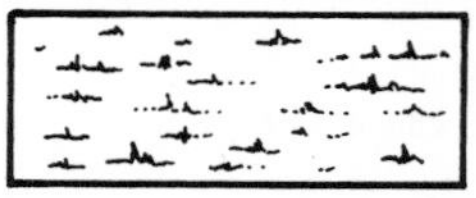	PROPERTY LINE	PR LN	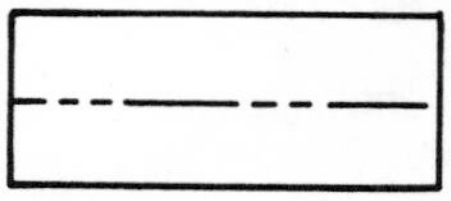
SUBMERGED MARSH	SUB MRS		SURVEYED CONTOUR LINE	SURV CON LN	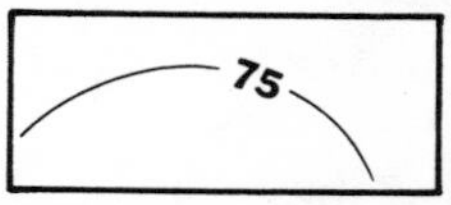
DENSE FOREST	DN FR		ESTIMATED CONTOUR	EST CON	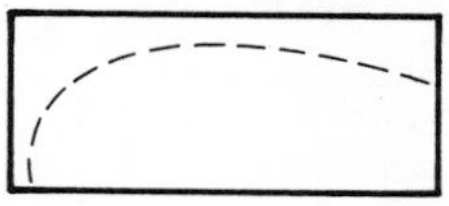
SPACED TREES	SP TR	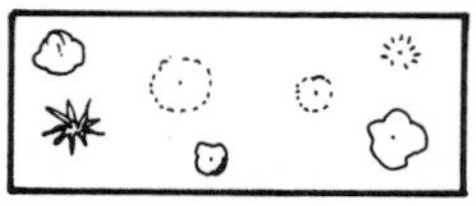	FENCE	FN	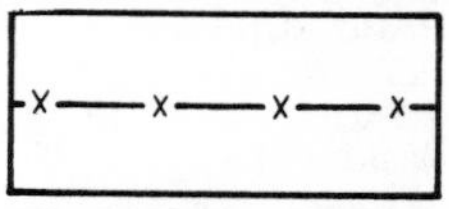
TALL GRASS	TL GRS	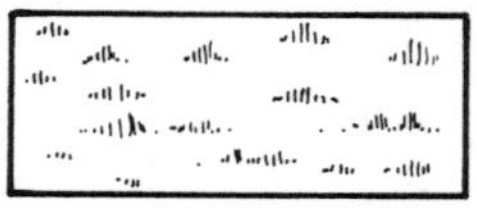	RAILROAD TRACKS	RR TRK	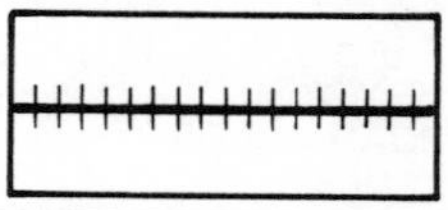
LARGE STONES	LRG ST		PAVED ROAD	PV RD	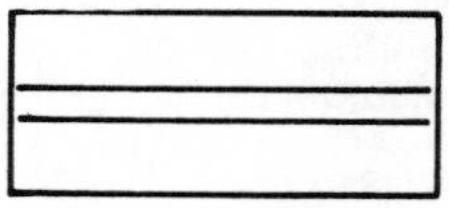
SAND	SND	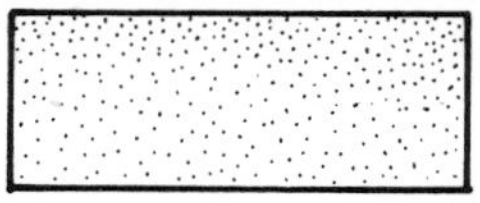	UNPAVED ROAD	UNPV RD	
DRY CRACKED CLAY	DRY CRK CLY	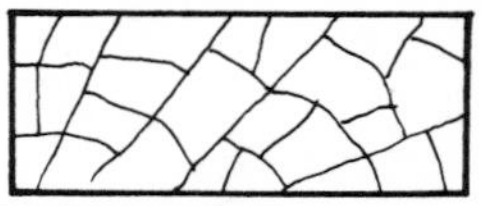	POWER LINE	POW LN	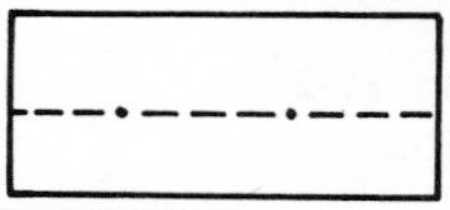

Fig. 10-1. Topographic symbols.

NAME	ABBREV	SYMBOL	NAME	ABBREV	SYMBOL
WATER LINE	WT LN	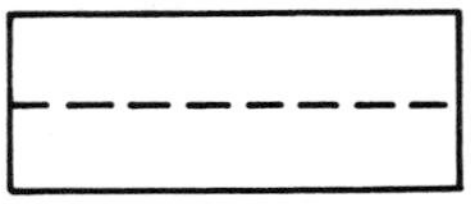	IMPROVED LIGHT DUTY ROAD	IMP LT DTY RD	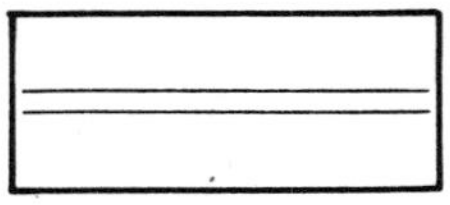
GAS LINE	G LN	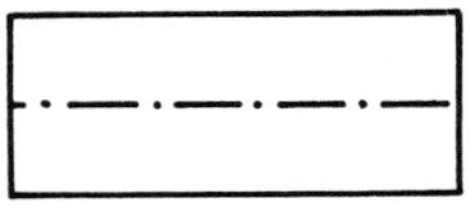	TRAIL UNIMPROVED DIRT ROAD	TRL UNIM DRT RD	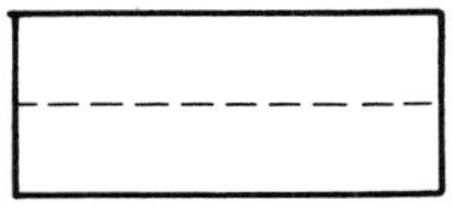
SANITARY SEWER	SAN SW	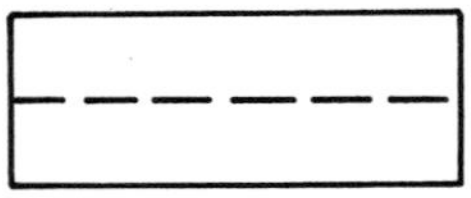	ROAD UNDER CONSTRUCTION	RD CONST	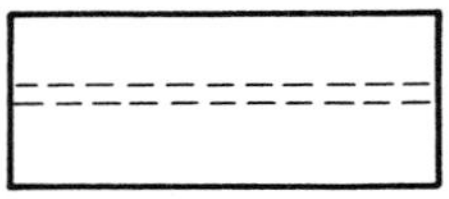
SEWER TILE	SW TL	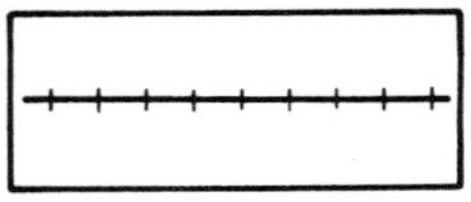	BRIDGE OVER ROAD	BRG OV RD	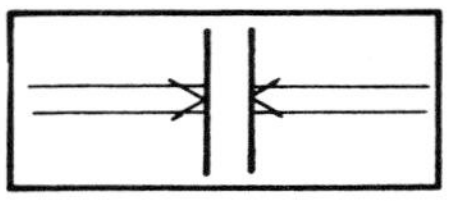
SEPTIC-FIELD LEACH LINE	SP FLD LCH LN	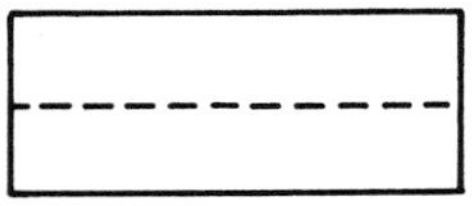	RAILROAD TUNNEL	RR TUN	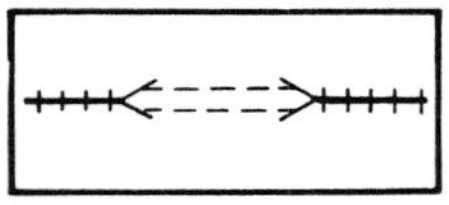
PROPERTY CORNER WITH ELEVATION	PROP CR EL	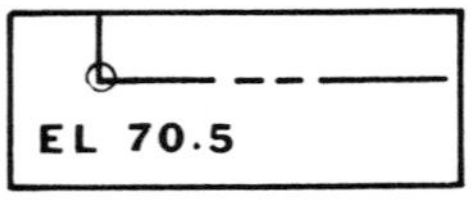 EL 70.5	ROAD OVERPASS	RD OVP	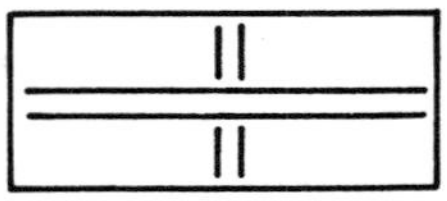
SPOT ELEVATION	SP EL	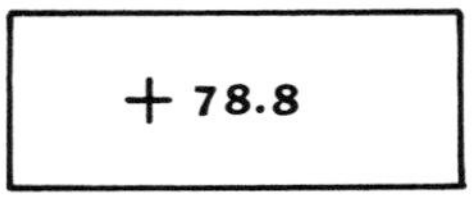 + 78.8	ROAD UNDERPASS	RD UNP	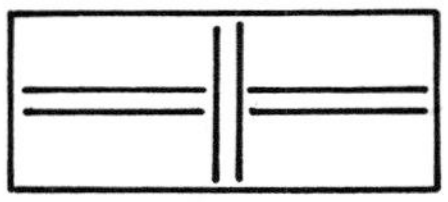
WATER ELEVATION	WT EL	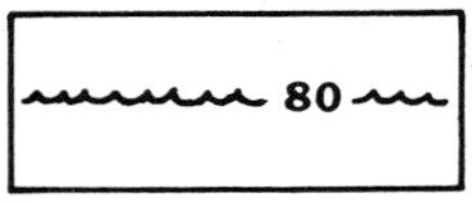 80	SMALL DAM	SM DA	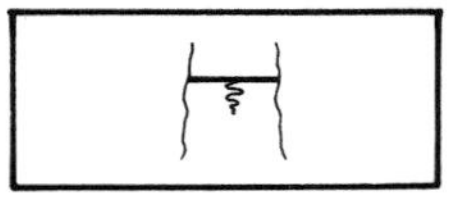
BENCH MARKS WITH ELEVATIONS	BM/EL	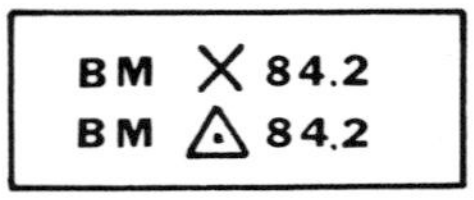 BM X 84.2 BM △ 84.2	LARGE DAM WITH LOCK	LRG DM LK	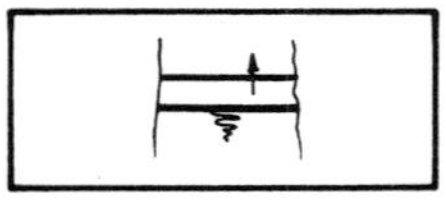
HARD-SURFACE HEAVY DUTY ROAD — FOUR OR MORE LANES	HRD SUR HY DTY RD		BUILDINGS	BLDGS	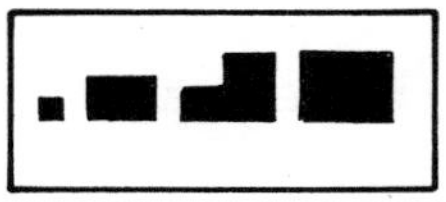
HARD-SURFACE HEAVY DUTY ROAD — 2 OR 3 LANES	HRD SUR HY DTY RD	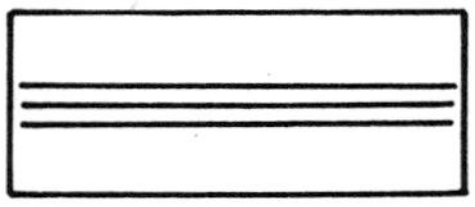	SCHOOL	SCH	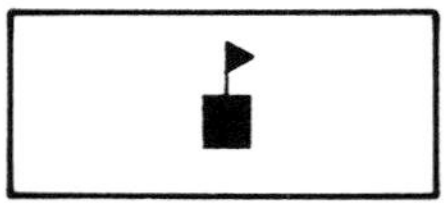
HARD-SURFACE MEDIUM DUTY ROAD — FOUR OR MORE LANES	HRD SUR MED DTY RD	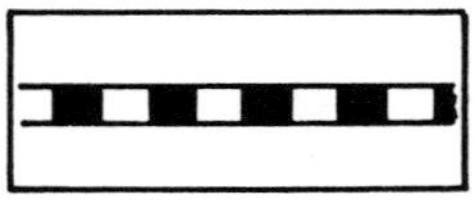	CHURCH	CH	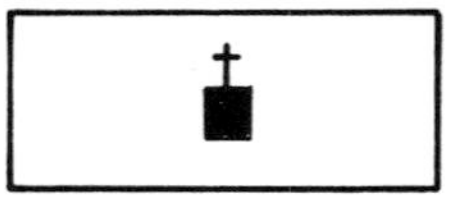
HARD-SURFACE MEDIUM DUTY ROAD — 2 OR 3 LANES	HRD SUR MED DTY RD	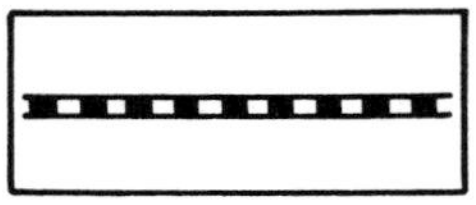	CEMETARY	CEM	 † CEM

Fig. 10-2. Topographic symbols.

NAME	ABBREV	SYMBOL	NAME	ABBREV	SYMBOL
POWER TRANSMISSION LINE	PW TR LN	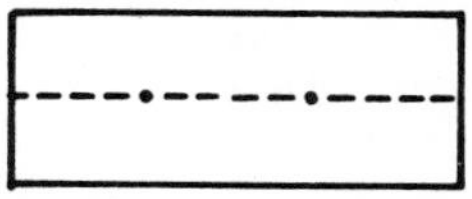	BOUNDRY, LAND GRANT	BND LD GR	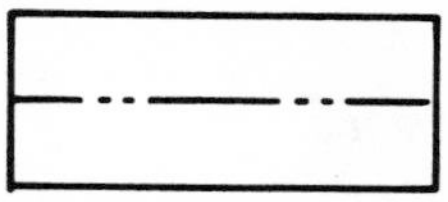
GENERAL LINE LABEL TYPE	GN LN	 oil line	BOUNDRY, U.S. LAND SURVEY TOWNSHIP	BND US LD SUR TWN	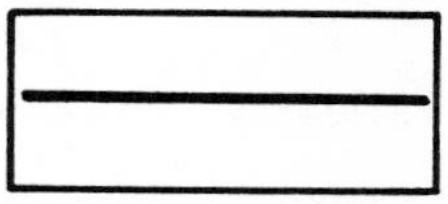
WELL LABEL TYPE	WL	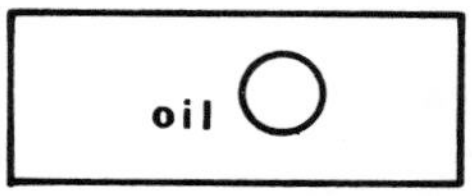 oil ○	BOUNDRY, TOWNSHIP APPROXIMATED	BND TWN	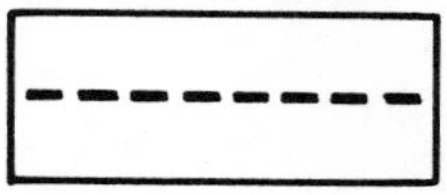
TANK LABEL TYPE	TK	 water ●	BOUNDRY, SECTION LINE U.S. LAND SURVEY	BND SEC LN US LD SUR	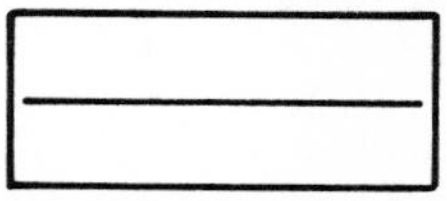
MINING AREA	MIN AR		BOUNDRY, SECTION LINE APPROXIMATED	BND SEC LN	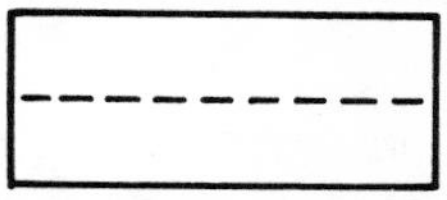
SHAFT	SHF	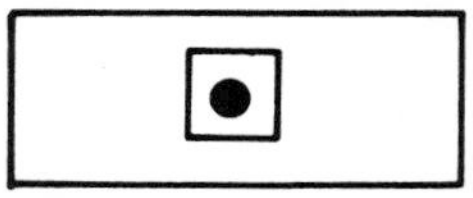	BOUNDRY, TOWNSHIP NOT U.S. LAND SURVEY	BND TWN	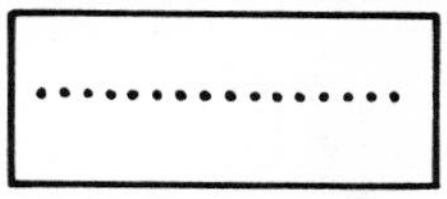
TUNNEL ENTRANCE	TUN ENT		INDICATION CORNER SECTION	COR SEC	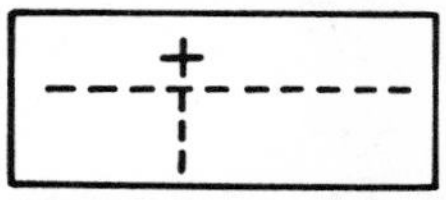
BOUNDRY, STATE	BND ST	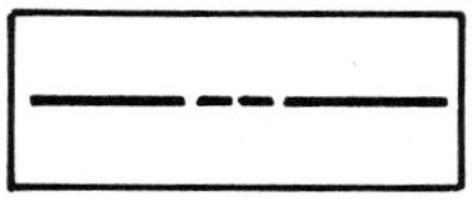	BOUNDRY MONUMENT	BND MON	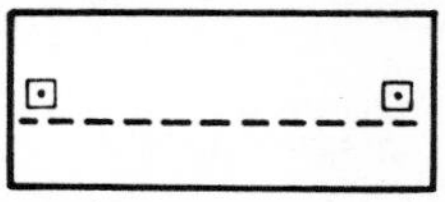
BOUNDRY, COUNTY	BND CNTY	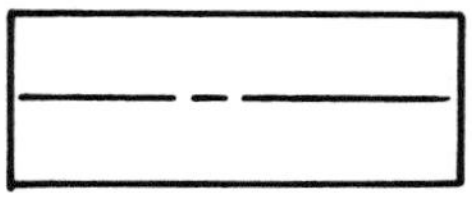	U.S. MINERAL OR LOCATION MONUMENT	U.S. MIN MON	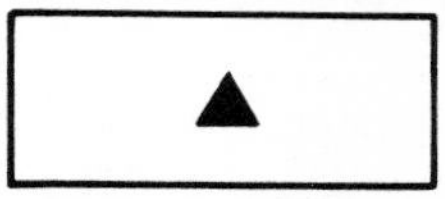 ▲
BOUNDRY, TOWN	BND TWN	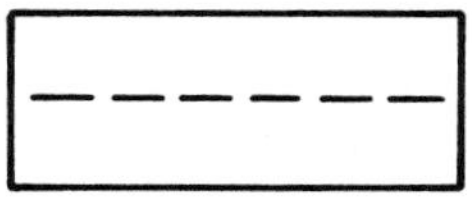	DEPRESSION CONTOURS	DEP CONT	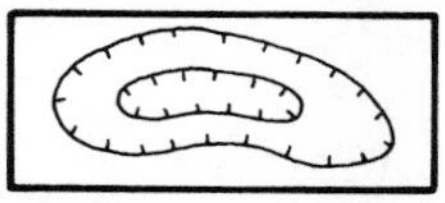
BOUNDRY, CITY INCORPORATED	BND CTY	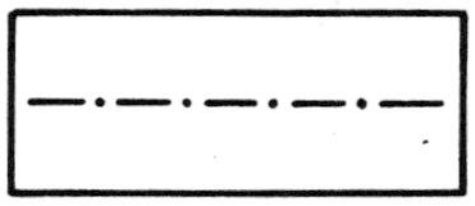	FILL	FL	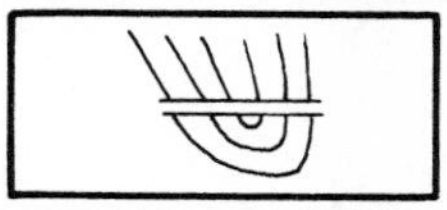
BOUNDRY, NATIONAL OR STATE RESERVATION	BND NAT OR ST RES	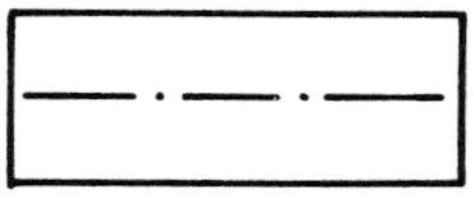	CUT	CT	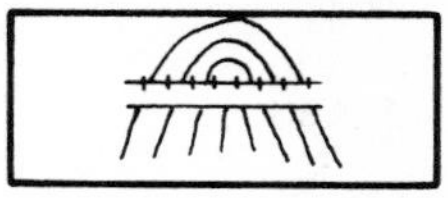
BOUNDRY, SMALL AREAS: PARKS, AIRPORTS, ETC.	BND	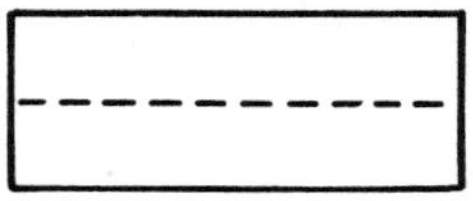	LEVEE	LEV	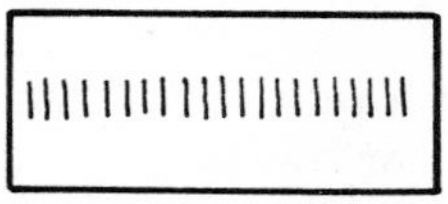

Fig. 10-3. Topographic symbols.

NAME	ABBREV	SYMBOL	NAME	ABBREV	SYMBOL
LEVEE WITH ROAD	LV RD	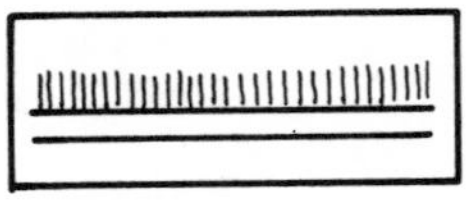	LAKE, INTERMITTEN	LK INT	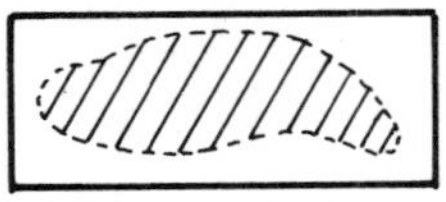
MINE DUMP	MN DP	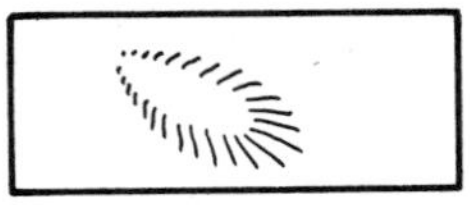	LAKE, DRY	LK DRY	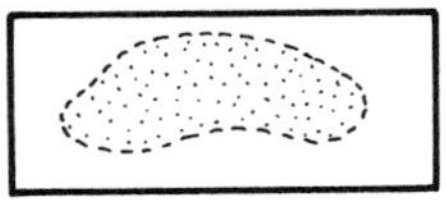
RIVER	RV	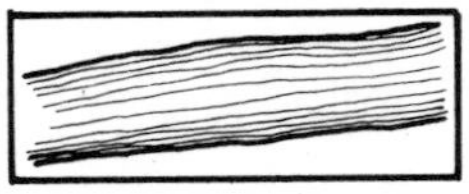	REEF	RF	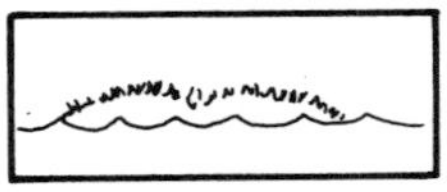
STREAM PRENNIAL	ST PRE	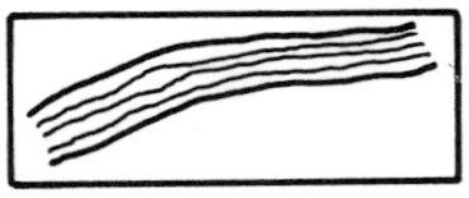	SOUNDING WATER DEPTH CURVE	SND WT DPT CUR	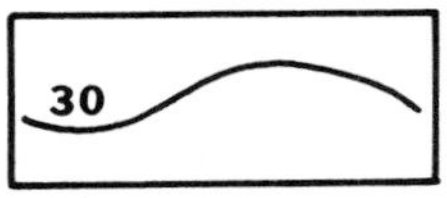 30
STREAM INTERMITTENT	ST INT	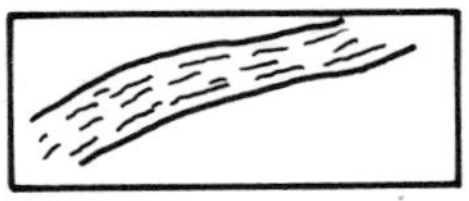	EXPOSED WRECK	EX WRK	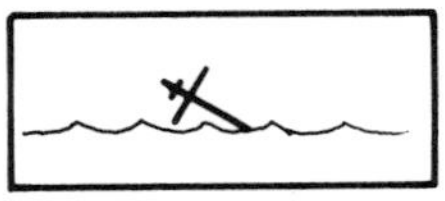
AQUEDUCT ELEVATED	AQ EL	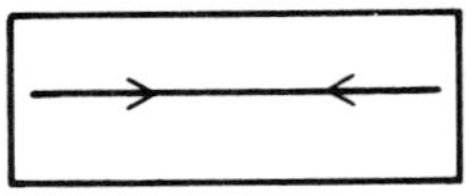	SUNKEN WRECK	SUN WRK	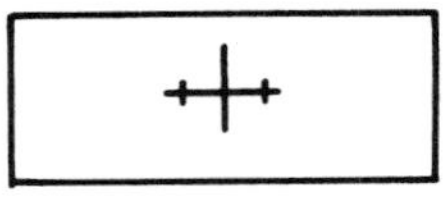
AQUEDUCT TUNNEL	AQ TUN	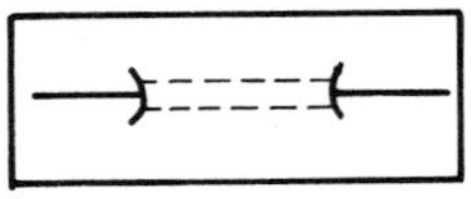	EXPOSED ROCK IN WATER	EX RK WK	
STREAM DISAPPEARING	ST DIS	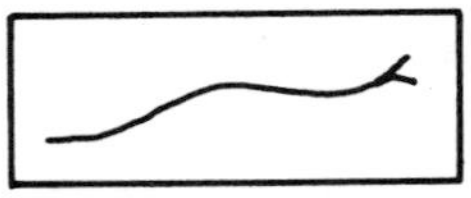	EXPOSED ROCK NAVIGATION DANGER	EX RK NAV DAN	
SMALL RAPIDS	SM RP	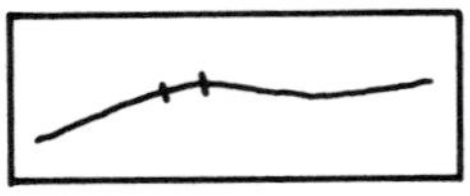	SPRING	SP	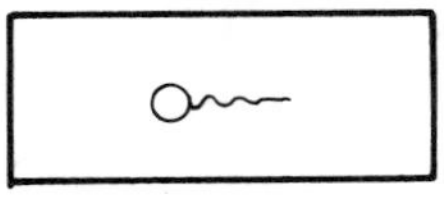
SMALL WATER FALL	SM WT FL	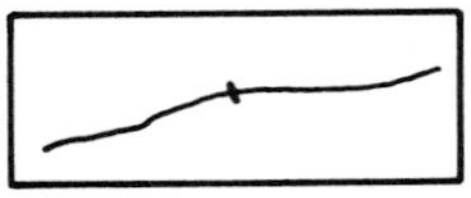	PILINGS	PLG	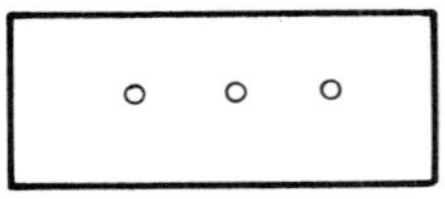
LARGE RAPIDS	LRG RP	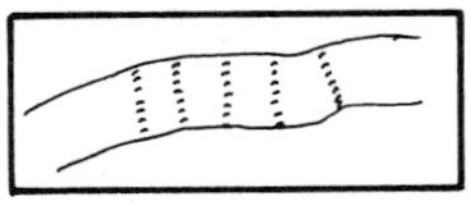	CANAL WITH LOCK	CN LK	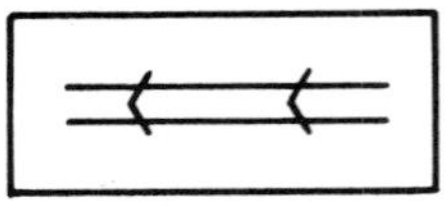
WASH	WSH	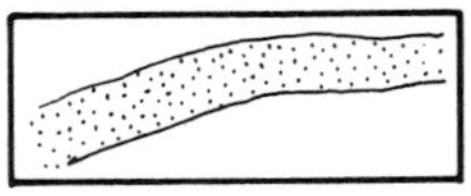	SWAMP	SWP	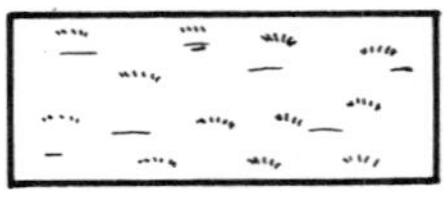
LARGE WATER FALL	LRG WT FL	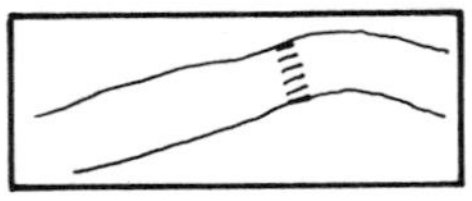	SHORELINE	SH LN	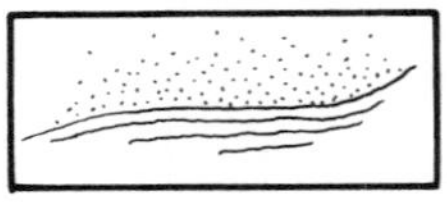

Fig. 10-4. Topographic symbols.

Unit 11
Reference Callout Symbols

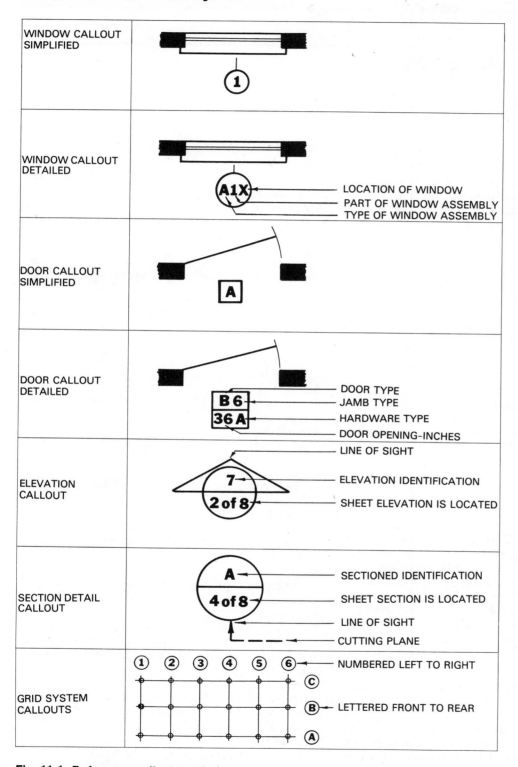

Fig. 11-1. Reference callout symbols.

CHAPTER 3

Using the Written Language

Although the graphic language is the primary tool of communication in the construction field, the written language is also of vital importance as part of this communication system. A drawing without labels, notes, and dimensions lacks real meaning. The written language is needed to describe the function of areas, the size of materials and components, and the kinds of materials used. Without written language, graphic language would be much more complex and many more construction details would be necessary. For example, a construction note about the kind and thickness of flooring on a basement plan eliminates the need for a detail drawing of that portion of the plan. Thus it is important that builders become familiar with the following construction terms and their abbreviations and synonyms. Abbreviations are necessary to fit as much information on a drawing as possible, and synonyms are valuable since some materials are referred to by different names in different parts of the world, and by different builders.

Unit 12

Construction Terms This unit contains definitions of construction terms found on most construction drawings. The same terms may be used differently in other geographic areas. Refer to Unit 14 for synonyms of construction terms. Refer also to Unit 13 for abbreviations of these terms, since abbreviations are more commonly used on most drawings than the full form of the term.

Abstract of title
A summary of all deeds, wills, and legal actions to show ownership.

Acoustics
The science of sound. In housing, acoustical materials used to keep down noise within a room or to prevent it from passing through walls.

Adobe construction
Construction using sun-dried units of adobe soil for walls; usually found in the southwestern United States.

Air conditioner
An apparatus that can heat, cool, clean, and circulate air.

Air-dried lumber
Lumber that is left in the open to dry rather than being dried by a kiln.

Air duct
A pipe usually made of sheet metal that conducts air to rooms from a central source.

Air trap
A U-shaped pipe filled with water and located beneath plumbing fixtures to form a seal against the passage of gases and odors.

Alcove
A recessed space connected at the side of a larger room.

Alteration
A change in, or addition to, an existing building.

Amortization
An installment payment of a loan, usually monthly for a home loan.

Ampere
The unit used in the measure of the rate of flow of electricity.

Anchor bolt
A threaded rod inserted in masonry construction for anchoring the sill plate to the foundation.

Angle iron
A structural piece of rolled steel shaped to form a 90° angle.

Appraisal
The estimated price of a house which a buyer would pay and the seller accept for a property. An appraisal is a detailed evaluation of the property.

Apron
The finish board immediately below a window sill. Also the part of the driveway that leads directly into the garage.

Arcade
A series of arches supported by a row of columns.

Arch
A curved structure that will support itself by mutual pressure and the weight above its curved opening.

Architect
A person who plans and designs buildings and oversees their construction.

Area wall
A wall surrounding an areaway to admit light and air to a basement.

Areaway
A recessed area below grade around the foundation to allow light and ventilation into a basement window or doorway.

Asbestos
A soft, fibrous, fireproof mineral fiber used in fireproofing building materials.

Asbestos board
A fire-resistant sheet made from asbestos fiber and portland cement.

Ashlar
A facing of squared stones.

Ashpit
The area below the hearth of a fireplace which collects the ashes.

Asphalt
Bituminous sandstones used for paving streets and waterproofing asbestos roof and wall coverings.

Asphalt shingles
Composition roof shingles made from asphalt-impregnated felt covered with mineral granules.

Assessed value
A value set by governmental assessors to determine tax assessments.

Atrium
An open court within a building.

Attic
The space between the roof and the ceiling.

Awning window
An out-swinging window hinged at the top.

Backfill
Earth used to fill in areas around exterior foundation walls.

Backhearth
The part of the hearth inside the fireplace.

Baffle
A partial blocking against a flow of wind or sound.

Balcony
A deck projecting from the wall of a building above the ground.

Balloon framing
The building-frame construction in which each of the studs is one piece from the foundation to the roof.

Balustrade
A series of balusters or posts connected by a rail, generally used for porches and balconies.

Banister
A handrail.

Base
The finish of a room at the junction of the walls and floor.

Baseboard
The finish board covering the interior wall where the wall and floor meet.

Base course
The lowest part of masonry construction.

Baseline
A located line for reference-control purposes.

Basement
The lowest story of a building, partially or entirely below ground.

Base plate
A plate, usually of steel, upon which a column rests.

Base shoe
A molding used next to the floor in interior baseboards.

Batt
A blanket insulation material usually made of mineral fibers and designed to be installed between framing members.

Batten
A narrow strip of board used to cover cracks between the boards in board and batten siding.

Batter
Sloping a masonry or concrete wall upward and backward from the perpendicular.

Batter boards
Boards at exact elevations nailed to posts just outside the corners of a proposed building. Strings are stretched across the boards to locate the outline of the foundation.

Bay window
A window projecting out from the wall of a building to form a recess in the room.

Beam
A horizontal structural member that carries a load.

Beam ceiling
A ceiling in which the ceiling beams are exposed to view.

Bearing plate
A plate that provides support for a structural member.

Bearing wall or partition
A wall supporting any vertical load other than its own weight.

Bench mark
A metal or stone marker placed in the ground by a surveyor with the elevation on it. This is the reference point to determine lines, grades, and elevations in the area.

Bending moment
A measure of the forces that break a beam by bending.

Bent
A frame consisting of two supporting columns and a girder or truss used in a vertical position in a structure.

Bevel siding
Shingles or other siding board thicker on one edge than the other. The thick edge overlaps the thin edge of the next board.

Bib
A threaded faucet allowing a hose to be attached.

Bill of material
A parts list of material accompanying a structural drawing.

Blanket insulation
Insulation in a rolled-sheet form and often backed by treated paper which forms a vapor barrier.

Blocking
Small wood framing members that fill in the open space between the floor and ceiling joists to add stiffness to the floors and ceiling.

Blueprint
An architectural drawing used by workers as a guide for building. The original drawing is transferred to a sensitized paper that turns blue with white lines when printed.

Board measure
A system of lumber measurement having as a unit a board foot (bd ft). One board foot is the equivalent of 1 foot square by 1 inch thick.

Brace
Any stiffening member of a framework.

Braced framing
Frame construction with posts and braces used for stiffening. More rigid than balloon framing.

Breezeway
A roofed walkway with open sides. It connects the house and garage. If large enough it can be used as a patio.

Broker
An agent in buying and selling property.

Btu
Abbreviation for British thermal unit, a standard unit for measuring heat gain or loss.

Buck
Frame for a door, usually made of metal, into which the finished door fits.

Building code
A collection of legal requirements for buildings, designed to protect the safety, health, and general welfare of people who work and live in them.

Building line
An imaginary line on a plot beyond which the building cannot extend.

Building paper
A heavy, waterproof paper used over sheathing and subfloors to prevent passage of air and water.

Building permit
A permit issued by a municipal government authorizing the construction of a building or structure.

Built-up beam
A beam constructed of smaller members fastened together.

Built-up roof
A roofing material composed of several layers of felt and asphalt.

Butterfly roof
A roof with two sides sloping down toward the interior of the house.

Butt joint
A joint formed by placing the end of one member against another member.

Buttress
A mass of masonry projecting beyond a wall to take thrust or pressure. A projection from a wall to create additional strength and support.

BX cable
Armored electric cable wrapped in rubber and protected by a flexible steel covering.

Cabinet work
The finish interior woodwork.

Canopy
A projection over windows and doors to protect them from the weather.

Cantilever
A projecting member supported only at one end.

Cant strip
An angular board used to eliminate a sharp right angle on roofs or flashing.

Carport
An automobile shelter not fully enclosed.

Carriage
The horizontal part of the stringers of a stair that supports the treads.

Casement window
A hinged window that opens out, usually made of metal.

Casing
A metal or wooden member around door and window openings to give a finished appearance.

Catch basin
An underground structure for drainage into which the water from a roof or floor will drain. It is connected to a sewer or drain.

Caulking
A waterproof material used to seal cracks.

Cavity wall
A hollow wall usually made up of two brick walls built a few inches apart and joined together with brick or metal ties.

Cedar shingles
Roofing and siding shingles made from western red cedar.

Cement
A masonry adhesive material purchased in the form of pulverized powder. Any substance used in its soft state to join other materials together and which afterward dries and hardens.

Central heating
A single source of heat which is distributed by pipes or ducts.

Certificate of title
A document given to the home buyer with the deed, stating that the title to the property named in the deed is clearly established.

Cesspool

A pit or cistern to hold sewage.

Chalk line

A string that is heavily chalked, held tight, then plucked to make a straight guideline against boards or other surfaces.

Chase

A vertical space within a building for ducts, pipes, or wires.

Checks

Splits or cracks in a board, ordinarily caused by seasoning.

Check valve

A valve that permits passage through a pipe in only one direction.

Chimney

A vertical flue for passing smoke and gases outside a building.

Chimney stack

A group of flues in the same chimney.

Chord

The principal members of a roof or bridge truss. The upper members are indicated by the term "upper chord." The lower members are identified by the term "lower chord."

Cinder block

A building block made of cement and cinder.

Circuit

The path of an electric current. The closed loop of wire in which an electric current can flow.

Circuit breaker

A device used to open and close an electrical circuit.

Cistern

A tank or other reservoir to store rain which has run off the roof.

Clapboard

A board, thicker on one side than the other, used to overlap an adjacent board to make house siding.

Clearance

A clear space to allow passage.

Clerestory

A set of high windows often placed above a roof line.

Clinch

To bend over the protruding end of a nail.

Clip

A small connecting angle used for fastening various members of a structure.

Collar beam

A horizontal member fastening opposing rafters below the ridge in roof framing.

Column

In architecture, a perpendicular supporting member, circular in section. In engineering, a vertical structural member supporting loads acting on or near and in the direction of its longitudinal axis.

Common wall

A wall that serves two dwelling units.

Compression

A force which tends to make a member fail because of crushing.

Concrete block

Precast hollow or solid blocks of concrete.

Condemn

To legally declare unfit for use.

Condensation

The formation of frost or drops of water on inside walls when warm vapor inside a room meets a cold wall or window.

Conductor

In architecture, a drain pipe leading from the roof. In electricity, anything that permits the passage of an electric current.

Conductor pipe

A round, square, or rectangular metal pipe used to lead water from the roof to the sewer.

Conduit

A channel built to convey water or other fluids. A drain or sewer. In electrical work, a channel that carries wires for protection and for safety.

Construction loan

A mortgage loan to be used to pay for labor and materials going into the house. Money is usually advanced to the builder as construction progresses and is repaid when the house is completed and sold.

Continuous beam

A beam that has three or more supports.

Contractor

A person offering to build for a specified sum of money.

Convector

A heat-transfer surface which uses convection currents to transfer heat.

Coping

The top course of a masonry wall which projects to protect the wall from the weather.

Corbel

A projection in a masonry wall made by setting courses beyond the lower ones.

Corner bead

A metal molding built into plaster corners to prevent the accidental breaking off of the plaster.

Cornice

The part of a roof that projects out from the wall.

Counterflashing

A flashing used under the regular flashing.

Course

A continuous row of stone or brick of uniform height.

Court

An open space surrounded partly or entirely by a building.

Crawl space

The shallow space below the floor of a house built above the ground. It is surrounded by the foundation wall.

Cricket
A roof device used at intersections to divert water.

Cripple
A structural member that is cut less than full length, such as a studding piece above a window or door.

Cross bracing
Boards nailed diagonally across studs or other boards to make framework rigid.

Cross bridging
Bracing between floor joists to add stiffness to the floors.

Crosshatch
Lines drawn closely together at an angle of 45°, to show a sectional cut.

Cull
Building material rejected as below standard grade.

Culvert
A passage for water below ground level.

Cupola
A small structure built on top of a roof.

Curb
A very low wall.

Cure
To allow concrete to dry slowly by keeping it moist to allow maximum strength.

Curtain wall
An exterior wall which provides no structural support.

Damp course
A layer of waterproof material.

Damper
A movable plate which regulates the draft of a stove, fireplace, or furnace.

Datum
A reference point of starting elevations used in mapping and surveying.

Deadening
Construction intended to prevent the passage of sound.

Dead load
All the weight in a structure made up of unmovable materials. *See also* Loads.

Decay
The disintegration of wood through the action of fungi.

Dehumidify
To reduce the moisture content in the air.

Density
The number of people living in a calculated area of land such as 1 sq mi or 1 km².

Depreciation
Loss of value.

Designer
A person who designs houses but is not a registered architect.

Detail
To provide specific instruction with a drawing: dimensions, notes, or specifications.

Dimension building material
Building material which has been precut to specific sizes.

Dimension line
A line with arrowheads at either end to show the distance between two points.

Dome
A hemispherical roof form.

Doorstop
The strips on the doorjambs against which the door closes.

Dormer
A structure projecting from a sloping roof to accommodate a window.

Double glazing
A pane of two pieces of glass with air space between and sealed to provide insulation.

Double header
Two or more timbers joined for strength.

Double hung
A window having top and bottom sashes each capable of movement up and down.

Downspout
A pipe for carrying rainwater from the roof to the ground.

Drain
A pipe for carrying waste water.

Dressed lumber
Lumber machined and smoothed at the mill. Usually 1/2 in. less than nominal (rough) size.

Drip
A projecting construction member or groove below the member to prevent rainwater from running down the face of a wall or to protect the bottom of a door or window from leakage.

Dry rot
A term applied to many types of decay, especially an advanced stage when the wood can be easily crushed to a dry powder. The term is actually inaccurate because all fungi require considerable moisture for growth.

Dry-wall construction
Interior wall covering other than plaster, usually referred to as gypsum-board surfacing.

Dry well
A pit located on porous ground walled up with rock which allows water to seep through the pit. Used for the disposal of rainwater or the effluent from a septic tank.

Ducts
Sheet-metal conductors for warm- and cold-air distribution.

Easement
The right to use land owned by another, such as a utility company's right-of-way.

Eave
That part of a roof that projects over a wall.

Efflorescence
Whitish powder that forms on the surface of bricks or stone walls due to evaporation of moisture-containing salts.

Effluent
The liquid discharge from a septic tank after bacterial treatment.

Elastic limit
The limit to which a material may be bent or pulled out of shape and still return to its former shape and dimensions.

Elbow
An L-shaped pipe fitting.

Elevation
The drawings of the front, side, or rear face of a building.

Ell
An extension or wing of a building at right angles to the main section.

Embellish
To add decoration.

Eminent domain
The right of the local government to condemn for public use.

Enamel
Paint with a considerable amount of varnish. It produces a hard, glossy surface.

Equity
The interest in or value of real estate the owner has in excess of the mortgage indebtedness.

Escutcheon
The hardware on a door to accommodate the knob and keyhole.

Excavation
A cavity or pit produced by digging the earth in preparation for construction.

Fabrication
Work done on parts of a structure at the factory before delivery to the building site.

Facade
The face or front elevation of a building.

Face brick
A brick used on the outside face of a wall.

Facing
A finish material used to cover another.

Fascia
A vertical board nailed on the ends of the rafters. It is part of the cornice.

Fatigue
A weakening of structural members.

Federal Housing Administration (FHA)
A government agency that ensures loans made by regular lending institutions.

Felt papers
Papers, sometimes tar-impregnated, used on roofs and side walls to give protection against dampness and leaks.

Fenestration
The arrangement of windows.

Fiberboard
A building board made with fibrous material—used as an insulating board.

Filled insulation
A loose insulating material poured from bags or blown by machines into walls.

Finish lumber
Dressed wood used for building trim.

Firebrick
A brick that is especially hard and heat-resistant. Used in fireplaces.

Fireclay
A grade of clay that can withstand a large quantity of heat. Used for firebrick.

Fire cut
The angular cut at the end of a joist designed to rest on a brick wall.

Fire door
A door that will resist fire.

Fire partition
A partition designed to restrict the spread of fire.

Fire stop
Obstruction across air passages in buildings to prevent the spread of hot gases and flames. A horizontal blocking between wall studs.

Fished
A splice strengthened by metal pieces on the sides.

Fixed light
A permanently sealed window.

Fixture
A piece of electrical or plumbing equipment.

Flagging
Cut stone, slate, or marble used on floors.

Flagstone
Flat stone used for floors, steps, walks, or walls.

Flashing
The material used for, and the process of, making watertight the roof intersections and other exposed places on the outside of the house.

Flat roof
A roof with just enough pitch to let water drain.

Flitch beam
A built-up beam formed by a metal plate sandwiched between two wood members and bolted together for additional strength.

Floating
Spreading plaster, stucco, or cement on walls with use of a tool called a float.

Floor plan
The top view of a building at a specified floor level. A floor plan includes all vertical details at or above windowsill levels.

Floor plug
An electrical outlet flush with the floor.

Flue
The opening in a chimney through which smoke passes.

Flue lining
Terra-cotta pipe used for the inner lining of chimneys.

Flush surface
A continuous surface without an angle.

Footing
An enlargement at the lower end of a wall, pier, or column, to distribute the load into the ground.

Footing form
A wooden or steel structure placed around the footing that will hold the concrete to the desired shape and size.

Framing (western)
The wood skeleton of a building.

Frieze
The flat board of cornice trim which is fastened to the wall.

Frost line
The depth of frost penetration into the soil.

Fumigate
To destroy harmful insect or animal life with fumes.

Furring
Narrow strips of board nailed upon the wall and ceilings to form a straight surface for the purpose of attaching some wallboards or ceiling tile.

Fuse
A strip of soft metal inserted in an electric circuit and designed to melt and open the circuit should the current exceed a predetermined value.

Gable
The triangular end of an exterior wall above the eaves.

Gable roof
A roof which slopes from two sides only.

Galvanize
A lead and zinc bath treatment to prevent rusting.

Gambrel roof
A symmetrical roof with two different pitches or slopes on each side.

Garret
An attic.

Girder
A horizontal beam supporting the floor joists.

Glazing
Placing of glass in windows or doors.

Grade
The level of the ground around a building.

Gradient
The slant of a rod, piping, or the ground, expressed in percent.

Graphic symbols
Symbolic representations used in drawing which simplify presentations of complicated items.

Gravel stop
A strip of metal with a vertical lip used to retain the gravel around the edge of a built-in roof.

Green lumber
Lumber that still contains moisture or sap.

Grout
A thin cement mortar used for leveling and filling masonry holes.

Gusset
A plywood or metal plate used to strengthen the joints of a truss.

Gutter
A trough for carrying off water.

Gypsum board
A board made of plaster with a covering of paper.

Half timber
A frame construction of heavy timbers in which the spaces are filled in with masonry.

Hanger
An iron strap used to support a joist beam or pipe.

Hardpan
A compacted layer of soils.

Head
The upper frame on a door or window.

Header
The horizontal supporting member above openings, such as a lintel. Also, one or more pieces of lumber supporting ends of joists. Used in framing openings of stairs and chimneys.

Headroom
The clear space between floor line and ceiling, as in a stairway.

Hearth
That part of the floor directly in front of the fireplace, and the floor inside the fireplace on which the fire is built. It is made of fire-resistant masonry.

Heel plate
A plate at the ends of a truss.

Hip rafter
The diagonal rafter that extends from the plate to the ridge to form the hip.

Hip roof
A roof with four sloping sides.

House drain
Horizontal sewer piping within a building which receives wastes from the soil stacks.

House sewer
The watertight soil pipe extending from the exterior of the foundation wall to the public sewer.

Humidifier
A mechanical device which controls the amount of water vapor to be added to the atmosphere.

Humidistat
An instrument used for measuring and controlling moisture in the air.

I beam
A steel beam with an I-shaped cross section.

Indirect lighting
Artificial light that is bounced off ceiling and walls for general lighting.

Insulating board
Any board suitable for insulating purposes. Usually manufactured board made from vegetable fibers, such as fiber board.

Insulation
Materials for obstructing the passage of sound, heat, or cold from one surface to another.

Interior trim
General term for all the finish molding, casing, baseboard, etc.

Jack rafter
A short rafter, which is usually used on hip roofs.

Jalousie
A type of window consisting of a number of long, thin, hinged panels.

Jamb
The sides of a doorway or window opening.

Jerry built
Poor construction.

Joints
The meeting of two separate pieces of material for a common bond.

Joist
A horizontal structural member which supports the floor system or ceiling system.

Kalamein door
A fireproof door with a metal covering.

Keystone
The top, wedge-shaped stone of an arch.

Kiln
A heating chamber for drying lumber.

King post
In a roof truss, the central upright piece.

Knee brace
A corner brace, fastened at an angle from wall stud to rafter, stiffening a wood or steel frame to prevent angular movement.

Knee wall
Low wall resulting from 1½-story construction.

Knob and tube
Electric wiring through walls where insulated wires are supported with porcelain knobs and tubes when passing through wood construction members.

Lally column
A steel column used as a support for girders and beams.

Laminated beam
A beam made by bonding together several layers of material.

Landing
A platform in a flight of steps.

Landscape architect
A professional person who prepares designs for the aesthetic, functional, and ecological use of terrain.

Lap joint
A joint produced by lapping two pieces of material.

Lath (metal)
Sheet-metal screening used as a base for plastering.

Lath (wood)
A wooden strip nailed to studding and joists to which plaster is applied.

Lattice
A grille or open work made by crossing strips of wood or metal.

Lavatory
A washbasin or a room equipped with a washbasin.

Leaching bed field
A system of trenches that carries wastes from sewers. It is constructed in sandy soils or in earth filled with stones or gravel.

Leader
A vertical pipe or downspout that carries rainwater from the gutter to the ground.

Lean-to
A shed whose rafters lean against another building or other part of the same building.

Ledger
A wood strip nailed to the lower side of a girder to provide a bearing surface for joists.

Lessee
The one who leases.

Lessor
The owner of leased property.

Lien
A legal claim on a property, which may be exercised in default of payment of a debt.

Linear foot
A measurement of 1 ft along a straight line.

Lintel
A horizontal piece of wood, stone, or steel across the top of door and window openings to bear the weight of the walls above the opening.

Loads
Live load: the total of all moving and variable loads that may be placed upon a building. Dead load: the weight of all permanent, stationary construction, including a building.

Load-bearing walls
Walls that support weight from above as well as their own weight.

Loggia
A roofed, open passage along the front or side of a building. It is often at an upper level, and it often has a series of columns on either or both sides.

Lookout
A horizontal framing member extending from studs out to end of rafters.

Lot line
The line forming the legal boundary of a piece of property.

Louver
A set of fixed or movable slats adjusted to provide both shelter and ventilation.

Mansard roof
A roof with two slopes on each side, with the lower slope much steeper than the upper.

Mantel
A shelf over a fireplace.

Market price
What property can be sold for at a given time.

Market value
What property is worth at a given time.

Masonry
Anything built with stone, brick, tiles, or concrete.

Meeting rail
The horizontal rails of a double-hung sash that fit together when the window is closed.

Member
A single piece in a structure that is complete in itself.

Metal tie
A strip of metal used to fasten construction members together.

Metal wall ties
Strips of corrugated metal used to tie a brick-veneer wall to framework.

Mildew
A mold on wood caused by fungi.

Millwork
The finish woodwork in a building, such as cabinets and trim.

Mineral wool
An insulating material made into a fibrous form from mineral slag.

Modular construction
Construction in which the size of the building and the building materials are based on a common unit of measure.

Moisture barrier
A material such as specially treated paper that retards the passage of vapor or moisture into walls and prevents condensation within the walls.

Monolithic
Concrete construction poured and cast in one piece without joints.

Monument
A boundary marker set by surveyors to locate property lines.

Mortar
A mixture of cement, sand, and water, used by the mason to bind bricks and stone.

Mortgage
A pledging of property, conditional on payment of the debt in full.

Mortgagee
The lender of money to the mortgagor.

Mortgagor
The owner who mortages property in return for a loan.

Mosaic
Small colored tile, glass, stone, or similar material arranged on an adhesive ground to produce a decorative surface.

Mud room
A small room or entranceway where muddy overshoes and wet garments can be removed before entering other rooms.

Mullion
A vertical bar in a window separating two windows.

Muntin
A small bar separating the glass lights in a window.

Newel
A post supporting the handraft at the top or bottom of a stairway.

Nominal dimension
Dimensions for finished lumber in which the stated dimension is usually larger than the actual dimension. These dimensions are usually larger by an amount required to smooth a board.

Nonbearing wall
A dividing wall that does not support a vertical load other than its own weight.

Nonferrous metal
Metal containing no iron, such as copper, brass, or aluminum.

Nosing
The rounded edge of a stair tread.

Obscure glass
Sheet glass that is made translucent instead of transparent.

On center
Measurement from the center of one member to the center of another (noted "oc").

Open-end mortgage
A mortgage that permits the remaining amount of the loan to be increased, as for improvements, by mutual agreement of the lender and borrower, without rewriting the mortgage.

Orientation
The positioning of a house on a lot in relation to the sun, wind, view, and noise.

Outlet
Any kind of electrical box allowing current to be drawn from the electrical system for lighting or appliances.

Overhang
The horizontal distance that a roof projects beyond a wall.

Panelboard
The center for controlling electric circuits.

Parapet
A low wall or railing around the edge of a roof.

Parging
A thin coat of plaster applied to masonry surfaces for smoothing purposes.

Parquet flooring
Flooring, usually of wood, laid in an alternating or inlaid pattern to form various designs.

Partition
An interior wall that separates two rooms.

Party wall
A wall between two adjoining buildings which both owners share, such as a common wall between row houses.

Patio
An open court.

Pediment

The triangular space forming the gable end of a low-pitched roof. A similar form is often used as a decoration over doors in classic architecture.

Penny

A term for the length of a nail, abbreviated "d".

Periphery

The entire outside edge of an object.

Perspective

A drawing of an object in a three-dimensional form on a plane surface. An object drawn as it would appear to the eye.

Pier

A block of concrete supporting the floor of a building.

Pilaster

A portion of a square column, usually set within or against a wall for the purpose of strengthening the wall. Also, a decorative column attached to a wall.

Piles

Long posts driven into the soil in swampy locations, or whenever it is difficult to secure a firm foundation, upon which the foundation footing is laid.

Pillar

A column used for supporting parts of a structure.

Pinnacle

Projecting or ornamental cap on the high point of a roof.

Plan

A horizontal, graphic representational section of a building, showing the walls, doors, windows, stairs, chimneys, and surrounding objects, such as walks and landscape.

Planks

Material 2 or 3 in. (50 or 75 mm) thick and more than 4 in. (100 mm) wide, such as joists, flooring, and the like.

Plaster

A mortarlike composition used for covering walls and ceilings. Usually made of portland cement mixed with sand and water.

Plasterboard

A board made of plastering material which is covered on both sides with heavy paper. It is often used instead of plaster. Also called gypsum board.

Plaster ground

A nailer strip included in plaster walls to act as a gage for thickness of plaster and to give a nailing support for finish trim around openings and near the base of the wall.

Plat

A map or chart of an area showing boundaries of lots and other pieces of property.

Plate

The top horizontal member of a row of studs in a frame wall to carry the trusses of a roof or to carry the rafters directly. Also, a shoe or base member, as of a partition or other frame.

Plate cut

The cut in a rafter which rests upon the plate. It is also called the seat cut or birdmouth.

Plate glass

A high-quality sheet of glass used in large windows.

Plenum system

A system of heating or air conditioning in which the air is forced through a chamber connected to distributing ducts.

Plot

The land on which a building stands.

Plow

To cut a groove running in the same direction as the grain of the wood.

Plumb

Said of an object when it is in true vertical position as determined by a plumb bob.

Plywood

A piece of wood made of three or more layers of veneer joined with glue and usually laid with the grain of adjoining plies at right angles.

Porch

A covered area attached to a house at an entrance.

Portico

A roof supported by columns, whether attached to a building or wholly by itself.

Portland cement

A hydraulic cement, extremely hard, formed by burning silica, lime, and alumina together and then grinding up the mixture.

Post

A perpendicular supporting member.

Post and beam construction

Wall construction consisting of posts rather than studs.

Precast

Concrete shapes made separately before being used in a structure.

Prefabricated houses

Houses that are built in sections or component parts in a factory, and then assembled at the site.

Primary coat

The first coat of paint.

Principal

The original amount of money loaned.

Purlin

A structural member spanning from truss to truss and supporting the rafters.

Quad

An enclosed court.

Quarry tile

A machine-made, unglazed tile.

Quoins

Large squared stones set in the corners of a masonry building for appearance.

Radiant heating

A system using heating elements in the floors, ceilings, or walls to radiate heat into the room.

Rafters
Structural members used to frame a roof. Several types are common: hip, jack, valley, and cripple.

Raglin
The open joint in masonry to receive flashing.

Realtor
A real estate broker who is a member of a local chapter of the National Association of Real Estate Boards.

Register
The open end of a duct in a room for warm or cool air.

Reinforced concrete
Concrete in which steel bars or webbing has been embedded for strength.

Rendering
The art of shading or coloring a drawing.

Restoration
Rebuilding a structure so it will appear in its original form.

Restrictions
Limitations on the use of real estate as set by law or contained in a deed.

Retaining wall
A wall to hold back an earth embankment.

Rheostat
An instrument for regulating electric current.

Ribbon
A support for joists. A board set into studs that are cut to support joists.

Ridge
The top edge of the roof where two slopes meet.

Ridge cap
A wood or metal cap used over roofing at the ridge.

Riprap
Stones placed on a slope to prevent erosion. Also broken stone used for foundation fill.

Rise
The vertical height of a roof.

Riser
The vertical board in a stairway between two treads.

Rock wool
An insulating material that looks like wool but is composed of such substances as granite or silica.

Rodding
Stirring freshly poured concrete with a vibrator to remove air pockets.

Roll roofing
Roofing material of fiber and asphalt.

Rough floor
The subfloor on which the finished floor is laid.

Rough hardware
All the hardware used in a house, such as nails and bolts, that cannot be seen in the completed house.

Roughing in
Putting up the skeleton of the building.

Rough lumber
Lumber as it comes from the saw.

Rough opening
Any unfinished opening in the framing of a building.

Run
Stonework having irregular-shaped units and no indication of systematic course work. The horizontal distance covered by a flight of stairs. The length of a rafter.

Saddle
The ridge covering of a roof designed to carry water from the backs of chimneys. Also called a cricket or a threshold.

Safety factor
The ultimate strength of the material divided by the allowable working load. The element of safety needed to make certain that there will be no structural failures.

Sand finish
A final plaster coat. A skim coat.

Sap
All the fluids in a tree.

Sash
The movable framework in which window panes are set.

Scab
A small wood member used to join other members which is fastened on the outside face.

Scarfing
A joint between two pieces of wood which allows them to be spliced lengthwise.

Schedule
A list of parts or details.

Scratch coat
The first coat of plaster. It is scratched to provide a good bond for the next coat.

Screed
A guide for the correct thickness of plaster or concrete being placed on surfaces.

Scuttle
A small opening in a ceiling to provide access to an attic or roof.

Seasoning
Drying out of green lumber, either in an oven or kiln or by exposing it to air.

Second mortgage
A mortgage made by a home buyer to raise money for a down payment required under the first mortgage.

Section
The drawing of an object that is cut to show the interior. Also, a panel construction used in walls, floors, ceilings, or roofs.

Seepage pit
A pit or cesspool into which sewage drains from a septic tank, and which is so constructed that the liquid waste seeps through the sides of the pit into the ground.

Septic tank
A concrete or steel tank where sewage is reduced partially by bacterial action. About half the sewage solids become gases which escape back through the vent stack in the house. The

other solids and liquids flow from the tank into the ground through a tile bed.

Service connection
The electrical wires to the building from the outside power lines.

Set
The hardening of cement or plaster.

Set back
A zoning restriction on the location of the home on a lot.

Settlement
Compression of the soil or the members in a structure.

Shakes
Thick hand-cut shingles.

Sheathing
The structural covering of boards or wallboards, placed over exterior studding or rafters of a structure.

Sheathing paper
A paper barrier against wind and moisture applied between sheathing and outer wall covering.

Shed roof
A flat roof slanting in one direction.

Shim
A piece of material used to level or fill in the space between two surfaces.

Shingles
Thin pieces of wood or other materials which overlap each other in covering a roof. The number and kind needed depend on the steepness of the roof slope and other factors. Kinds of shingles include tile shingles, slate shingles, asbestos-cement shingles, and asphalt shingles.

Shiplap
Boards with lapped joints along their edges.

Shoe mold
The small mold against the baseboard at the floor.

Shoring
Lumber placed in a slanted position to support the structure of a building temporarily.

Siding
The outside boards of an exterior wall.

Sill
The horizontal exterior member below a window or door opening. Also the wood member placed directly on top of the foundation wall in wood-frame construction.

Skeleton construction
Construction where the frame carries all the weight.

Skylight
An opening in the roof for admitting light.

Slab foundation
A reinforced concrete floor and foundation system.

Sleepers
Strips of wood, usually 2 × 2's, laid over a slab floor to which finished wood flooring is nailed.

Smoke chamber
The portion of a chimney flue located directly over the fireplace.

Soffit
The undersurface of a projecting structure.

Softwood
Wood from trees having needles rather than broad leaves. The term does not necessarily refer to the softness of the wood.

Soil stack
The main vertical pipe which receives waste from all fixtures.

Solar heat
Heat from the sun's rays.

Sole
The horizontal framing member directly under the studs.

Spacing
The distance between structural members.

Spackle
To cover wallboard joints with plaster.

Span
The distance between structural supports.

Specification
The written or printed directions regarding the details of a building or other construction.

Spike
A large, heavy nail.

Splice
Joining of two similar members in a straight line.

Stack
A vertical pipe.

Stakeout
Marking the foundation layout with stakes.

Steel framing
Skeleton framing with structural steel beams.

Steening
Brickwork without mortar.

Stile
A vertical member of a door, window, or panel.

Stirrup
A metal, U-shaped strap used to support framing members.

Stock
Common sizes of building materials and equipment sold at most commercial industries.

Stool
An inside windowsill.

Stop
A small strip to hold a door or window sash in place.

Storm door or window
An extra door or extra window placed outside an ordinary door or window for added protection against cold.

Storm sewer
A sewer that is designed to carry away water from storms, but not sewage.

Stress
Any force acting upon a part or member used in construction.

Stress-cover construction
Construction consisting of panels of sections with wood frameworks to which plywood or other sheet material is bonded with glue in order that the covering carries a large part of the loads.

Stretcher course
A row of masonry in a wall with the long side of the units exposed to the exterior.

Stringer
The sides of a flight of stairs. The supporting member cut to receive the treads and risers.

Stripping
Removing of concrete forms from the hardened concrete.

Stucco
Any of various plasters used for covering walls, especially an exterior wall covering in which cement is used.

Studs
Upright beams in the framework of a building. Usually referred to as 2 × 4's, and spaced at 16 in. from center to center.

Subfloor
The rough flooring under the finish floor that rests on the floor joists.

Sump
A pit in a basement floor to collect water, into which a sump pump is placed to remove the water through sewer pipes.

Surfaced lumber
Lumber that is dressed by running it through a planer.

Surveyor
A person skilled in land measurement.

Swale
A drainage channel formed where two slopes meet.

Tamp
To ram and concentrate soil.

Tar
A dark, heavy oil used in roofing and roof surfacing.

Tempered
Thoroughly mixed cement or mortar.

Tensile strength
The greatest stretching stress a structural member can bear without breaking or cracking.

Termite shield
Sheet metal used to block the passage of termites.

Thermal conductor
A substance capable of transmitting heat.

Threshold
The beveled piece of stone, wood, or metal over which the door swings. It is sometimes called a carpet strip or a saddle.

Throat
A passage directly above the fireplace opening where a damper is set.

Tie
A structural member used to bind others together.

Timber
Lumber with a cross section larger than 4 × 6 in. (100 × 150 mm), for posts, sills, and girders.

Title insurance
An agreement to pay the buyer for losses in title of ownership.

Toe nail
To drive nails at an angle.

Tolerance
The acceptable variance of dimensions from a standard size.

Tongue
A projection on the edge of wood that joins with a similarly shaped groove.

Total run
The total of all the tread widths in a stair.

Tread
The step or horizontal member of a stair.

Trimmers
Single or double joists or rafters that run around an opening in framing construction.

Truss
A triangular-shaped unit for supporting roof loads over long spans.

Underpinning
A foundation replacement or reinforcement for temporary braced supports.

Undressed lumber
Lumber that is not squared or finished smooth.

Unit construction
Construction which includes two or more preassembled walls, together with floor and ceiling construction, for shipment to the building site.

Valley
The internal angle formed by the two slopes of a roof.

Valley jacks
Rafters that run from a ridge board to a valley rafter.

Valley rafter
The diagonal rafter forming the intersection of two sloping roofs.

Valve
Device regulating the flow of material in a pipe.

Vapor barrier
A watertight material used to prevent the passage of moisture or water vapor into and through walls.

Veneer
A thin covering of valuable material over a less expensive material.

Vent
A screened opening for ventilation.

Ventilation
Supplying and removing air by natural or mechanical means to or from any space.

Vent pipes
Small ventilating pipes extending from each fixture of a plumbing system to the vent stack.

Vent stack
The upper portion of a soil or waste stack above the highest fixture.

Vergeboard
The board which serves as the eaves finish on the gable end of a building.

Vestibule
A small lobby or entrance room.

Vitreous
Pertaining to a composition of materials that resemble glass.

Volume
The amount of space occupied by an object. Measured in cubic units.

Wainscot
Facing for the lower part of an interior wall.

Wallboard
Wood pulp, gypsum, or similar materials made into large, rigid sheets that may be fastened to the frame of a building to provide a surface finish.

Warp
Any change from a true or plane surface. Warping includes bow, crook, cup, and twist.

Warranty deed
A guarantee that the property is as promised.

Wash
The slant upon a sill, capping, etc., to allow the water to run off.

Waste stack
A vertical pipe in a plumbing system which carries the discharge from any fixture.

Waterproof
Material or construction which prevents the passage of water.

Water table
A projecting mold near the base on the outside of a building to turn the rainwater outward. Also the level of subterranean water.

Watt
A unit of electrical energy.

Weathering
The mechanical or chemical disintegration and discoloration of the surface of exterior building materials.

Weatherstrip
A strip of metal or fabric fastened along the edges of windows and doors to reduce drafts and heat loss.

Weep hole
An opening at the bottom of a wall to allow the drainage of water.

Well opening
A floor opening for a stairway.

Zoning
Building restrictions as to size, location, and type of structures to be built in specific areas.

Unit 13
Architectural Abbreviations

Since space on most construction drawings is extremely limited, abbreviations are used more often than actual terms. Therefore, understanding the following list of abbreviations is necessary for the accurate and consistent interpretation of construction drawings. In this list most abbreviations are written in capital letters; and in cases where an abbreviation may be confused with a whole word (e.g., COMB.), a period is added. Observe also that the same abbreviation is used for both singular and plural forms and that different abbreviations are very similar in form and must be distinguished from one another.

Access panel	AP
Acoustic	ACST
Actual	ACT.
Addition	ADD.
Adhesive	ADH
Aggregate	AGGR
Air condition	AIR COND
Alternating current	AC
Aluminum	AL
Ampere	AMP
Anchor bolt	AB
Apartment	APT.
Approved	APPD
Approximate	APPROX
Architectural	ARCH
Area	A
Asbestos	ASB
Asphalt	ASPH
At	@
Automatic	AUTO
Avenue	AVE
Average	AVG
Balcony	BALC
Basement	BSMT
Bathroom	B
Bathtub	BT
Beam	BM
Bearing	BRG
Bedroom	BR
Bench mark	BM
Between	BET.
Blocking	BLKG
Blower	BLO
Blueprint	BP
Board	BD
Boiler	BLR
Both sides	BS
Brick	BRK
British thermal units	Btu
Bronze	BRZ
Broom closet	BC
Building	BLDG
Building line	BL

Cabinet	CAB.	Equipment	EQUIP.
Caulking	CLKG	Estimate	EST
Cast concrete	C CONC	Excavate	EXC
Cast iron	CI	Existing	EXIST.
Catalog	CAT.	Exterior	EXT
Ceiling	CLG	**Fabricate**	FAB
Cement	CEM	Feet	(') FT
Center	CTR	Feet board measure	FBM
Centerline	CL	Finish	FIN.
Center to center	C to C	Fireproof	FPRF
Ceramic	CER	Fixture	FIX.
Circle	CIR	Flashing	FL
Circuit	CKT	Floor	FL
Circuit breaker	CIR BKR	Floor drain	FD
Circumference	CIRC	Flooring	FLG
Cleanout	CO	Fluorescent	FLUOR
Clear	CLR	Foot	(') FT
Closet	CL	Footcandle	FC
Coated	CTD	Footing	FTG
Column	COL	Foundation	FDN
Combination	COMB.	Full size	FS
Common	COM	Furred ceiling	FC
Composition	COMP	**Galvanize**	GALV
Concrete	CONC	Galvanized iron	GI
Conduit	CND	Garage	GAR
Construction	CONST	Gas	G
Continue	CONT	Gage	GA
Contractor	CONTR	Girder	G
Corrugate	CORR	Glass	GL
Courses	C	Grade	GR
Cross section	X-SECT	Grade line	GL
Cubic foot	CU FT	Gypsum	GYP
Cubic inch	CU IN.	**Hall**	H
Cubic yard	CU YD	Hardware	HDW
Damper	DMPR	Head	HD
Dampproofing	DP	Heater	HTR
Dead load	DL	Height	HT
Degree	(°) DEG	Horizontal	HOR
Design	DSGN	Hose bib	HB
Detail	DET	Hot water	HW
Diagonal	DIAG	House	HSE
Diagram	DIAG	Hundred	C
Diameter	DIA	**I beam**	I
Dimension	DIM	Impregnate	IMPG
Dining room	DR	Inch	(") IN.
Dishwasher	DW	Incinerator	INCIN
Ditto	DO.	Insulate	INS
Division	DIV	Intercommunication	INTERCOM
Door	DR	Interior	INT
Double	DBL	Iron	I
Double hung	DH	**Joint**	JT
Down	DN	Joist	JST
Downspout	DS	**Kilowatt**	KW
Drain	DR	Kilowatt hour	KWH
Drawing	DWG	Kip (1000 lb)	K
Dryer	D	Kitchen	KIT
East	E	**Laminate**	LAM
Electric	ELEC	Laundry	LAU
Elevation	EL	Lavatory	LAV
Enamel	ENAM	Left	L
Entrance	ENT	Length	LG
Equal	EQ	Length overall	LOA

Light	LT
Linear	LIN
Linen closet	L CL
Live load	LL
Living room	LR
Long	LG
Louver	LV
Lumber	LBR
Main	MN
Manhole	MH
Manual	MAN.
Manufacturing	MFG
Material	MATL
Maximum	MAX
Medicine cabinet	MC
Membrane	MEMB
Metal	MET.
Meter	M
Minimum	MIN
Minute	(') MIN
Miscellaneous	MISC
Mixture	MIX.
Model	MOD
Modular	MOD
Motor	MOT
Molding	MLDG
Natural	NAT
Nominal	NOM
North	N
Not to scale	NTS
Number	NO.
Obscure	OB
On center	OC
Opening	OPNG
Opposite	OPP
Overall	OA
Overhead	OVHD
Panel	PNL
Parallel	PAR.
Part	PT
Partition	PTN
Penny (nails)	d
Permanent	PERM
Perpendicular	PERP
Piece	PC
Plaster	PL
Plate	PL
Plumbing	PLMB
Pound	LB
Precast	PRCST
Prefabricated	PREFAB
Preferred	PFD
Quality	QUAL
Quantity	QTY
Radiator	RAD
Radius	R
Range	R
Receptacle	RECP
Reference	REF
Refrigerate	REF
Refrigerator	REF
Register	REG

Reinforce	REINF
Reproduce	REPRO
Required	REQD
Return	RET
Riser	R
Roof	RF
Tongue and groove	T & G
Total	TOT.
Tread	TR
Tubing	TUB.
Typical	TYP
Unfinished	UNFIN
Urinal	UR
Valve	V
Vapor proof	VAP PRF
Vent pipe	VP
Ventilate	VENT.
Vertical	VERT
Vitreous	VIT
Volt	V
Volume	VOL
Washing machine	WM
Water closet	WC
Water heater	WH
Waterproofing	WP
Watt	W
Weather stripping	WS
Weatherproof	WP
Weep hole	WH
Weight	WT
West	W
Width	W
Window	WDW
With	W/
Without	W/O
Wood	WD
Wrought iron	WI
Yard	YD

Unit 14
Architectural Synonyms
Architectural terms are standard. Nevertheless, architects, designers, and builders often use different terms for the same object. Geographic location can influence a person's word choice. For instance, what is referred to as a faucet in one locale is called a tap in another area. What one person calls an attic, another calls a garret, and still another a loft. A footing is often called a footer, a bannister, or a ballaster. In the list that follows, each entry is shown with a word or words which someone—somewhere—uses to refer to an object or building component.

Abutment: support
Acoustics: sound control
Adobe brick: firebrick, fireclay brick
Aeration: ventilation

Aggregate: cement matrix, concrete, mortar, plaster
Air-dried: air-seasoned, seasoned
Air-seasoned: air-dried, seasoned
Anchorage: footing, footer
Anchor bolt: securing bolt, sill bolt
Apartment: tenement, multiple dwelling, condominium
Arcade: corridor
Armored cable: conduit, tubing, metal casing, BX
Asbestos: fireproof material
Attic: garret, cot loft, half story, loft
Automatic heat control: thermostat
Auxiliary door: storm door
Auxiliary window: storm window
Awning: overhang, canopy

Backfill: fill, earth
Back plaster: parget
Baffle: screen
Baked clay: terra-cotta
Balcony: ledge, gallery, platform, veranda
Baseboard: mopboard, finish board, skirting
Basement: cellar, storm cellar, cyclone cellar, substructure
Base mold: shoe mold
Batten: cleat
Bead: thin molding
Beams: rafter, shaft, timber, girder, wood, spar, lumber
Bearing partition: support partition, bearing wall
Bearing plate: sill, load plate
Bearing soil: compact soil
Belvedere: gazebo, pavilion
Bermuda roof: hip roof
Beveled: mitered, chamfered
Bibs: faucets, taps
Birdmouth: plate cut, seat cut, seat of a rafter
Blanket insulation: sheet insulation
Blind: window shade
Blind nailing: secret nailing
Board insulation: heat barrier
Border: curb, parapet
Bracing: trussing
Breezeway: dogtrot
Brick: stone, masonry
Bridging: bracing, joining, cross supports, strutting
Buck: doorframe
Builder: contractor
Building area: setback, building lines
Building board: compo board, insulating board, dry wall, gypsum board, Sheetrock, wallboard, rocklath, plasterboard
Building code: building regulations
Building lines: setback, building area
Building paper: felt, tar paper, sheathing paper, construction paper, roll roofing
Building regulations: building code
Building steel: structural steel

Buttress: support, shield
BX: conduit, tubing, metal casing, armored cable

Candela: footcandle
Canopy: awning, overhang
Caps: coping
Carpenter's cloth: wire mesh, screen, wire cloth
Carport: car shed, open garage
Carriage: stringer
Carriage bolt: square bolt, threaded rod
Casement window: hinged window
Casing: window frame
Catch basin: cistern, dry well, reservoir
Caulking: sealer, oakum, pointing, masonry
Caulking compound: grout
Cavity wall: hollow wall
Ceiling clearance: headroom
Cellar: basement, storm cellar, cyclone cellar, substructure
Cement matrix: aggregate, concrete, mortar, plaster
Ceramic: porcelain, china
Cesspool: sewage basin, seepage pit
Chamber: bedroom
Chamfered: beveled, mitered
Chimney: flue, smokestack
Chimney pot: flue cap
China: ceramic, porcelain
Chute: trough
Cinder: rock, slag
Circuit box: fuse box, power panel, distribution panel
Circuit breaker: fuse
Cistern: catch basin, dry well, reservoir
Cleat: batten
Clipped ceiling: hung ceiling, drop ceiling
Closet: cloakroom, storage area
Colonial: Early American
Colonnade: portico
Column: post, pillar, cylinder, pile, spile
Column base: plinth, wall base
Comb board: cricket, saddle
Common wall: party wall
Compact soil: bearing soil
Compo board: insulating board, building board, dry wall, gypsum board, Sheetrock, wallboard, rocklath, plasterboard
Composition board: fiberboard, particle board
Concrete: cement matrix, aggregate, mortar, plaster
Condominium: apartment, tenement, multiple dwelling
Conductor: heat transmission, transmitter
Conduit: tubing, metal casing, BX, armored cable
Connectors: splice
Construction paper: building paper, felt, tar paper, sheathing paper, roll roofing
Contemporary: modern
Contractor: builder
Coping: caps
Corridor: arcade, hallway, passageway, lanai

Cot loft: attic, garret, half story, loft
Court: yard, patio, quad
Cover: escutcheon, shield, plate, hood
Cricket: saddle, threshold, doorsill
Cripple stud: short stud
Cross supports: joining, bracing, bridging
Culvert: gutter, channel, ditch, waste drain
Curb: border, parapet
Curtain wall: filler wall
Cyclone cellar: basement, storm cellar, substructure, cellar

Damper: flue control
Deck: landing, platform
Decking: floor
Decorative: ornamental
Deed: ownership document
Dehydration: evaporation
Den: library, reading room, quiet room, sitting room
Disposal system: leach lines, sewage line
Distribution panel: power panel, fuse box, circuit box
Ditch: culvert, gutter, channel, waste drain
Dogtrot: breezeway
Domicile: home, house, dwelling, residence
Doorframe: buck
Doorsill: saddle, threshold, cricket
Dormer: gable window, projected window, eyebrow
Double hung: double sashed
Double plate: top plate
Downspout: drainage pipe, rain drainage
Drainage hole: weep hole
Drainage pipe: downspout, rain drainage
Drain line: flow line
Drop ceiling: clipped ceiling, hung ceiling
Drop support: hanger, iron strap
Dry wall: gypsum board, Sheetrock, wallboard, building board, rocklath, plasterboard, compo board, insulating board
Dry well: cistern, catch basin, reservoir
Duct: pipeline, vent, raceway, plenum
Dumbwaiter: elevator, hoist, lift
Dwelling: domicile, home, house, residence

Early American: colonial
Earth: fill, backfill, topsoil
Easement: right-of-way
Eave overhang: roof projection, roof overhang
Egress: exit, outlet
Elevator: dumbwaiter, hoist, lift
Entrance: lobby, vestibule, stoop, porch, portal
Escalator: motor stairs
Escutcheon: shield, plate, cover
Evaporation: dehydration
Exit: egress, outlet
Exterior: facade, proscenium, facing
Exterior brick: face brick
Exterior wall: outside wall
Eyebrow: dormer

Facade: proscenium, facing, exterior
Face brick: exterior brick
Facing: facade, proscenium, exterior
Fan: blower
Faucets: taps, bibs
Felt: building paper, tar paper, sheathing paper, roll roofing, construction paper
Fiberboard: composition board, particle board
Fill: backfill
Fillers: shims
Filler stud: trimmer
Filler wall: curtain wall
Finish board: baseboard, mop board
Finish work: trim, millwork
Firebrick: adobe brick, fireclay brick
Fire door: resistance door
Fireplace: ingle
Fireproof material: asbestos
Flagstone: flagging
Flashing: vapor barrier
Flat roof: horizontal roof, shed roof, pent roof
Float valve: flush valve
Floor: decking
Flow line: drain line
Flue: chimney, smokestack
Flue cap: chimney pot
Flue control: damper
Flush plate: switch plate
Flush valve: float valve
Footcandle: candela
Footer: anchorage, footing
Footing: anchorage, footer
Foundation sill: mudsill
Framing: rough carpentry, skeleton
Fuse: circuit breaker
Fuse box: power panel

Gable window: dormer, projected window
Gallery: balcony, ledge, platform
Garret: attic, cot loft, half story, loft
Gazebo: pavilion, belvedere
Girder: beam, timber
Glazing bar: muntin, pane frame, sash bar
Glue: laminate
Grade: ground level, ground line, grade line
Granite: igneous rock, stone
Grate: spaced bars
Grease trap: U trap
Ground line: grade, grade line, ground level
Grout: caulking compound
Gutter: culvert, channel, ditch, waste drain
Gypsum: plaster
Gypsum board: dry wall, Sheetrock, wallboard, building board, insulating board, compo board, rocklath

Half story: loft, cot loft, garret, attic
Hallway: corridor, passageway, lanai
Handle: knob, pull
Handrail: newel
Hanger: drop support, iron strap

Hardboard: Masonite
Hatchway: opening, trapdoor, scuttle
Header: lintel
Headroom: ceiling clearance
Heat transmission: conductor, transmitter
Hinged window: casement window
Hip roof: Bermuda roof
Hoist: lift, elevator, dumbwaiter
Hollow-core door: veneer door
Hollow wall: cavity wall
Hood: cover
Horizontal roof: shed roof, flat roof, pent roof
Hung ceiling: drop ceiling, clipped ceiling

Igneous rock: stone, granite
Illumination level: light intensity
Ingle: fireplace
Insulating board: compo board, building board, dry wall, gypsum board, rocklath, Sheetrock, wallboard
Iron strap: drop support, hanger

Jalousies: louvers
Joining: bridging, bracing, cross supports

Knob: handle, pull

Lacing: lattice bars
Lambert: light unit, lumen
Laminate: glue
Laminated wood: plywood
Lanai: passageway, corridor, hallway
Landing: platform, deck
Larder: pantry
Lattice bars: lacing
Laundry: utility room, service porch
Laundry tray: slop sink, work sink
Lavatory: sink
Leach lines: disposal system, sewage line
Ledge: gallery, platform, balcony
Library: den, reading room, quiet room, sitting room
Lift: dumbwaiter, elevator, hoist
Light intensity: illumination level
Light unit: lumen, lambert
Lintel: header
Live load: moving load
Load: weight
Load plate: bearing plate, sill
Lobby: vestibule, stoop, porch, portal, entrance
Loft: half story, cot loft, garret, attic
Lot: plot, property, site
Louvers: jalousies
Lumen: light unit, lambert

Mantel: shelf
Masonite: hardboard
Masonry: stone, brick, pointing, caulking
Membrane: sheet, sisalkraft
Metal casing: tubing, conduit, BX, armored cable
Millwork: trim, finish work
Mitered: beveled, chamfered

Modern: contemporary
Module: standard unit
Moisture barrier: vapor barrier
Mop board: baseboard, finish board
Mortar: aggregate, cement matrix, concrete, plaster
Motor stairs: escalator
Moving load: live load
Mudsill: foundation sill
Mullion: window divider
Multiple dwelling: apartment, tenement, condominium
Muntin: glazing bar, pane frame, sash bar

Newel: handrail

Opening: trapdoor, scuttle, hatchway
Outside wall: exterior wall
Overhang: awning, canopy
Ownership document: deed

Pane frame: sash bar, glazing bar, muntin
Pantry: larder
Parapet: curb, border
Parget: back plaster
Particle board: composition board, fiberboard
Partition: wall
Party wall: common wall
Passageway: lanai, corridor, hallway
Patio: quad, court, yard, terrace
Pavement: paving, road, surface, sidewalk
Pavilion: gazebo, belvedere
Paving: pavement, road surface
Pent roof: shed roof, flat roof
Pier: support, abutment
Pilaster: wall column
Pillar: post, column, pile, spile, cylinder
Pipeline: vent, raceway, duct, plenum
Pitch: slant, slope
Plank and beam: post and beam, post and lintel
Plaster: mortar, gypsum, concrete, cement matrix, aggregate
Plasterboard: dry wall, Sheetrock
Plastic membrane: vapor barrier, sisalkraft
Plate: escutcheon, shield, cover, shoe, scantling, sole
Plate cut: birdmouth, seat cut, seat of a rafter
Platform: balcony, ledge, gallery
Platform framing: western framing
Plenum: pipeline, vent, raceway, duct
Plinth: column base, wall base
Plot: lot, property, site
Plywood: laminated wood, veneer
Pointing: caulking, masonry
Porcelain: ceramic, china
Porch: stoop, lobby, entry, ingress, entrance, portal, gallery, lanai, terrace, veranda, vestibule
Portico: colonnade
Post: column, pillar, cylinder, pile, spile
Post and lintel: post and beam, plank and beam
Power and panel: distribution panel, fuse box, circuit box

Pressure: stress
Property: plot, lot, site
Proscenium: facade, facing, exterior
Pull: handle, knob

Quad: patio, court
Quiet room: sitting room, reading room, library, den
Quoins: stone coping

Raceway: duct, pipeline, plenum, vent
Rafter: beam, shaft, timber
Rain drainage: drainage pipe, downspout
Reading room: quiet room, sitting room, library, den
Recessed fluorescent fixture: troffer
Reservoir: catch basin, cistern, dry well
Residence: domicile, home, house, dwelling
Resistance door: fire door
Ridge: roof peak, ridge pole, ridge board
Right-of-way: easement
Rock: slab, cinder
Rocklath: compo board, insulating board, building board, dry wall, gypsum board, Sheetrock, wallboard
Roll roofing: construction paper, sheathing paper, tar paper, felt, building paper
Roof overhang: roof projection, eave overhang
Roof peak: ridge, ridge board, ridge pole
Rough carpentry: framing
Rough floor: subfloor
Rough lumber: undressed lumber

Saddle: threshold, doorsill, cricket
Sanitary sewer: storm sewer
Sash bar: pane frames, glazing bar, muntin
Scaffold: staging
Scantling: shoe, sole, plate
Screen: wire mesh, wire cloth, carpenter's cloth, baffle
Screw stairs: spiral stairs, winding stairs
Scupper: wall drain
Scuttle: hatchway, opening, trapdoor
Sealer: oakum, caulking
Seasoned: air-dried
Seat of a rafter: seat cut, plate cut, birdmouth
Secret nailing: blind nailing
Securing bolt: anchor bolt, sill bolt
Seepage pit: sewage basin, cesspool
Service porch: laundry, utility room
Setback: building lines, building area
Sewage basin: cesspool, seepage pit
Sewage line: leach lines, disposal system
Shaft: beam, rafter, timber
Sheathing paper: building paper, felt, tar paper, construction paper, roll roofing
Shed roof: flat roof, horizontal roof, pent roof
Sheet: sisalkraft, membrane
Sheet insulation: blanket insulation
Sheetrock: compo board, insulating board, building board, dry wall, gypsum board, wallboard, rocklath, plasterboard

Shelf: mantel
Shield: plate cover, escutcheon
Shims: fillers
Shoe: plate, scantling, sole
Shoe mold: base mold
Shoring: supporting, timber brace
Short stud: cripple stud
Sidewalk: pavement
Sill: bearing plate, load plate
Sill bolt: securing bolt, anchor bolt
Sink: lavatory
Sisalkraft: vapor barrier, plastic membrane, membrane, sheet
Site: plot, property, lot
Sitting room: quiet room, reading room, library, den
Skeleton: framing
Skirting: baseboard
Slag: cinder, rock
Slope: slant, pitch
Slop sink: work sink, laundry tray
Smokestack: flue, chimney
Soffit: underside
Solar energy: sun energy
Sole: shoe, plate, scantling
Sound control: acoustics
Spaced bars: grate
Spar: lumber, beam, wood, timber
Spile: pile, column, pillar, post
Spiral stairs: screw stairs, winding stairs
Splice: connectors
Square bolt: carriage bolt, threaded rod
Staging: scaffold
Standard unit: module
Steel connector: strap
Step: tread
Stiffener: tie
Stone: granite, igneous rock, brick, masonry
Stone coping: quoins
Stoop: porch, portal, vestibule, lobby, entrance
Storage area: cloakroom, closet
Storm cellar: basement, cyclone cellar, substructure, cellar
Storm door: auxiliary door
Storm sewer: sanitary sewer
Storm window: auxiliary window
Strap: steel connector
Strengthen: reinforce
Stress: pressure
Stringer: carriage
Strip insulation: weather stripping
Structural steel: building steel
Strutting: bridging
Subfloor: rough floor
Substructure: storm cellar, basement, cyclone cellar, cellar
Sun energy: solar energy
Support: shield, buttress, abutment, pier, plate cover
Support partition: bearing partition
Surface drainage: surface flow
Switch plate: flush plate

Taps: faucets, bibs
Tar paper: building paper, felt, sheathing paper, construction paper, roll roofing
Tenement: apartment, multiple dwelling, condominium
Terrace: patio, passageway, corridor, hallway
Terra-cotta: baked clay
Thermostat: automatic temperature control
Thin molding: bead
Threaded rod: square bolt, carriage bolt
Threshold: saddle, doorsill, cricket
Tie: stiffener
Timber: shaft, rafter, beam, girder, lumber, wood
Timber brace: shoring, supporting
Top plate: double plate
Topsoil: earth
Tower: turret
Trapdoor: scuttle, hatchway, opening
Transmitter: conductor, heat transmission
Tread: step
Trim: finish work, millwork
Trimmer: filler stud
Troffer: recessed fluorescent fixture
Trough: chute
Trussing: bracing
Tubes: ducts, channels, air pipes
Tubing: BX, conduit, metal casing, armored cable

Underside: soffit
Undressed lumber: rough lumber
Utility room: service porch, laundry
U trap: grease trap

Vapor barrier: flashing, sisalkraft, plastic membrane, moisture barrier
Veneer: plywood
Veneer door: hollow-core door
Vent: duct, opening
Ventilation: aeration
Veranda: passageway, balcony
Vestibule: stoop, porch, portal, lobby, entrance

Wall: partition
Wall base: column base, plinth
Wallboard: compo board, insulating board, building board, dry wall, gypsum board, Sheetrock
Wall column: pilaster
Wall drain: scupper
Waste drain: gutter, culvert, channel, ditch
Water closet: toilet, WC
Water table: water level
WC: water closet, toilet
Weather stripping: strip insulation
Weep hole: drainage hole
Weight: load
Western framing: platform framing
Winding stairs: spiral stairs, screw stairs
Window divider: mullion
Window frame: casing
Window shade: blind
Wire cloth: screen, wire mesh, carpenter's cloth
Wood: beam, spar, lumber, timber
Work sink: laundry tray, slop sink

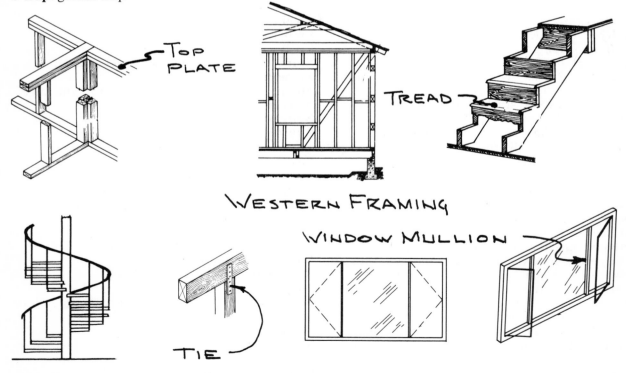

Figure 14-1. Drawings of some building parts.

CHAPTER 4

Reading Scaled Drawings

In ancient times simple structures were built without detailed architectural plans and even without established dimensions. The outline of the structure and the position of each room could be determined experimentally by "pacing off" approximate distances. The builder could then erect the structure using existing materials by adjusting sizes and dimensions as necessary during the building process. In this case the builder played the role of the architect, designer, contractor, carpenter, mason, and, perhaps, the manufacturer of materials and components. Today, building materials are so varied, construction methods so complex, and design requirements so demanding, that a complete dimensioned set of drawings is absolutely necessary to complete any building exactly as designed.

Unit 15
Size Description
Dimensions show the size of materials and exactly where they are to be located. Dimensions show the builder the width, height, and length of the building and subdivisions of the building. They show the location of doors, windows, stairs, fireplace, and planters. The number of dimensions included on an architectural plan depends largely on how much freedom of interpretation the designer wants to give the builder. If complete dimensions are shown on a plan, the builder cannot deviate greatly from the original design. However, if only a few dimensions are shown, then the builder must determine the sizes of many areas, fixtures, and details. When this occurs the builder must provide the dimensions and is placed in the position of designer.

Dimensions, or distances between points, are shown on architectural drawings by the use of dimension lines, extension lines, arrowheads, and numerals, as shown in Fig. 15-1. Dimensions are placed on a line that represents the distance

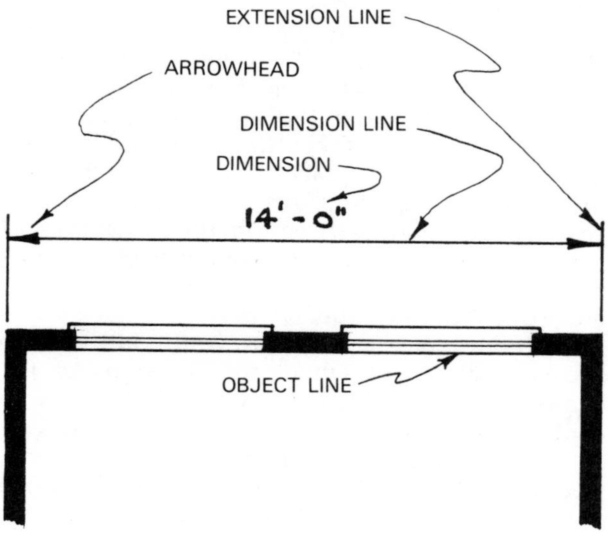

EXTENSION LINE

ARROWHEAD

DIMENSION LINE

DIMENSION

14'-0"

OBJECT LINE

Fig. 15-1. Dimensions show the distance between arrowheads.

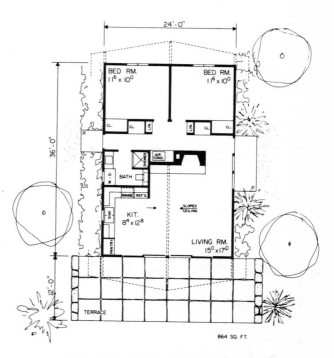

864 SQ. FT.

Fig. 15-2. Abbreviated dimensions are often used on simple drawings. (Home Planners, Inc.)

between two arrowheads. All completely dimensioned architectural plans are shown in this manner. Abbreviated architectural plans such as the one shown in Fig. 15-2 show only approximate dimensions of areas. In general, more detailed plans and complex designs include more detailed dimensioning. Although arrowheads, as shown in the top line in Fig. 15-3, are the accepted standard, many architectural drawings include alternate symbols to describe the length of an area, as shown in the bottom portion of Fig. 15-3. Information concerning the reading of dimensions on specific drawings is covered in the appropriate unit.

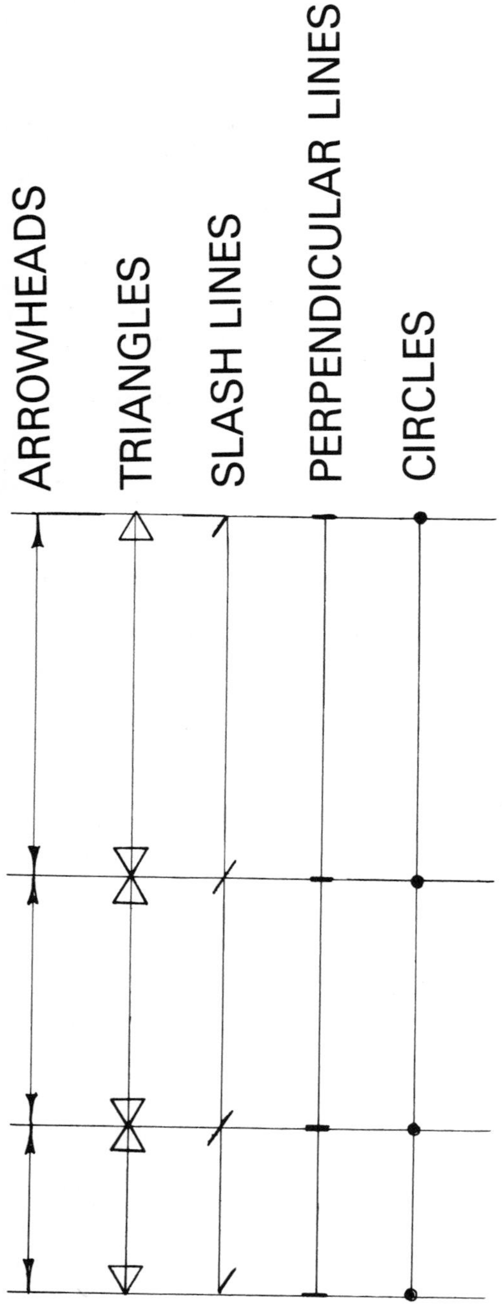

Unit 16
Scales

When a drawing of an object is exactly the same scale (size) as the actual object, this is full-size scale (1:1). However, architectural structures are obviously too large to be drawn full size. They must, therefore, be reduced in scale to fit on sheets of paper. Three types of measuring scales are commonly used for scaled construction drawings: the architect's scale, the engineer's scale, and the metric scale.

The main function of any scale is to enable the designer to draw a building at a convenient size and to enable the builder to think in relation to the actual size of the structure as he uses the drawing to construct the building. When a drawing is prepared to a reduced scale, 1 ft (12 in.) may actually be drawn $1/4$ in. long. On the reduced scale, the builder does not think of this $1/4$-in. line as actually representing $1/4$ in. but thinks of it as being 1 ft long. If the area to be represented is very large, the reduction ratio will be very large. A very large plot plan may be reduced as much as 200 times, while a detailed drawing may be reduced only one or two times its actual size. Figures 16-1 and 16-2 show the common scales used to reduce construction drawings.

Architect's Scale

Architect's scales are either the bevel type shown in Fig. 16-1 or the triangular type shown in Fig. 16-3. The triangular scale has 6 sides and 11 different scales: a full scale of 12 in. graduated into 16 parts to an inch and 10 open-divided scales with ratios of $3/32$, $1/8$, $3/16$, $1/4$, $3/8$, $1/2$, $3/4$, 1, $1 1/2$, and 3. Two scales are located on each face. One scale reads from left to right. The other scale, which is twice as large, reads from right to left. For example, the $1/4$ scale and half of this, the $1/8$ scale, are placed on the same face. Similarly, the $3/4$ scale and the $3/8$ scale are placed on the same face but are read from different directions. If the scale is read from the wrong direction, the measurement could be wrong, since the second row of numbers reads from the opposite side of the scale at half scale or twice the value.

The architect's scale is most commonly used to measure distances where the divisions of the scale equal 1 ft or 1 in. For example, in the $1/4$-in. scale shown in Fig. 16-3, $1/4$ in. can equal either 1 in. or 1 ft. Since buildings are very large, most major architectural drawings use a scale which relates the parts of the scale to a foot. Architectural details such as cabinet construction and joints often use these same parts of the scale to represent 1 in.

ARROWHEADS TRIANGLES SLASH LINES PERPENDICULAR LINES CIRCLES

Fig. 15-3. Dimension lines with alternate symbols.

Fig. 16-1. Construction drawings are prepared with either an architect's, metric, or engineer's scale. (Sterling Scales.)

SCALE SELECTIONS:			
DRAWING	ARCHITECT	METRIC	ENGINEER
LARGE PLOT PLAN	NOT USED	1:200	1″ = 20′
SMALL PLOT PLAN	1/8″ = 1′ – 0″	1:100	1″ = 10′
FLOOR PLAN	1/4″ = 1′ – 0″	1:50	NOT USED
DETAIL DRAWINGS	1/2″ = 1′ – 0″	1:20	NOT USED
	3/4″ = 1′ – 0″	1:10	
	1″ = 1′ – 0″	1:5	

Fig. 16-2. The scale of a drawing depends on the amount of reduction from object to drawing.

Architect's scales are either open-divided or fully divided. In fully divided scales, each main unit on the scale is fully subdivided throughout the scale. On open-divided scales, only the main units of the scale are graduated with a fully subdivided extra unit at each end, as shown in Fig. 16-4. Figure 16-5 is an open-divided scale showing that the $^3/_{16}$ scale reads from right to left and is twice as large as the $^3/_{32}$ scale, which reads from left to right.

In reading an open-divided scale, the section at the end of the scale is not part of the numerical scale. When measuring with the scale, start with the zero line, *not* with the outside end line of the fully divided section. Always start with the number of feet or inches you wish to measure and then add the additional inches (or feet) in the subdivided area. For example, in Fig. 16-6 the distance 4 ft 11 in. is derived by measuring from the line 4 to 0, then 11 in. past zero, since each of the lines in the subdivision equals 1 in. On smaller scales, the lines in the fully divided portion may equal 2 in.

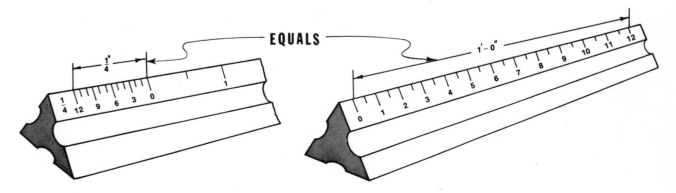

Fig. 16-3. ¼″ = 1′-0″ on this architect's scale.

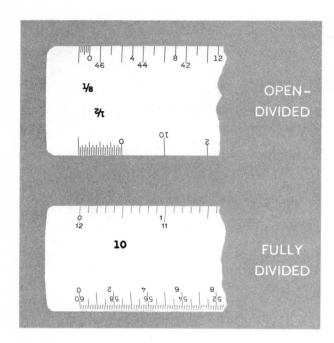

Fig. 16-4. Two types of architect's scale divisions.

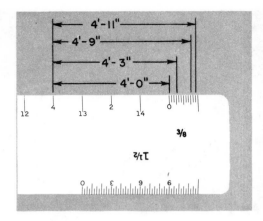

Fig. 16-6. Subdivisions at the end of an open-divided scale used for inch measurements.

On larger scales, they may equal ¹/₂ in. Figure 16-7 shows this principle applied to a ¹/₂″ = 1'-0″ scale where ¹/₂ in. equals 1 ft. And Fig. 16-8 shows the use of this same scale in a ¹/₂″ = 1″ ratio.

Figure 16-9 shows a further application of the use of the architect's scale. Notice the dimensioned distance of 8'-0″ extends from the 8 to the 0 on the scale and the 6-in. wall is shown as ¹/₂ ft on the subdivided foot on the end of the scale.

When the scale of the drawing changes, the length of each line increases or decreases and the width of various areas also increases or decreases. The actual appearance of a typical corner wall at a scale of ¹/₁₆″ = 1'-0″, ¹/₈″ = 1'-0″, ¹/₄″ = 1'-0″, and ¹/₂″ = 1'-0″ is shown in Fig. 16-10. You can see that the wall drawn to the scale of ¹/₁₆″ = 1'-0″ is small and that a great amount of

detail would be impossible. The ¹/₂″ = 1'-0″ wall would probably cover too large an area on a drawing if the building is very large. Therefore, the ¹/₄″ = 1'-0″ and ¹/₈″ = 1'-0″ scales are the most popular for most basic architectural drawings.

Figure 16-11 shows the comparative distances used to measure 1 ft 9 in. as it appears on various architect's scales. The colored bar on each of these scales represents 1 ft 9 in. The same comparison would exist if the scales were related to 1 in. rather than 1 ft. In this case, a distance of 1³/₄ in. would have the same line length as 1 ft 9 in. on the foot representation. The ³/₃₂-in., ³/₁₆-in., ¹/₈-in., ¹/₄-in., ³/₈-in., ¹/₂-in., and ³/₄-in. scales represent a distance smaller than full size (1³/₄ in.). The 1¹/₂-in. and 3-in. scales represent a distance 1¹/₂ times and 3 times as large as the full scale. Figure 16-12 shows the same comparison using a distance of 5 ft. 6 in. on the foot equivalent scale. When an inch equivalent (1″ = 1'-0″) scale is used, the distance shown is 5¹/₂ in.

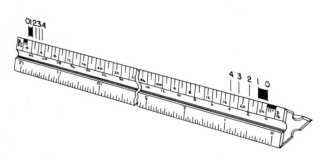

Fig. 16-5. The ³/₁₆″ scale reads right to left, and the ³/₃₂″ scale reads from left to right.

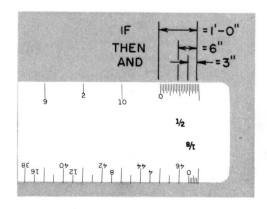

Fig. 16-7. If ¹/₂″ = 1'-0″, then ¹/₄″ = 6″, and ¹/₈″ = 3″.

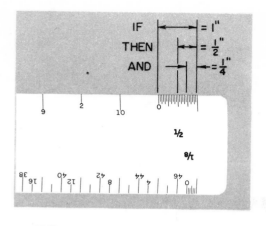

Fig. 16-8. If $1/2'' = 1''$, then $1/4'' = 1/2''$, and $1/8'' = 1/4''$.

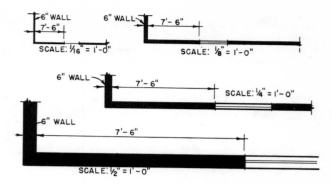

Fig. 16-10. Comparison of similar walls drawn to different scales.

Engineer's Scale

The civil engineer's scale is often used for plot plans, surveys, and other large land tract plans. Each scale divides the inch into decimal parts. These parts, as shown in Fig. 16-13, are 10, 20, 30, 40, 50, and 60 parts per inch. Each one of these units can represent any linear value, such as an inch, foot, yard, or mile, depending on the drawing size. For example, on the ten ratio scale, 10 units can equal 1 in., 1 ft, 1 yd, or 1 mi; and the subdivisions of that unit are in tenths of an inch, foot, yard, or mile. Subdivisions of tenths, as shown in the enlargement in Fig. 16-13, show each unit subdivided into hundredths. Thus all units on a civil engineer's scale are divisible or multipliable by units of 10. Figure 16-15 shows the use of a 30 ratio scale using the same graduation to depict from 5.1 in. up to 5100 in., depending on whether the scale of $1'' = 3''$ is used or the scale of $1'' = 3000'$ is used.

Although rarely used, Fig. 16-16 shows the most dramatic range possible using the engineer's scale to depict from inches to miles.

Metric Scale

The basic units of measure in the metric system are the meter (m) for distance, the kilogram (kg) for mass (weight), and the liter (L) for volume. Since most measurements used on architectural drawings are linear distances, multiples or subdivisions of the meter are the metric units most commonly used. The meter, as shown in Fig. 16-17, represents a basic unit of one. To eliminate the use of many zeros, prefixes are used to change the base (m) to larger or smaller amounts by units of ten.

Prefixes which represent multiples of meters are deka, hecto, and kilo. A dekameter (dam) equals 10 meters. A hectometer (hm) equals 100 meters. A kilometer (km) equals 1000 meters.

Prefixes which represent subdivisions of meters are deci, centi, and milli. A decimeter (dm) equals one-tenth (0.1) of a meter. A centimeter (cm) equals one one-hundredth (0.01) of a meter. A millimeter (mm) equals one-thousandth (0.001) of a meter. The most useful subdivision of a meter for construction drawings is the millimeter. Figure 16-18 shows a portion of a meter scale in which the numbers on the scale denote millimeters and every tenth line represents centimeters. Each line on this scale represents a milli-

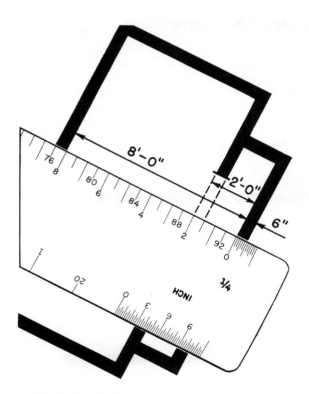

Fig. 16-9. Subdivisions on the architect's scale can be used to indicate overall dimensions and subdimensions.

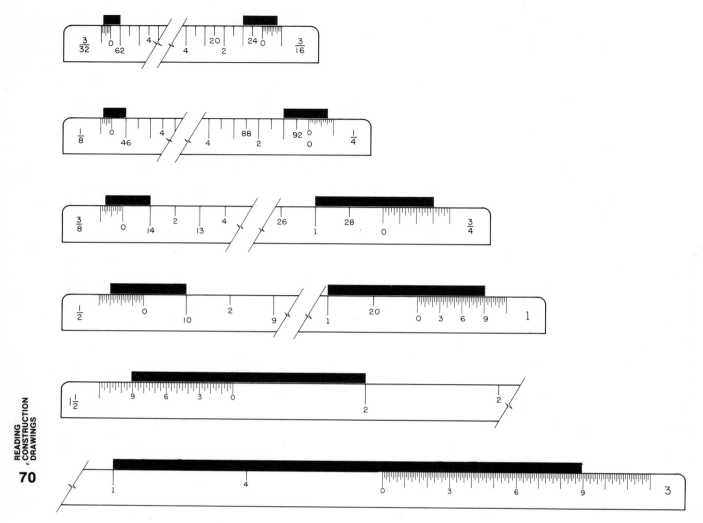

Fig. 16-11. 1′-9″ shown on different scales.

READING
CONSTRUCTION
DRAWINGS

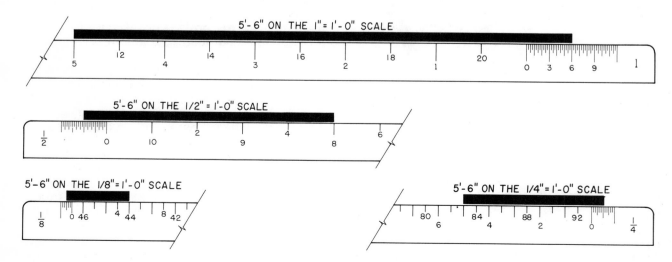

Fig. 16-12. 5′-6″ shown on different scales.

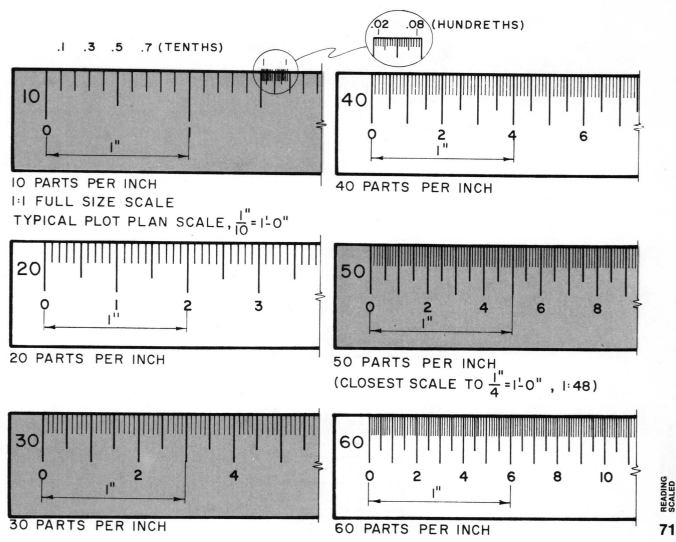

.I .3 .5 .7 (TENTHS)

.02 .08 (HUNDRETHS)

10 PARTS PER INCH
1:1 FULL SIZE SCALE
TYPICAL PLOT PLAN SCALE, $\frac{1''}{10}$=1'-0"

20 PARTS PER INCH

30 PARTS PER INCH

40 PARTS PER INCH

50 PARTS PER INCH
(CLOSEST SCALE TO $\frac{1''}{4}$=1'-0" , 1:48)

60 PARTS PER INCH

Fig. 16-13. Civil engineer's scale.

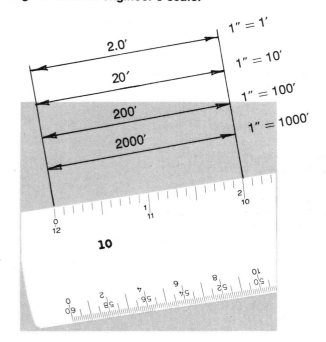

1" = 1'
2.0'
1" = 10'
20'
1" = 100'
200'
1" = 1000'
2000'

Fig. 16-14. The civil engineer's scale divided by tenths.

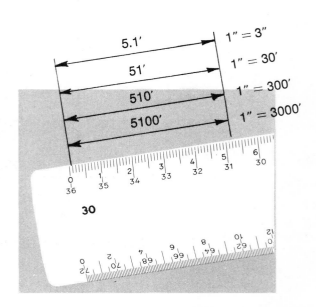

1" = 3"
5.1'
1" = 30'
51'
1" = 300'
510'
1" = 3000'
5100'

Fig. 16-15. The civil engineer's scale divided by thirtieths of an inch.

meter. Notice that there are 10 millimeters between each centimeter. Figure 16-19 shows the most useful metric prefixes and the relationship of these prefixes to the meter. This consistent use of prefixes makes the metric system much easier to use than customary units.

METRIC DIMENSIONS Linear metric sizes used on basic architectural drawings such as floor plans and elevations are found in meters; or, if less than 1 m, in millimeters, as shown in Fig. 16-20. Using millimeters exclusively is the preferred standard. When meter dimensions are

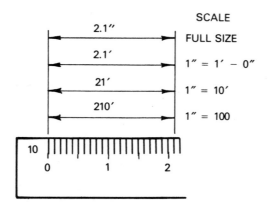

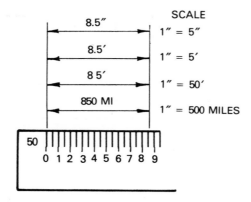

Fig. 16-16. Different scales possible on engineer's scale.

1 YARD

1 METER

1 YARD = 0.9 METER

1 METER = 1.1 YARDS

Fig. 16-17. A meter is slightly larger than a yard.

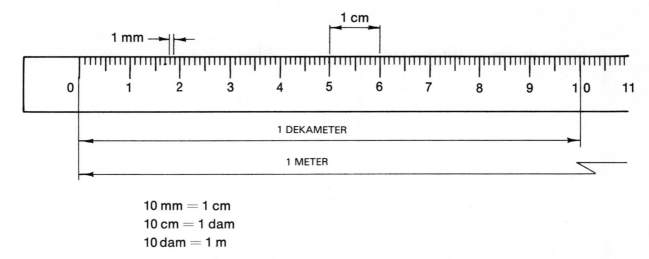

10 mm = 1 cm
10 cm = 1 dam
10 dam = 1 m

Fig. 16-18. A meter scale subdivided into millimeters.

PREFIX SYMBOL + METER =				
$1000 = 10^3$	KILO	k	KILOMETER	km
$100 = 10^2$	HECTO	h	HECTOMETER	hm
$10 = 10^1$	DEKA	da	DEKAMETER	dam
$0.1 = 10^{-1}$	DECI	d	DECIMETER	dm
$0.01 = 10^{-2}$	CENTI	c	CENTIMETER	cm
$0.001 = 10^{-3}$	MILLI	m	MILLIMETER	mm

Fig. 16-19. Prefixes change the base unit by increments of ten.

used on plans, they are carried to three decimal points, as shown in Fig. 16-21. In detail drawings such as the one shown in Fig. 16-22, millimeters are used exclusively.

Metric scales such as those shown in Fig. 16-23 are used in the same manner as the architect's scale is used: to scale reduced-size drawings. Metric scales, however, use ratios in increments of 10 rather than fractional ratios of 12 used in archi-

tect's scales. Just as with fractional scales, the ratio chosen in a metric scale depends on the size of the drawing compared to the full size of the object. Figure 16-24 shows some common metric ratios used in various types of architectural drawings. Scales representing some of these ratios are shown in Fig. 16-25.

METRIC CONVERSION Designers attempt to design exclusively in either U.S. customary units or in metric units. But often conversion is necessary because some components are not available in metric sizes, even though the basic design is prepared metrically. In renovation work, drawings of existing structures or structural components may be in customary units, and proposed additions may be prepared in metric units. When it is necessary to convert U.S. customary architectural dimensions and distances to metric dimensions and distances, the information found in Fig. 16-26 can be used for that pur-

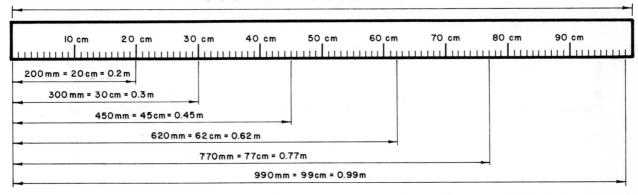

Fig. 16-20. All architectural dimensions are in millimeters or meters.

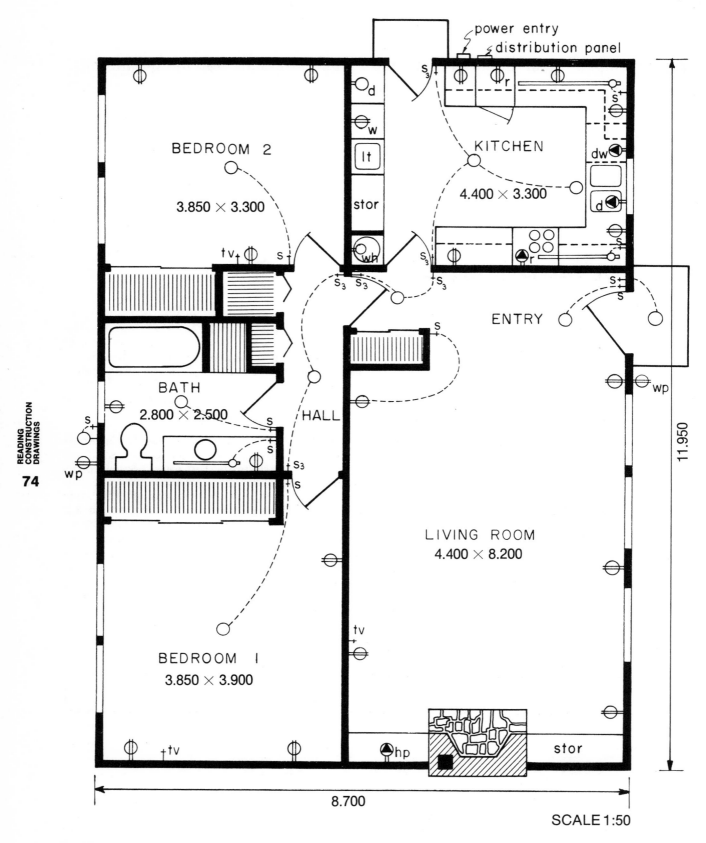

power entry
distribution panel

BEDROOM 2

3.850 × 3.300

d

w

lt

stor

wh

KITCHEN

4.400 × 3.300

r

dw

d

r

t v

s

s₃

s₃

s₃

s₃

s₃

s

s

ENTRY

s

s

wp

BATH

2.800 × 2.500

s

s

s

s₃

s

HALL

s

LIVING ROOM

4.400 × 8.200

wp

tv

BEDROOM 1

3.850 × 3.900

tv

hp

stor

11.950

8.700

SCALE 1:50

Fig. 16-21. Room sizes shown in meters to three decimal points. (3.750 m can also be read 3750 mm.)

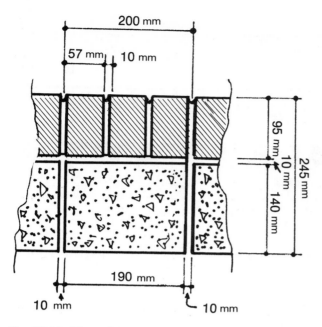

Fig. 16-22. **The millimeter is used for small detail dimensions.**

USE	RATIO	COMPARISON TO 1 METER
CITY MAP	1:2500	(0.4 mm EQUALS 1 m)
	1:1250	(0.8 mm EQUALS 1 m)
PLAT PLANS	1:500	(2 mm EQUALS 1 m)
	1:200	(5 mm EQUALS 1 m)
PLOT PLANS	1:100	(10 mm EQUALS 1 m)
	1:80	(12.5 mm EQUALS 1 m)
FLOOR PLANS	1:75	(13.3 mm EQUALS 1 m)
	1:50	(20 mm EQUALS 1 m)
	1:40	(25 mm EQUALS 1 m)
DETAILS	1:20	(50 mm EQUALS 1 m)
	1:10	(100 mm EQUALS 1 m)
	1:5	(200 mm EQUALS 1 m)

Fig. 16-24. **Architectural use of metric ratios.**

pose. These are approximate metric units compared to similar U.S. customary units. When very accurate conversion from customary units to metric units is necessary, a metric conversion handbook or ANSI Z210.1-1976 Standard for Metric Practice should be used.

As metrication becomes more widely accepted, more building materials will be manufactured in metric sizes, greatly simplifying the use of metric-prepared drawings. Figures 16-27, 16-28, and 16-29 show probable standard lengths of lumber in metric sizes.

When a drawing contains both metric and customary dimensions, two kinds of dimensions may be provided. Dual dimensions may be included in the drawing, as shown in Fig. 16-30; or conversion tables may be included with the drawing, as shown in Fig. 16-31. Metric conversion tables are keyed to dimensions on the drawing that are either customary or metric. The equivalent is then read from the chart. Charts are sometimes master charts showing all equivalents or may include only the equivalents of the dimensions actually used on the drawing.

Knowing how to read the architect's, engineer's, and metric scales helps the builder read and understand scaled drawings. Use of scales can help find missing dimensions; however, be extremely careful in scaling prints of drawings. Print paper stretches in the process of printing, creating distances greater than those found on the original drawing. Measure several known dimensions on a print to determine the amount of stretch before using a scale to determine unknown dimensions.

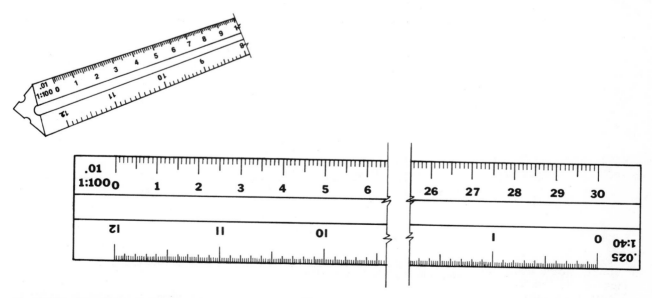

Fig. 16-23. **A metric scale showing two ratios: 1:100 and 1:40.**

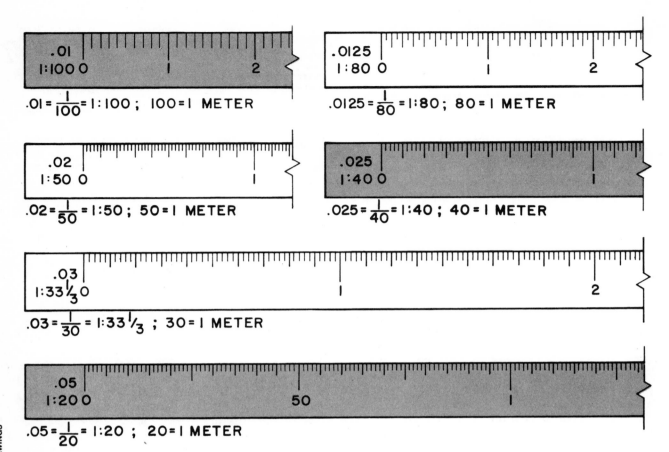

.01 = $\frac{1}{100}$ = 1:100 ; 100 = 1 METER

.0125 = $\frac{1}{80}$ = 1:80 ; 80 = 1 METER

.02 = $\frac{1}{50}$ = 1:50 ; 50 = 1 METER

.025 = $\frac{1}{40}$ = 1:40 ; 40 = 1 METER

.03 = $\frac{1}{30}$ = 1:33 $\frac{1}{3}$; 30 = 1 METER

.05 = $\frac{1}{20}$ = 1:20 ; 20 = 1 METER

Fig. 16-25. Typical metric ratios.

Fig. 16-26. Metric units compared to similar U.S. customary units.

	WHEN YOU KNOW	YOU CAN FIND:	IF YOU MULTIPLY BY:
LENGTH	INCHES	MILLIMETERS	25.0
	FEET	CENTIMETERS	30.0
	YARDS	METERS	0.9
	MILES	KILOMETERS	1.6
	MILLIMETERS	INCHES	0.04
	CENTIMETERS	INCHES	0.4
	METERS	YARDS	1.1
	KILOMETERS	MILES	0.6
AREA	SQUARE INCHES	SQUARE CENTIMETERS	6.5
	SQUARE FEET	SQUARE METERS	0.09
	SQUARE YARDS	SQUARE METERS	0.8
	SQUARE MILES	SQUARE KILOMETERS	2.6
	ACRES	SQUARE HECTOMETERS (HECTARES)	0.4
	SQUARE CENTIMETERS	SQUARE INCHES	0.16
	SQUARE METERS	SQUARE YARDS	1.2
	SQUARE KILOMETERS	SQUARE MILES	0.4
MASS	OUNCES	GRAMS	28.0
	POUNDS	KILOGRAMS	0.45
	SHORT TONS	MEGAGRAMS (METRIC TONS)	0.9
	GRAMS	OUNCES	0.035
	KILOGRAMS	POUNDS	2.2
	MEGAGRAMS (METRIC TONS)	SHORT TONS	1.1
LIQUID VOLUME	OUNCES	MILLILITERS	30.0
	PINTS	LITERS	0.47
	QUARTS	LITERS	0.95
	GALLONS	LITERS	3.8
	MILLILITERS	OUNCES	0.034
	LITERS	PINTS	2.1
	LITERS	QUARTS	1.06
	LITERS	GALLONS	0.26
TEMPERATURE	DEGREES FAHRENHEIT	DEGREES CELSIUS	5/9 (AFTER SUBTRACTING 32)
	DEGREES CELSIUS	DEGREES FAHRENHEIT	9/5 (THEN ADD 32)

1.8 m	3.0 m	4.2 m	5.4 m
2.1 m	3.3 m	4.5 m	5.7 m
2.4 m	3.6 m	4.8 m	6.0 m
2.7 m	3.9 m	5.1 m	6.3 m

Fig. 16-27. Probable standard length of construction lumber in meters.

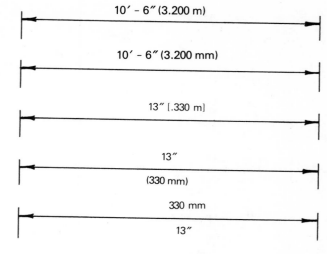

Fig. 16-30. Dual dimensioning methods.

SIZE (mm)

1800 × 1200
2400 × 1200
2700 × 1200
3000 × 1200
3600 × 1200
2400 × 900

Fig. 16-28. Probable standard lumber sheet sizes in millimeters.

THICKNESS, mm	WIDTH, mm								
16x	75	100	125	150					
19x	75	100	125	150					
22x	75	100	125	150					
25x	75	100	125	150	175	200	225	250	300
32x	75	100	125	150	175	200	225	250	300
36x	75	100	125	150					
38x	75	100	125	150	175	200	225		
40x	75	100	125	150	175	200	225		
44x	75	100	125	150	175	200	225	250	300
50x	75	100	125	150	175	200	225	250	300
63x		100	125	150	175	200	225		
75x		100	125	150	175	200	225	250	300
100x		100		150		200		250	300
150x				150		200			300
200x						200			
250x								250	
300x									300

Fig. 16-29. Probable standard sizes of construction lumber in millimeters.

78

Feet   Inches

Meters and millimeters

Feet	0	1	2	3	4	5	6	7	8	9	10	11
0	—	25	51	76	102	127	152	178	203	229	254	279
1	305	330	356	381	406	432	457	483	508	533	559	584
2	610	635	660	686	711	737	762	787	813	838	864	889
3	914	940	965	991	1.016	1.041	1.067	1.092	1.118	1.143	1.168	1.194
4	1.219	1.245	1.270	1.295	1.321	1.346	1.372	1.397	1.422	1.448	1.473	1.499
5	1.524	1.549	1.575	1.600	1.626	1.651	1.678	1.702	1.727	1.753	1.778	1.803
6	1.829	1.854	1.880	1.905	1.930	1.956	1.981	2.007	2.032	2.057	2.083	2.108
7	2.134	2.159	2.184	2.210	2.235	2.261	2.286	2.311	2.337	2.362	2.388	2.413
8	2.438	2.464	2.489	2.515	2.540	2.565	2.591	2.616	2.642	2.667	2.692	2.718
9	2.743	2.769	2.794	2.819	2.845	2.870	2.896	2.921	2.946	2.972	2.997	3.023
10	3.048	3.073	3.099	3.124	3.150	3.175	3.200	3.226	3.251	3.277	3.302	3.327
11	3.353	3.378	3.404	3.429	3.454	3.480	3.505	3.531	3.556	3.581	3.607	3,632
12	3.658	3.683	3.708	3.734	3.759	3.785	3.810	3.835	3.861	3.886	3.912	3.937
13	3.962	3.988	4.013	4.039	4.064	4.089	4.115	4.140	4.168	4.191	4.216	4.242
14	4.267	4.293	4.318	4.343	4.369	4.394	4.420	4.445	4.470	4.496	4.521	4.547
15	4.572	4.597	4.623	4.648	4.674	4.699	4.724	4.750	4.775	4.801	4.826	4.851
16	4.877	4.902	4.928	4.953	4.978	5.004	5.029	5.055	5.080	5.105	5.131	5.156
17	5.182	5.207	5.232	5.258	5.283	5.309	5.334	5.359	5.385	5.410	5.436	5.461
18	5.486	5.512	5.537	5.563	5.588	5.613	5.639	5.664	5.690	5.715	5.740	5.766
19	5.791	5.817	5.842	5.867	5.893	5.918	5.944	5.969	5.994	6.020	6.045	6.071
20	6.096	6.121	6.147	6.172	6.198	6.223	6.248	6.274	6.299	6.325	6.350	6.375
21	6.401	6.426	6.452	6.477	6.502	6.528	6.553	6.579	6.604	6.629	6.655	6.680
22	6.706	6.731	6.756	6.782	6.807	6.833	6.858	6.883	6.909	6.934	6.960	6.985
23	7.010	7.036	7.061	7.087	7.112	7.137	7.163	7.188	7.214	7.239	7.264	7.290
24	7.315	7.341	7.366	7.391	7.417	7.442	7.468	7.493	7.518	7.544	7.569	7.595
25	7.620	70645	7.671	7.696	7.722	7.747	7.772	7.798	7.823	7.849	7.874	7.899
26	7.925	7.950	7.976	8.001	8.026	8.052	8.077	8.103	8.128	8.158	8.179	8.204
27	8.230	8.255	8.280	8.306	8.331	8.357	8.382	8.407	8.433	8.458	8.484	8.509
28	8.634	8.560	8.585	8.611	8.636	8.661	8.687	8.712	8.738	8.763	8.788	8.814
29	8.839	8.865	8.890	8.915	8.941	8.966	8.992	9.017	9.042	9.068	9.093	9.119
30	9.144	9.169	9.195	9.220	9.246	9.271	9.296	9.322	9.347	9.373	9.398	9.423
31	9.449	9.474	9.500	9.525	9.550	9.576	9.601	9.627	9.652	9.677	9.703	9.728
32	9.754	9.779	9.804	9.830	9.855	9.881	9.906	9.931	9.957	9.982	10.008	10.033
33	10.058	10.084	10.109	10.135	10.160	10.185	10.211	10.236	10.262	10.287	10.312	10.338
34	10.363	10.389	10.414	10.439	10.465	10.490	10.516	10.541	10.566	10.592	10.617	10.643
35	10.668	10.693	10.719	10.744	10.770	10.795	10.820	10.846	10.871	10.897	10.922	10.947
36	10.973	10.998	11.024	11.049	11.074	11.100	11.125	11.151	11.176	11.201	11.227	11.252
37	11.278	11.303	11.328	11.354	11.379	11.405	11.430	11.455	11.481	11.506	11.532	11.557
38	11.582	11.608	11.633	11.659	11.684	11.709	11.735	11.760	11.786	11.811	11.836	11.862
39	11.887	11.913	11.938	11.963	11.989	12.014	12.040	12.065	12.090	12.116	12.141	12.167
40	12.192	12.217	12.243	12.268	12.294	12.319	12.344	12.370	12.395	12.421	12.446	12.471
41	12.497	12.522	12.548	12.573	12.598	12.624	12.649	12.675	12.700	12.725	12.751	12.776
42	12.802	12.827	12.852	12.878	12.903	12.929	12.954	12.979	13.005	13.030	13.058	13.081
43	13.106	13.132	13.157	13.183	13.208	13.233	13.259	13.284	13.310	13.335	13.360	13.386
44	13.411	13.437	13.462	13.487	13.513	13.538	13.564	13.589	13.614	13.640	13.665	13.691
45	13.716	13.741	13.767	13.792	13.818	13.843	13.868	13.894	13.919	13.945	13.970	13.995
46	14.021	14.046	14.072	14.097	14.122	14.148	14.173	14.199	14.224	14.249	14.275	14.300
47	14.326	14.351	14.376	14.402	14.427	14.453	14.478	14.503	14.529	14.554	14.580	14.605
48	14.630	14.656	14.681	14.707	14.732	14.757	14.783	14.808	14.834	14.859	14.884	14.910
49	14.935	14.961	14.986	15.011	15.037	15.062	15.088	15.113	15.138	15.164	15.189	15.215
50	15.240	15.265	15.291	15.316	15.342	15.367	15.392	15.418	15.443	15.469	15.494	15.519
51	15.545	15.570	15.596	15.621	15.646	15.672	15.697	15.723	15.748	15.773	15.799	15.824
52	15.850	15.875	15.900	15.926	15.951	15.977	16.002	16.027	16.053	16.078	16.104	16.129
53	16.154	16.180	16.205	16.231	16.256	16.281	16.307	16.332	16.358	16.383	16.408	16.434
54	16.459	16.485	16.510	16.535	16.561	16.586	16.612	16.637	16.662	16.688	16.713	16.739
55	16.764	16.789	16.815	16.840	16.866	16.891	16.916	16.942	16.967	16.993	17.018	17.043
56	17.069	17.094	17.120	17.145	17.170	17.196	17.221	17.247	17.272	17.297	17.323	17.348
57	17.374	17.399	17.424	17.450	17.475	17.501	17.526	17.551	17.577	17.602	17.628	17.653
58	17.678	17.704	17.729	17.755	17.780	17.805	17.830	17.856	17.882	17.907	17.932	17.958
59	17.983	18.009	19.034	18.059	18.085	18.110	18.136	18.161	18.186	18.212	18.237	18.263
60	18.288	18.313	18.339	18.364	18.390	18.415	18.440	18.466	18.491	18.517	18.542	18.567
61	18.593	18.618	18.644	18.669	18.694	18.720	18.745	18.771	18.796	18.821	18.847	18.872
62	18.893	18.923	18.948	18.974	18.999	19.025	19.050	19.075	19.101	19.126	19.152	19.177
63	19.202	19.228	19.253	19.279	19.304	19.329	19.355	19.308	19.406	19.431	19.456	19.482
64	19.507	19.533	19.558	19.583	19.609	19.634	19.660	19.685	19.710	19.736	19.761	19.787
65	19.812	19.837	19.863	19.888	19.914	19.939	19.964	19.990	20.015	20.041	20.068	20.091
66	20.117	20.142	20.168	20.193	20.218	20.244	20.269	20.295	20.320	20.345	20.371	20.386
67	20.422	20.447	20.472	20.498	20.523	20.549	20.574	20.599	20.625	20.650	20.678	20.701
68	20.720	20.752	20.777	20.803	20.828	20.853	20.879	20.904	20.930	20.955	20.980	21.006
69	21.031	21.057	21.082	21.107	21.133	21.158	21.184	21.209	21.234	21.260	21.285	21.311

Fig. 16-31.   **Metric-customary conversion table.**

CHAPTER 5

Reading Floor Plans

Orthographic views of a structure show the exterior sides (elevations), top (roof), and bottom of a structure. The bottom is never used in architectural work and the top (roof) view is usually shown only on location plans. However, none of these views shows interior details. A floor plan is a top orthographic view as seen if the building is cut (sectioned) horizontally about four feet above the floor line, as shown in Fig. 17-1. The floor plan is a basic plan showing the interior layout of the design.

The floor plan is the most significant of all architectural working drawings since it contains more information pertaining to the design and construction of the structure than any other single plan.

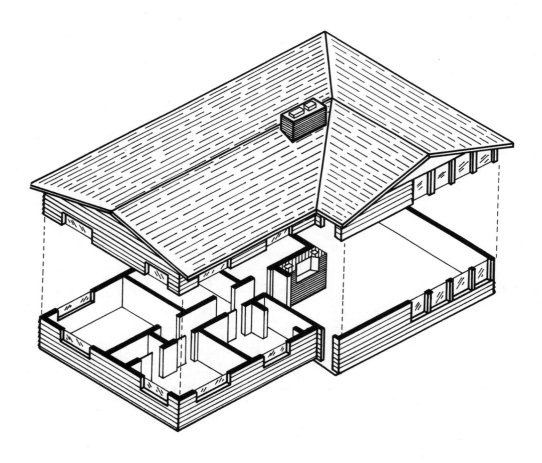

Fig. 17-1. A floor plan is a top view cut through the building approximately 4 ft (1.2 m) above the floor line.

Unit 17
Types of Floor Plans

There are two basic types of floor plans: general-design floor plans and working-drawing floor plans.

General-Design Floor Plans

General-design floor plans are used only to show the basic layout, arrangement of areas, and de-

sign features. They do not include much detail nor are they fully dimensioned. General design plans are also called abbreviated plans. Plans of this type range from very simple floor-plan sketches (Fig. 17-2), which show only the approximate position of walls and partitions, to single-line floor plans (Fig. 17-3), which are more precise but which do not show specific wall thicknesses or detail dimensions. Pictorial floor plans such as the one shown in Fig. 17-4 are also used for general design interpretation purposes. These plans are easier read by the layman because more depth is added to exterior walls and internal partitions, giving the plan a third dimension. The most popular abbreviated floor plan is the type shown in Fig. 17-5. These types of plans are prepared primarily for sales purposes. They do not include sufficient information for construction, but they are drawn to scale and do include approximate internal dimensions for each room and total width and length dimensions for the structure. These plans are usually drawn or reduced to a scale too small to include many construction details and symbols.

Often abbreviated floor plans of specific rooms, such as the ones shown in Figs. 17-6a and b and

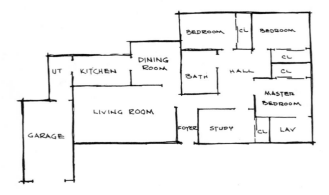

Fig. 17-2. A floor-plan sketch.

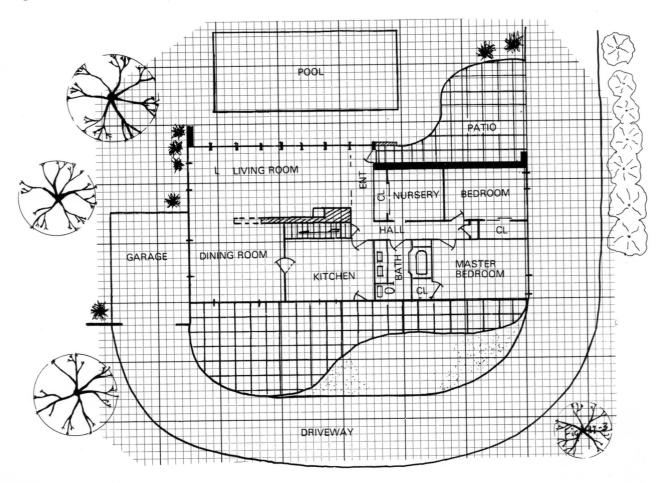

Fig. 17-3. Single-line floor plan.

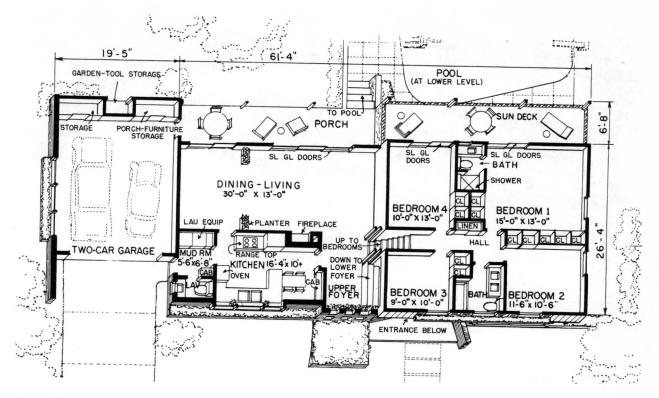

Fig. 17-4. A pictorial floor plan.

17-7*a* and *b*, are used to show the positioning of appliances and fixtures for sales purposes.

Abbreviated plans are often reversed as shown in Fig. 17-8 to show the option of building a structure exactly opposite from the way it was designed. Any plan can be reversed simply by running the original drawing through a copy machine upside down.

Working-Drawing Floor Plans

Abbreviated floor plans are sufficient to show rough layouts and preliminary design for presentation and sales purposes but are not accurate or complete enough to be used as working drawings for building purposes. An accurate floor plan complete with dimensions and materials sym-

bols, such as the one shown in Fig. 17-9, is necessary for construction purposes. Using a plan of this type, a contractor can accurately interpret the desires of the designer. The prime function of a working-drawing floor plan is to communicate design information to the contractor and construction technicians. A complete floor plan eliminates misunderstandings between the designer and the builder. The builder's judgment must be used to fill in omitted details if the floor plan is inaccurate, incomplete or abbreviated. The floor plan shown in Fig. 17-9 is a complete working-drawing floor plan prepared for construction purposes. In some cases electrical, air-conditioning, and plumbing symbols are added to the floor plan rather than preparing separate specialized floor plans.

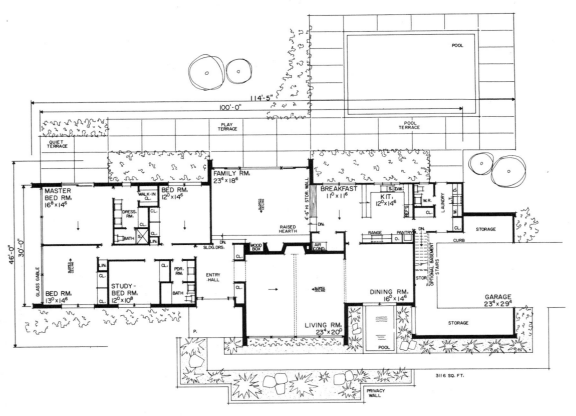

Fig. 17-5. Abbreviated floor plan with a related pictorial drawing. (Home Planners, Inc.)

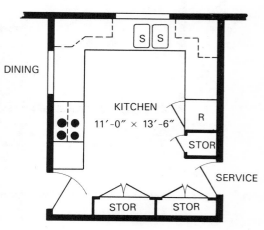

Fig. 17-6. Floor plan of a bath.

Fig. 17-7. A floor plan of a kitchen. (Home Planners, Inc.)

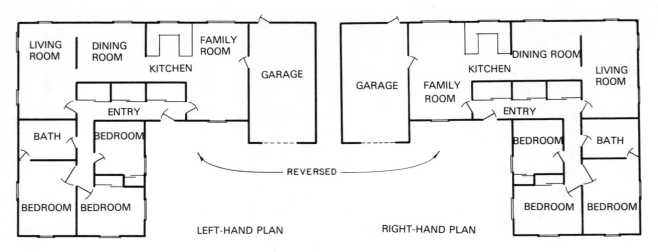

Fig. 17-8. A reversed floor plan.

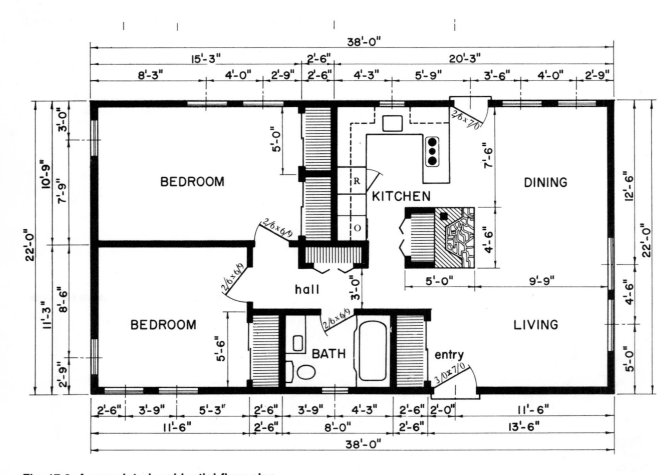

Fig. 17-9. A completed residential floor plan.

Unit 18
Floor-Plan Symbols
Architects substitute symbols for materials and fixtures just as stenographers substitute shorthand symbols for words. It is obviously more convenient and time-saving to draw a symbol of the material than to repeat a description every time the material is used. It is also impractical to describe all construction materials shown on floor plans such as fixtures, doors, windows, stairs, and partitions without the use of symbols.

In addition to the use of symbols to represent materials and fixtures on a floor plan, different

LINE CONVENTIONS

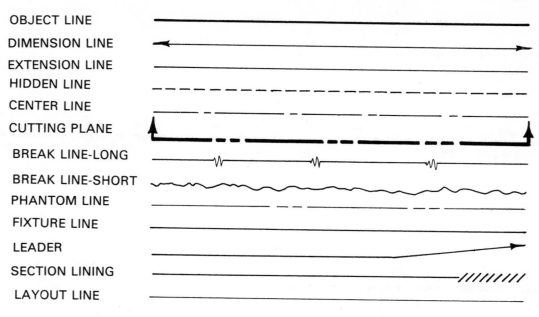

OBJECT LINE

DIMENSION LINE

EXTENSION LINE

HIDDEN LINE

CENTER LINE

CUTTING PLANE

BREAK LINE-LONG

BREAK LINE-SHORT

PHANTOM LINE

FIXTURE LINE

LEADER

SECTION LINING

LAYOUT LINE

Fig. 18-1. Lines found on floor plans.

types of lines are used on floor plans as symbols. Figure 18-1 shows lines used on floor plans. Lines are actually symbols, since different lines convey different meanings.

Object lines are used to show the main outline of the building, including exterior walls, interior partitions, porches, patios, and driveways. These lines are the most outstanding on the drawing.

Dimension lines are thinner unbroken lines upon which building dimensions are placed.

Extension lines extend from the building to permit dimensioning outside the plan.

Hidden lines are used to show areas which are not visible on the surface but which exist below the plane of projection. Hidden lines are also used in floor plans to show objects above the floor-plan section line, such as wall cabinets, arches, and beams.

Center lines are used to identify and locate the center of symmetrical objects such as exterior doors and windows. These lines are usually necessary for dimensioning purposes.

Cutting-plane lines are very heavy lines used to denote a sectioned area. In some cases, this line is drawn completely through the plan; in other cases, this line is broken to eliminate interference with details of the plan.

Break lines are used when an area is not drawn in its entirety. They are used where a feature remains exactly the same or is repeated in a pattern

over a long distance. Breaks are used to shorten lines without enlarging the scale of a drawing. A ruled line with freehand breaks is used for long, straight breaks. A wavy, uneven freehand line is used for smaller, irregular breaks.

Phantom lines are used to indicate alternative positions of fixtures, movable partitions, or future construction additions. Existing structures to be removed are also shown with phantom lines.

Fixture lines outline the shape of kitchen, laundry, bathroom fixtures, or built-in furniture which is part of the structure.

Leaders connect a note or dimension with the object it represents. They are very light and sometimes are curved to eliminate confusion with other lines.

Section lines are used to show section lining in sectional drawings. A different symbol is used for each building material. Section-lining lines are lighter than object lines.

Figures 18-2a through 18-2f show the application of symbols to a floor plan. Figures 18-3, 18-5, and 18-6 show typical construction details represented by symbols on floor-plan drawings. Although most architectural symbols are standardized, some variations of symbols are used in different parts of the country by architects, designers, and draftsmen. For example, Fig. 18-4 shows several methods used for drawing outside

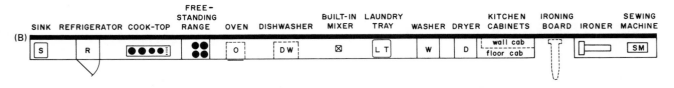

Fig. 18-2a. Built-in component floor-plan symbols.

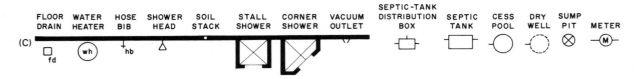

Fig. 18-2b. Kitchen and laundry floor-plan symbols.

Fig. 18-2c. Sanitation facility floor-plan symbols.

Fig. 18-2d. Heating and air-conditioning floor-plan symbols.

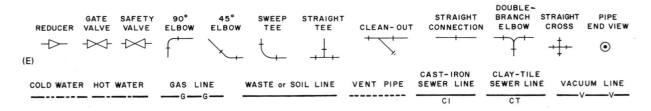

Fig. 18-2e. Plumbing floor-plan symbols.

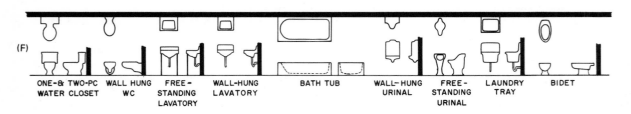

Fig. 18-2f. Lavatory-bath fixture symbols.

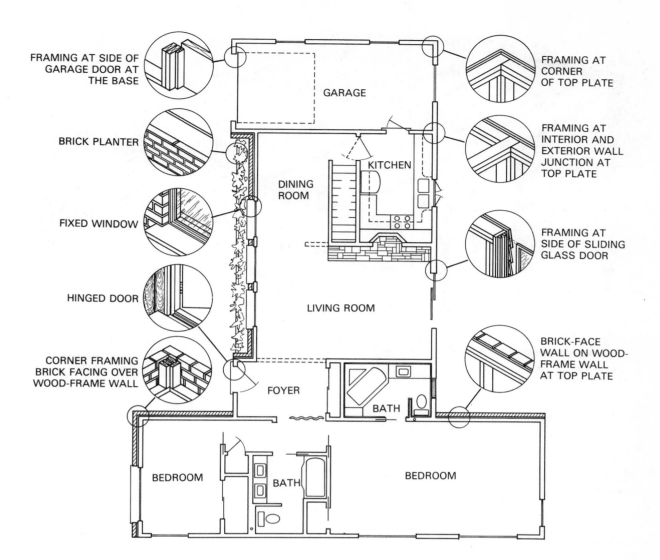

FRAMING AT SIDE OF
GARAGE DOOR AT
THE BASE

BRICK PLANTER

FIXED WINDOW

HINGED DOOR

CORNER FRAMING
BRICK FACING OVER
WOOD-FRAME WALL

GARAGE

DINING
ROOM

KITCHEN

LIVING ROOM

FOYER

BATH

BEDROOM

BATH

BEDROOM

FRAMING AT
CORNER
OF TOP PLATE

FRAMING AT
INTERIOR AND
EXTERIOR WALL
JUNCTION AT
TOP PLATE

FRAMING AT
SIDE OF SLIDING
GLASS DOOR

BRICK-FACE
WALL ON WOOD-
FRAME WALL
AT TOP PLATE

Fig. 18-3. Floor-plan symbols are abbreviated to represent details.

wall symbols for frame buildings. These four alternative methods represent the same type of construction. Learning and remembering floor-plan symbols will be easier if you associate each symbol with the actual material or fixture it represents.

Although floor-plan symbols represent an approximation of the exact appearance of the floor-plan section as viewed from above, sometimes a complete representation is not possible. Many floor-plan symbols are too intricate to be drawn to $1/4'' = 1'\text{-}0''$ or $1/8'' = 1'\text{-}0''$ scale. Therefore, many details are not shown on floor plans, but instead are shown on separate detail drawings.

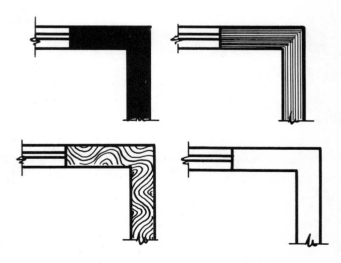

Fig. 18-4. Framing details shown on floor plans.

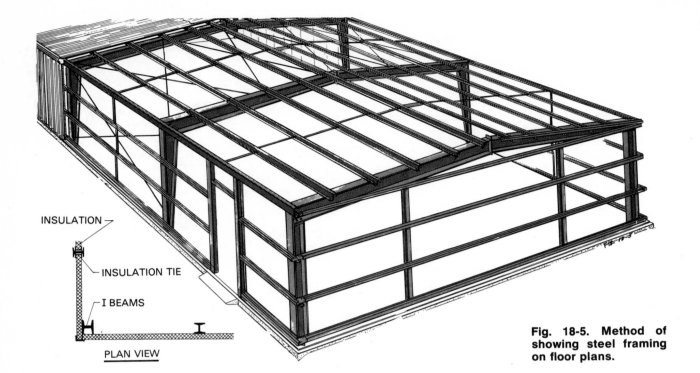

INSULATION

INSULATION TIE

I BEAMS

PLAN VIEW

Fig. 18-5. Method of showing steel framing on floor plans.

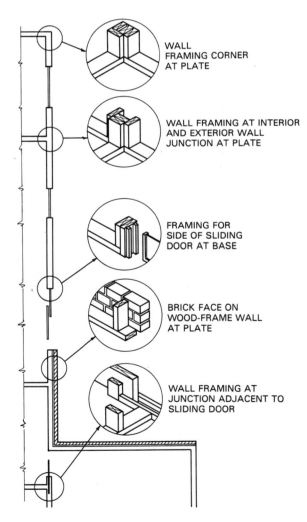

WALL FRAMING CORNER AT PLATE

WALL FRAMING AT INTERIOR AND EXTERIOR WALL JUNCTION AT PLATE

FRAMING FOR SIDE OF SLIDING DOOR AT BASE

BRICK FACE ON WOOD-FRAME WALL AT PLATE

WALL FRAMING AT JUNCTION ADJACENT TO SLIDING DOOR

Fig. 18-6. Different methods of showing frame construction.

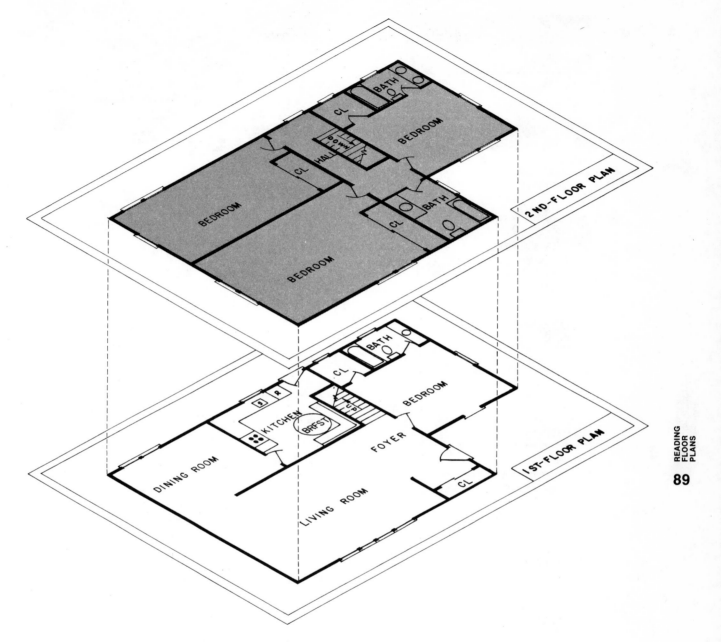

Inside the figure:
- BATH
- CL
- BEDROOM
- DOWN
- HALL
- CL
- BEDROOM
- CL
- BATH
- BEDROOM
- 2ND-FLOOR PLAN
- BATH
- CL
- BEDROOM
- KITCHEN
- BRFST
- DINING ROOM
- UP
- FOYER
- LIVING ROOM
- CL
- 1ST-FLOOR PLAN

Fig. 19-1. Relationship of second-floor plan to first-floor plan.

Unit 19
Level Designations
Structures with 2 or more stories, 1½ stories, and split-levels require a separate floor plan for each level. Figure 19-1 shows the relationship of a second-floor plan to a first-floor plan. Each successive floor plan shows the alignment of exterior walls, interior bearing partitions, fireplaces, stairwells, walls containing piping, wiring, and duct work that extend vertically. Figure 19-2 shows the basic types of multiple-story floor plans. Full two-or-more-story plans have the same overall dimensions as the first-floor plan. In Fig. 19-3 notice how the position of the fireplace and stairwell on the first level coincides with the position of the fireplace on the second level. In split-level structures, like the one shown in Fig. 19-4, the left portion and the upper right portion are combined on the first-floor plan. The lower level is shown in a separate plan. Thus the right portion of both plans align, but the stairwell is shown on both plans: solid on the first-floor plan and dotted on the lower-level plan. Also, notice that the stairwell area is labeled "up" on the first-floor plan to further indicate

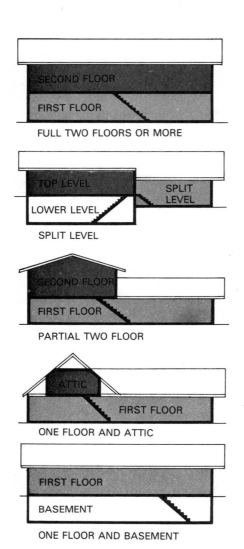

FULL TWO FLOORS OR MORE

SPLIT LEVEL

PARTIAL TWO FLOOR

ONE FLOOR AND ATTIC

ONE FLOOR AND BASEMENT

Fig. 19-2. Basic styles of multiple-story dwellings.

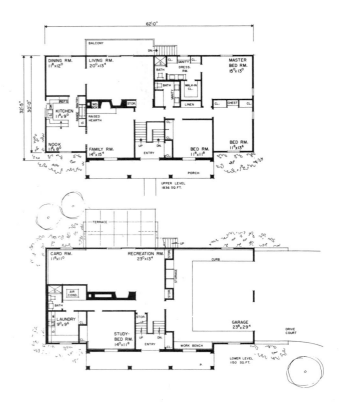

Fig. 19-3. Full two-story plan. (Home Planners, Inc.)

which level is highest and in which direction the stairs lead.

Many structures are designed with a second story (level) over only part of the first floor. In this case, the roof line representing the first-floor area is drawn in order to locate the second-floor plan correctly over the appropriate area of the first-floor plan, as shown in Fig. 19-5. When the first-floor roof outline is not included (Fig. 19-6), it is more difficult to align the second-and first-floor plans. However, using key points such as stairwells and fireplaces will usually reveal the position to the builder.

One-and-one-half story plans are sometimes difficult to interpret since part of the roof line extends through interior partitions. The story-and-a-half plan shown in Fig. 19-7, in addition to revealing the second-floor plan, shows the outline of the roof. In this plan, lines are used to show the outlines of the building under the roof. In some story-and-a-half plans (Fig. 19-8), the outline of

the roof and dormers are also shown in order to orient the design of the second floor with the positioning of dormers.

Interpreting multilevel floor plans with three or more levels is similar to interpreting the difference between a first-floor plan and a second-floor plan. Each floor must be interpreted and related to the floor directly underneath. This means relating the second floor to the first, the third floor to the second, the fourth floor to the third, and so forth, as shown in Fig. 19-9.

Movement from one level to another is designed into a structure either through the use of elevators, ramps, or stairs. Stairs are the most commonly used device in one-, two-, and three-

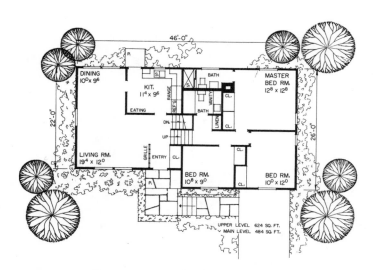

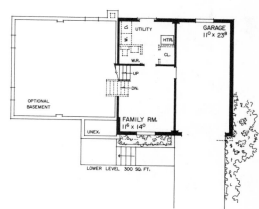

Fig. 19-4. Split-level plan. (Home Planners, Inc.)

story structures. For buildings over three stories, elevators are most commonly used.

Stairs are of two types: static and movable. Most structures under four stories use static (fixed) stairs. Figure 19-10 shows the basic types of static stairs. Stairs on floor plans show the number of risers and the direction of the stairs from the plan level. Stairs shown on first-floor plans with a notation "down" indicate the stairs go down to the basement. Stairs on a first-floor plan with an indication "up" show the stairs go

up to the second floor. To fully understand the stairwell symbols as shown on a floor plan, the following terminology used in describing stairways (Fig. 19-11) must be understood: A *tread* is the walking surface of a stairway. *Risers* are the vertical members between treads. There is always one less tread than riser in any stair system. *Rise* is the total vertical distance from one floor to the next. *Run* is the total horizontal distance of the stairwell. *Stringers* are the side support of a stairwell system. *Nosing* is the projection tread

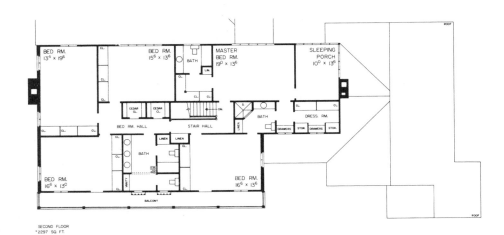

SECOND FLOOR
2297 SQ. FT.

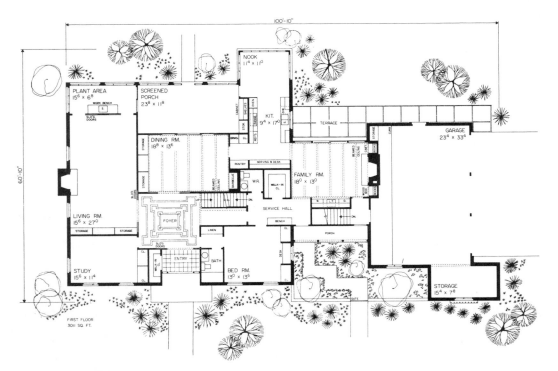

FIRST FLOOR
3011 SQ. FT.

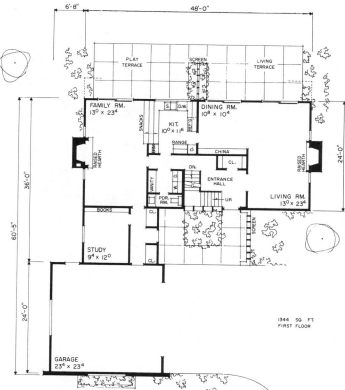

Fig. 19-6. Partial two-story plan without first-level roof outline. (Home Planners, Inc.)

beyond the riser. *Headroom* is the vertical distance from the outside edge nosing (nosing line) of a tread to the nearest ceiling line.

A *breakline* is often used to indicate that the stairs continue past that point on the floor plan to the next level, but are beyond the plane of projection.

The simplest and most common stair type is the straight-run stair, shown in Fig. 19-12. However, when the length of the rise is excessive, or when the run would interfere with other design features, landings are incorporated into the stair system in order to condense and reduce the length of the run. The L stairs shown in Fig. 19-13 is one of the most popular designs used in reduc-

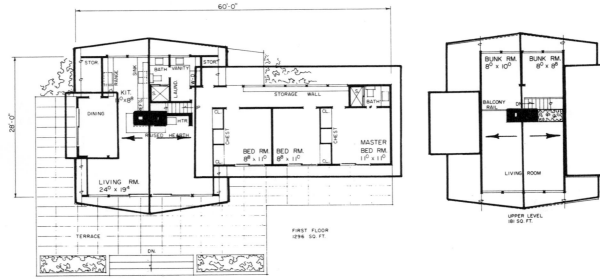

Fig. 19-7. One-and-one-half story plan with full roof outline shown. (Home Planners, Inc.)

ing the length of the run. Variations of this type include: the long L (Fig. 19-14), the wide L (Fig. 19-15), and the double L, shown in Fig. 19-16. Study the plan view of each of these stair systems and relate it to the pictorial drawing shown in each illustration.

When space is even more restrictive, winder stairs (Fig. 19-17) are used to further conserve horizontal space since three stairs take the place of one landing.

The use of the U-stair design (Fig. 19-18) also reduces the horizontal run as does the L type, but the U also conserves and condenses space by

paralleling two sets of stairs connected to one landing. Figure 19-19 shows another variation of the U-stair design.

The most efficient stair system in the conservation of floor-plan space is the spiral (circular) stairs (Fig. 19-20), which is a system of radial winders attached to a central pole. The plan view of a spiral staircase includes the direction from the floor plan in either a clockwise or counterclockwise direction.

As you study each of these stair systems, think of them in relation to the stairwell-opening outlines on the floor plan, as shown in Fig. 19-21.

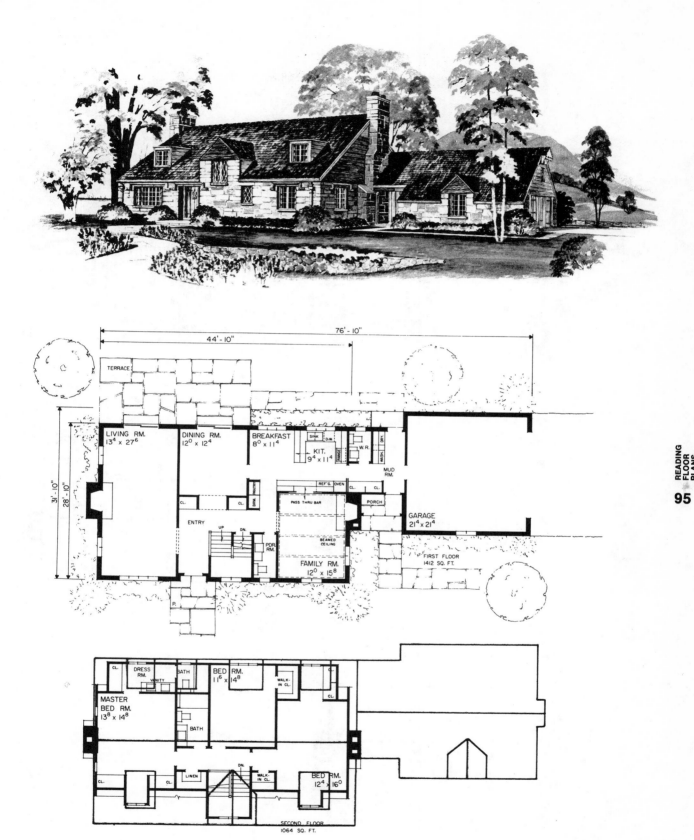

Fig. 19-8. One-and-one-half story plan with gable end, shed dormers, and wall positions shown. (Home Planners, Inc.)

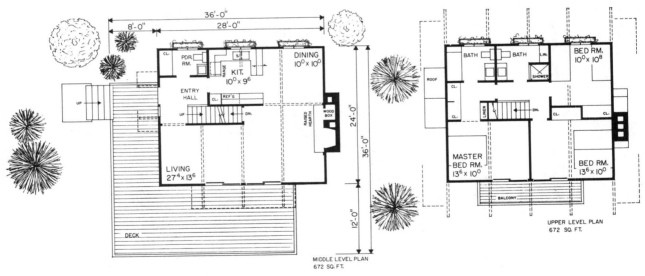

MIDDLE LEVEL PLAN
672 SQ. FT.

36'-0"
8'-0"
28'-0"
24'-0"
36'-0"
12'-0"

UP

CL.
PDR. RM.
RANGE
KIT.
10⁰ x 9⁶
DINING
10⁰ x 10⁰

ENTRY HALL
UP
CL. REF'G
DN.
RAISED HEARTH
WOOD BOX

LIVING
27⁴ x 13⁶

DECK

UPPER LEVEL PLAN
672 SQ. FT.

ROOF
BATH
BATH
LIN.
SHOWER
BED RM.
10⁰ x 10⁸
CL.
CL.
LINEN
DN.
CL.
CL.

MASTER
BED RM.
13⁶ x 10⁰
BED RM.
13⁶ x 10⁰

BALCONY

LOWER LEVEL PLAN
672 SQ. FT.

UP

CL.
LAUND.
W.
D.
BATH
SHOWER
WORK RM.
11⁰ x 9⁰
AIR COND.
WOOD BOX
RAISED HEARTH

UP
CL.

STORAGE

WET HALL

SKI-LOUNGE
19⁶ x 13⁴

COVERED TERRACE

**Fig. 19-9. Multiple-level plans.
(Home Planners, Inc.)**

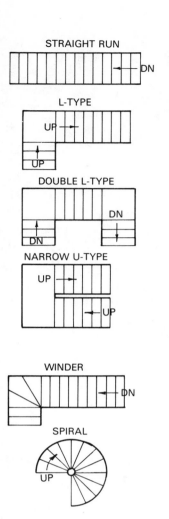

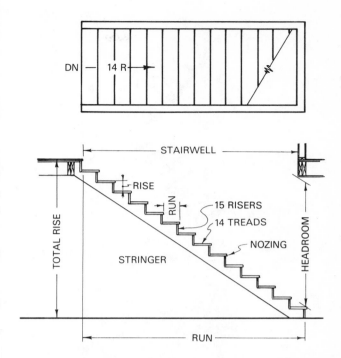

Fig. 19-11. Method of showing stairwell opening and stair direction on floor plan.

Fig. 19-10. Basic types of static stairs.

Fig. 19-12. Straight-run stairs.

PICTORIAL OF STRAIGHT RUN STAIRS

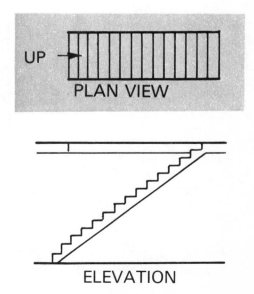

UP

PLAN VIEW

ELEVATION

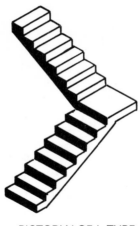

PICTORIAL OF L-TYPE
STAIRS

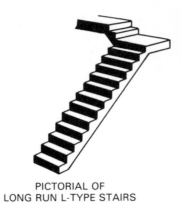

PICTORIAL OF
LONG RUN L-TYPE STAIRS

Fig. 19-14. Long L stairs.

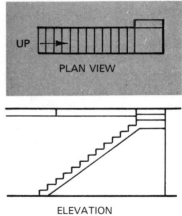

UP

PLAN VIEW

ELEVATION

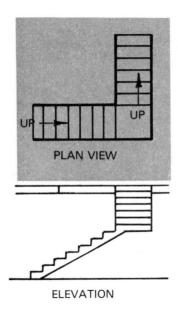

UP

UP

PLAN VIEW

ELEVATION

Fig. 19-13. L stairs.

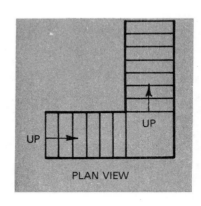

UP

UP

PLAN VIEW

ELEVATION

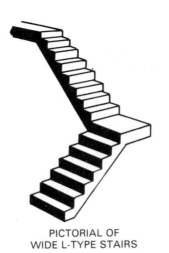

PICTORIAL OF
WIDE L-TYPE STAIRS

Fig. 19-15. Wide L stairs.

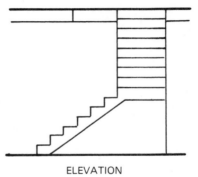

Fig. 19-16.
Double L stairs.

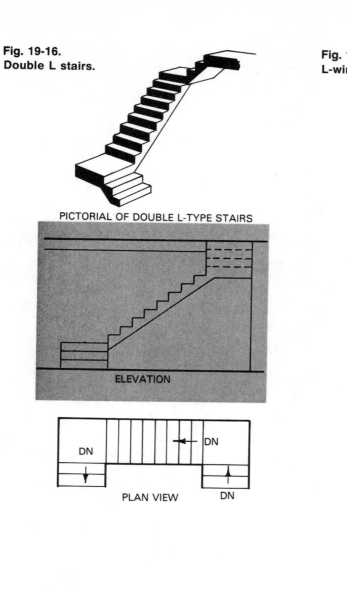

PICTORIAL OF DOUBLE L-TYPE STAIRS

ELEVATION

DN DN

DN

PLAN VIEW

Fig. 19-17.
L-winder stairs.

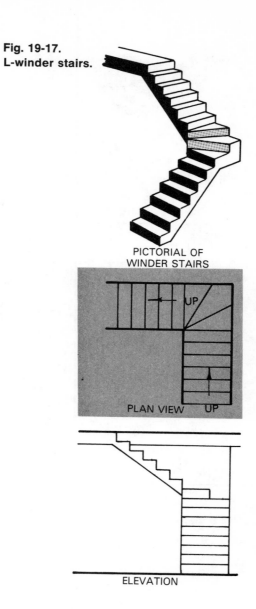

PICTORIAL OF
WINDER STAIRS

UP

PLAN VIEW UP

ELEVATION

Fig. 19-18. U stairs.

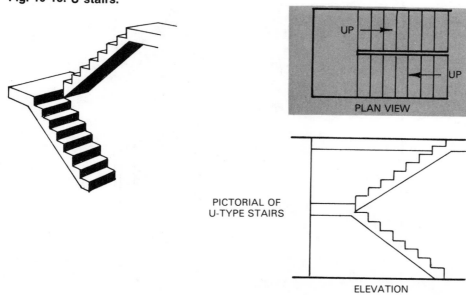

UP

UP

PLAN VIEW

PICTORIAL OF
U-TYPE STAIRS

ELEVATION

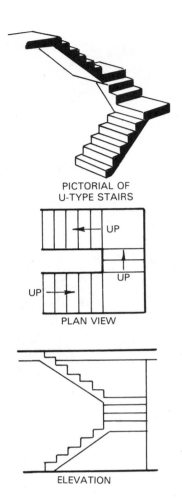

PICTORIAL OF
U-TYPE STAIRS

PLAN VIEW

ELEVATION

Fig. 19-19. Wide U stairs.

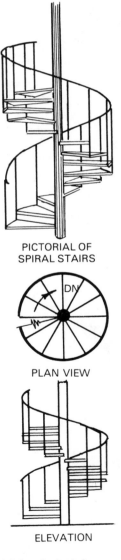

PICTORIAL OF
SPIRAL STAIRS

PLAN VIEW

ELEVATION

Fig. 19-20. Spiral (circular) stairs.

Fig. 19-21. Framing of stairwell opening.

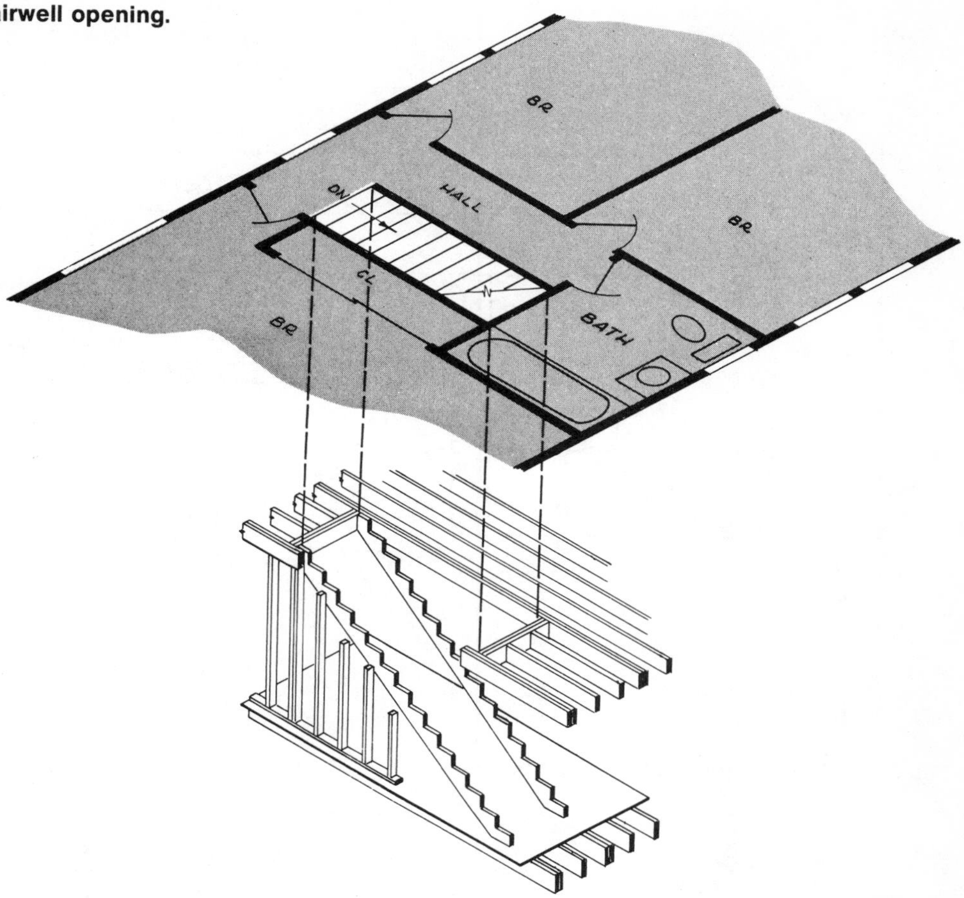

Unit 20
Reading Floor-Plan Dimensions

Floor-plan dimensions show the builder the width and length of the building. They show the location of doors, windows, stairs, and fireplaces. They show the width and length of each room, closet, and hall. There are several kinds of dimensions found on floor plans, as shown in Fig. 20-1. *Overall dimensions* show the total length of a building. *Subdimensions* show the length of subdivisions of a building. *Positioning dimensions* show the position of features such as doors, windows, and fixtures. *Size dimensions* indicate the size (width, length and/or height) of a material or components.

Some floor plans which are not used for construction purposes (abbreviated plans) include only bare minimum dimensions. Such a plan is shown in Fig. 20-2. If a floor plan is not to be used for construction purposes, many of the detail dimensions are unnecessary and clutter the floor plan, making reading the plan difficult. For this reason, abbreviated plans show only overall dimensions and sizes of rooms. Dimensions are not positioned on dimension lines. In this type of plan, the limited dimensions shown are sufficient to summarize the relative sizes of the building and its rooms. But these dimensions are not sufficient for building purposes.

A floor plan must be completely dimensioned as shown in Fig. 20-3 to ensure that the building will be constructed precisely as designed. These dimensions convey the exact wishes of the designer to the builder, and the contractor is given little tolerance in interpreting the size and position of the various features of the plan.

The number of dimensions included on a floor plan depends largely on how much freedom of interpretation the architect wants to give the builder. If complete dimensions are shown on the plan, the builder cannot deviate greatly from the original design. However, if only a few dimensions are shown, then the builder must determine many exact locations and sizes of many areas and components.

Rules for Reading Floor-Plan Dimensions

Many construction mistakes result from errors in either preparing or reading architectural drawings incorrectly. Most of these errors result from

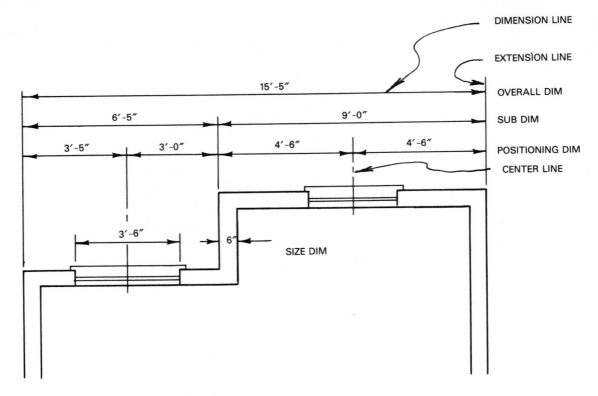

Fig. 20-1. Kinds of architectural dimensions.

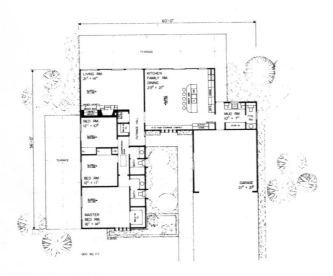

Fig. 20-2. Minimum floor-plan dimensions. (Home Planners, Inc.)

mistakes in misreading dimensions. Reading dimensions incorrectly is, therefore, costly in time, efficiency, and money. Familiarization with the following rules for reading floor-plan dimensions will eliminate much confusion and error. (These rules are illustrated by the numbered arrows in Fig. 20-3.)

1. Dimension lines are unbroken lines with dimensions placed centered above the line.

2. Foot and/or inch marks are used on all architectural dimensions unless otherwise noted. On metric drawings m is used to denote meters and mm to denote millimeters.

3. Dimensions over 1 ft are expressed in feet and inches.

4. Dimensions less than 1 ft are shown in inches.

5. A slash is often used with fractional dimensions to conserve vertical space. Do not confuse a fraction numerator with a whole number.

6. Dimensions read from the right or from the bottom of the drawing.

7. Overall building dimensions are found outside the other dimensions.

8. Line and arrowhead weights for architectural dimensions are lighter than object lines.

9. Room sizes are shown by stating width and length if dimension lines are omitted.

10. When an area to be dimensioned is too small for the numeral, the dimension is placed outside the extension line.

11. Rooms are sometimes dimensioned from the center line of partitions, as shown on the left side of Fig. 20-4. However, rule 13 is usually practiced.

12. Window and door sizes are shown directly on the door or window symbol or may be indexed to the door or window schedule (Fig. 20-5).

13. Rooms are dimensioned from wall to wall exclusive of thicknesses of wall covering materials shown on the right side of Fig. 20-4.

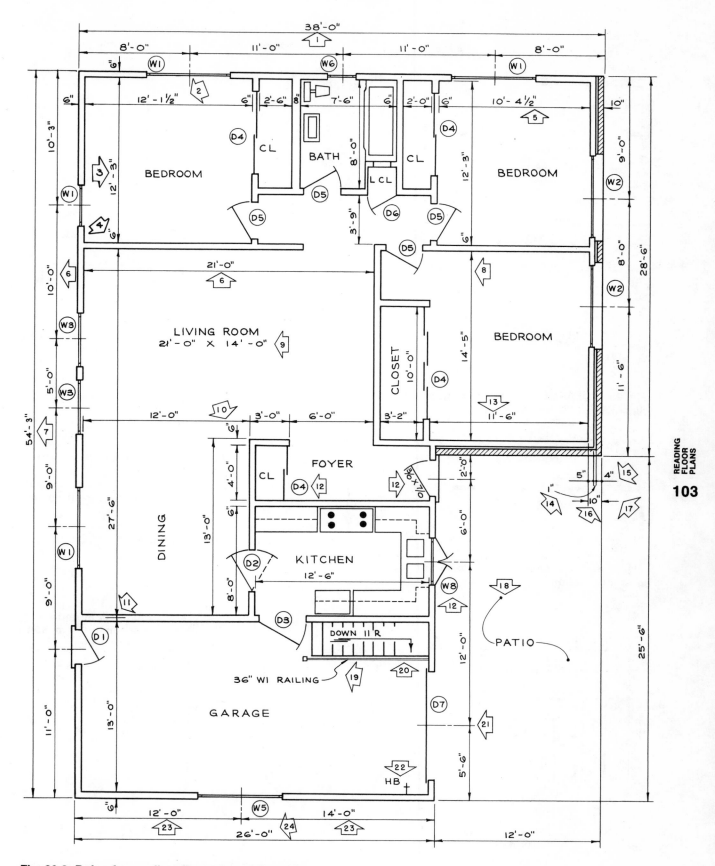

Fig. 20-3. Rules for reading dimensioned floor plans.

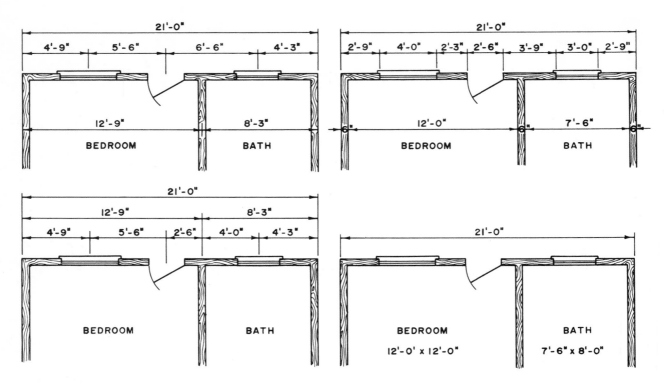

Fig. 20-4. Different methods of positioning floor-plan dimensions.

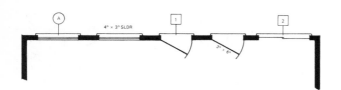

Fig. 20-5. Use of code symbols to show door and window information.

14. Curved leaders are often used to eliminate confusion with other dimension lines.
15. When areas are too small for arrowheads, dots or slashes may be used to indicate dimension limits.
16. The dimensions of brick and stone veneer are added to framing dimensions (Fig. 20-6).

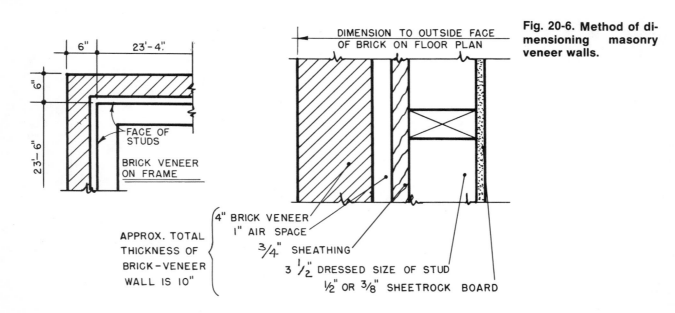

Fig. 20-6. Method of dimensioning masonry veneer walls.

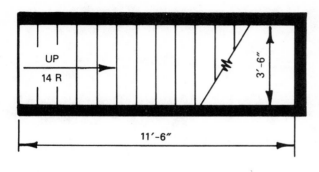

Fig. 20-7*a*. Method of dimensioning stairs.

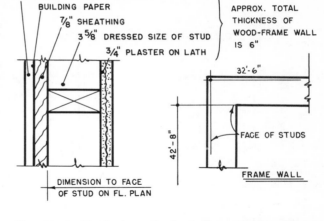

Fig. 20-7*b*. Typical method of dimensioning frame walls.

17. When space is small, arrowheads may be found outside the extension lines.
18. A dot with a leader refers to the large area noted.
19. Dimensions that cannot be seen on the floor plan or those too small to be placed on the object are placed on leaders for easier reading.
20. In dimensioning stairs, the number of risers is shown with an arrow indicating the direction up or down (Fig. 20-7*a*).
21. Windows, doors, pilasters, beams, and areaways are dimensioned to their centers with center lines.
22. Terms or abbreviations are used when graphic symbols do not show clearly what is intended.
23. Subdimensions must add up to overall dimensions. For example, 14'-0" + 12'-0" = 26'-0".
24. Floor-plan dimensions always refer to the actual size of the building regardless of the scale of the drawing. The dimensions of the building shown in Fig. 20-3 are 38 ft by 54 ft.
25. When framing dimensions are desirable, room dimensions are shown by distances between studs, as shown in Fig 20-7*a*.
26. Since building materials vary in thickness, check the thickness of each component in the wall and partitions, such as furring, panel, plaster stud, brick, and tile thicknesses (Fig. 20-7*a*). Add these thicknesses together to establish the total wall thickness if not dimensioned.

27. Masonry plan walls are dimensioned using the exact thickness of the masonry excluding exterior and interior wall covering materials, as shown in Fig. 20-8.
28. Refer to notes on floor plans for sizes and direction of ceiling joist, beams, headers, floor coverings, and room or area labels.
29. A north arrow on the plan indicates the compass orientation of the plan.
30. Scale of the plan refers to the ratio of the actual building to the drawing.
31. Avoid scaling a printed drawing since plans stretch in the reproduction process.

Reading Metric Dimensions

Metric dimensions are read in exactly the same manner as conventionally dimensioned floor plans except that all dimensions are in meters and millimeters. Figure 20-9 shows a conventionally dimensioned floor plan with metric equivalents shown under each dimension. This type of dual dimensioning is normally avoided but sometimes necessary when conventional and metric components are combined on the same plan.

Reading Modular Dimensions

Buildings to be erected with modular components are designed within modular limits. Dimensions used in modular system buildings are

Fig. 20-8. Method of dimensioning masonry walls.

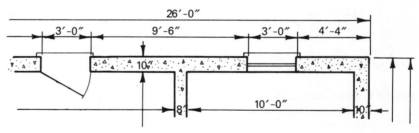

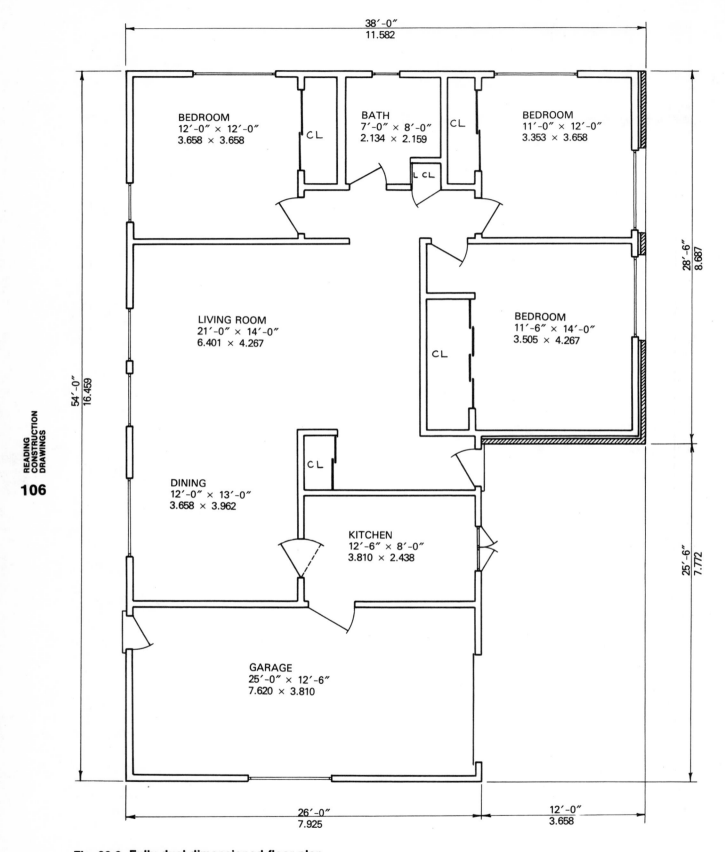

Fig. 20-9. Fully dual dimensioned floor plan.

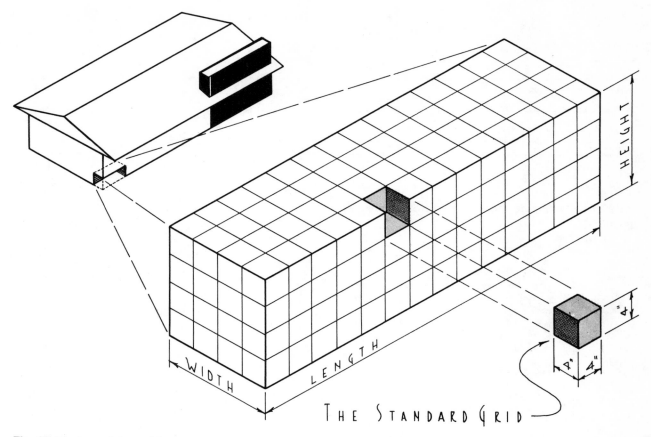

Fig. 20-10. A modular grid.

expressed in standard sizes. This procedure ensures the proper fitting of various components. The planning of rooms to accommodate standard materials also saves considerable labor, time,

and material. The modular system of coordination is based on a standard grid placed on the width, length, and height of a building, as shown in Fig. 20-10.

In modularly designed buildings, all building dimensions—width, length, and height—fall on some 4-in. module. Figure 20-11 shows a modular component grid which represents multiples of 4 in. Dimensions and components precisely designed on the 16-, 24-, and 48-in. spaces of the modular grid assure an accurate and less troublesome fitting of components. Building materials are selected to conform to this module. As new building materials are developed, their sizes are established to conform to modular sizes. However, many building materials do not conform to the 4-in. grid, and, therefore, the dimensioning procedure is adjusted accordingly. Dimensions that align with the 4-in. module are known as *grid dimensions*. Dimensions that do not align with the 4-in. grid are known as *nongrid dimensions*. Figure 20-12 shows the two methods of indicating grid dimensions and nongrid dimensions. Grid dimensions are shown by conventional arrowheads, and nongrid dimensions are shown by dots (instead of arrowheads) on the dimension lines.

The primary difference between reading dimensions on modular plans and reading conven-

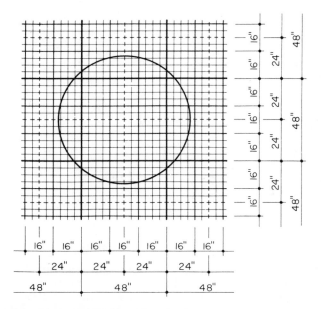

Fig. 20-11. Modular component grid.

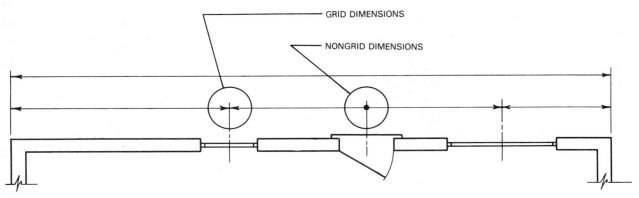

Fig. 20-12. **Combining modular and nonmodular dimensions.**

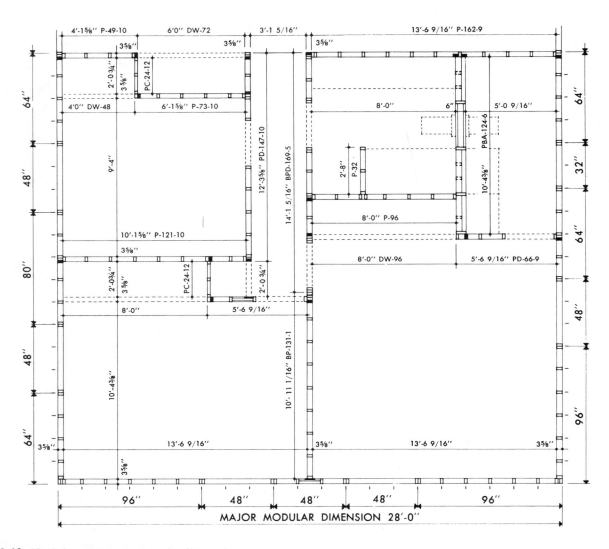

Fig. 20-13. **Modular component code dimensions.**

tionally dimensioned floor plans is the use of a component identification system. Figure 20-13 shows the use of component codes in place of dimensions on a modular component floor plan. In reading this type of plan, the builder need not be concerned with the actual detail dimensions but with the selection of the correct component, which has been precut and preassembled. Exact dimensional tolerances are maintained in the construction of components. Consequently, when components are assembled as designed, the assembled components will produce the building exactly as designed.

When preassembled modular components are used in conjunction with components framed in the field, a combination of dimensions and component coding is used.

Interpreting Sequence

In reading a floor plan, (step 1), look at the position of walls and partitions. Concentrate on these as though the drawing were an abbreviated floor plan, blocking out all other symbols, notes, and details. Memorize the plan in this form. Next, (step 2) study the position of the major components—doors, windows, stairs, fireplaces—and building equipment such as bathroom and kitchen fixtures. Now (step 3) carefully read the dimensions to get an understanding of the general size of the structure and its components. Start with the outside overall dimensions and then read the inside dimensions. Next (step 4) read over the construction notes. If a combined set is provided, next read the electrical, air-conditioning, and plumbing symbols. Figure 20-14 shows examples of these four basic steps.

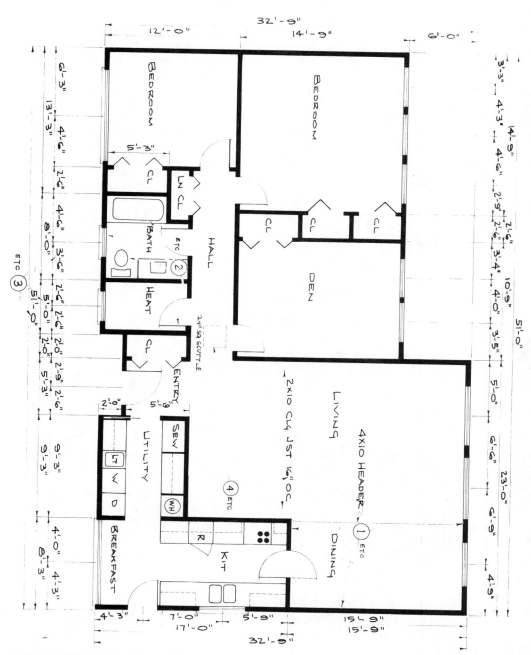

Figure 20-14. Steps in interpreting architectural plans.

CHAPTER 6

Reading Elevation Drawings

Elevation drawings represent the exterior view of a structure. The exterior sides (elevations) of a building are the same as the front, right-side, left-side, and rear views in an orthographic drawing.

Unit 21
Projection and Orientation
Elevation drawings are projected from architectural floor plans just as side views are projected from the top view of an orthographic drawing (Fig. 21-1). Figure 21-2 shows the four basic elevation planes of projection and their relationship when elevations are projected in the same plane (flat) as the floor plan. Trace the position of chimneys, doors, windows, overhangs, and building corners directly from the floor plan outward to the elevation plane. Trace with your eye or with a pencil the projection of the chimney from the floor plan to each of the elevations shown in Fig. 21-3. You will see the relative position of the chimney as you view the structure from four different directions which correspond to the four elevations. Think of the four elevations as they relate to each side of the floor plan. Also, think of the left and right elevations as hinging from the front elevation.

Elevations are classified according to their functions. They are called the front elevation, the rear elevation, the right elevation, and the left

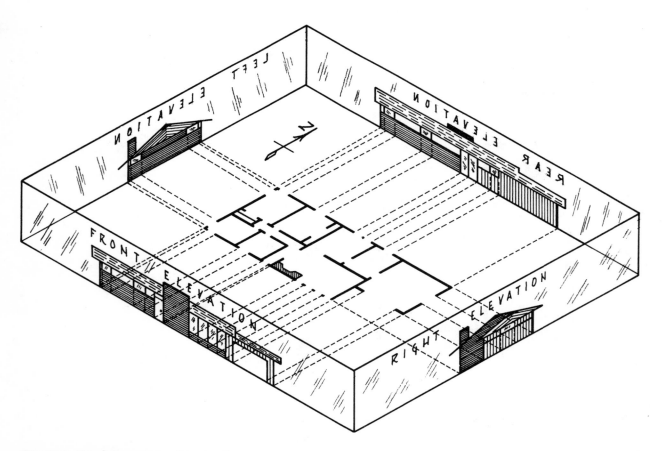

Fig. 21-1. Elevation plane of projection.

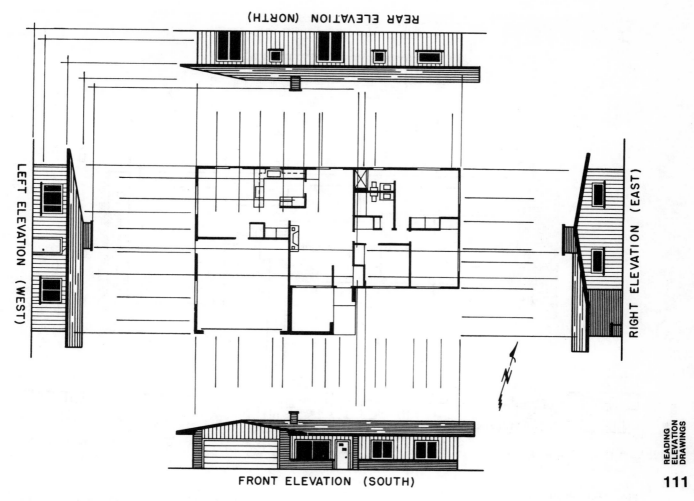

Fig. 21-2. Relationship of elevation to floor plan.

REAR ELEVATION (NORTH)

LEFT ELEVATION (WEST)

RIGHT ELEVATION (EAST)

FRONT ELEVATION (SOUTH)

READING ELEVATION DRAWINGS

111

elevation. The front view of a structure is known as the front elevation. The view projected from the rear of the floor plan is known as the rear elevation. The view projected from the right side of the front is known as the right elevation, and the view projected from the left side of the front is known as the left elevation. When these elevations are projected on the same drawing sheet as the floor plan, the rear elevation appears to be upside down, and the right and left elevations appear to rest on their sides. Because most elevation drawings are prepared at a large scale, and because of the desirability of seeing drawings as we normally view them, each elevation is usually shown with the ground line on the bottom and the roof line on the top, as shown in Fig. 21-4.

Compass Orientation

The north, east, south, and west compass points are often used by designers to describe and label elevation drawings. This method is preferred when there is no actual front or rear view of a structure. When this method is used, the north arrow on the floor plan is the key to the title designation of each elevation. For example, in Fig. 21-4 the rear elevation is facing north. Therefore, the rear elevation could also be called the north elevation. Here the front elevation is also the south elevation and the left elevation is also the west elevation. Compass directions are not used if the final positioning of the building on the terrain has not been determined.

Auxiliary Elevations

When a floor plan has more than four sides or if any of the sides deviate from the normal 90° projection, an auxiliary elevation shows the view of any side of a structure perpendicular to the side of the floor plan to which it is related. Figure 21-5 shows an auxiliary elevation view of one portion of a wall. If only a front, rear, left, and right view were drawn on this plan, walls that extend at

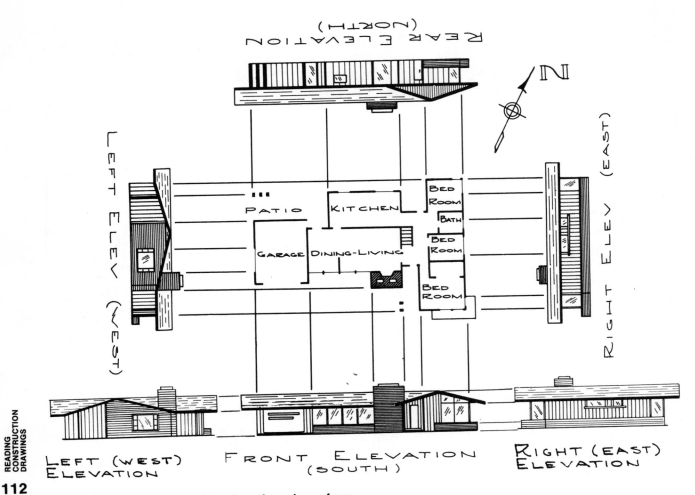

Fig. 21-3. Projection of side elevation views from front elevation.

other than a 90° angle would have foreshortened lines, and the elevation would not reflect true horizontal distances. There are five walls on the plan in Fig. 21-5 which would require auxiliary projections to show their true dimensions.

Elevation Design Factors

Reading elevation drawings accurately is extremely important since a variation of design factors such as heights of the ground line, siding materials, chimney proportion, roof pitch, roof overhang, door and window heights and style, and the positioning of landscaping features significantly affects the appearance of a structure. Figure 21-6, for example, shows two elevation designs projected from the same floor plan. By altering the pitch and style of the roof (Fig. 21-7), the type of siding, heights for windows, and grade line, the structure looks significantly different although both elevations relate to the same floor plan.

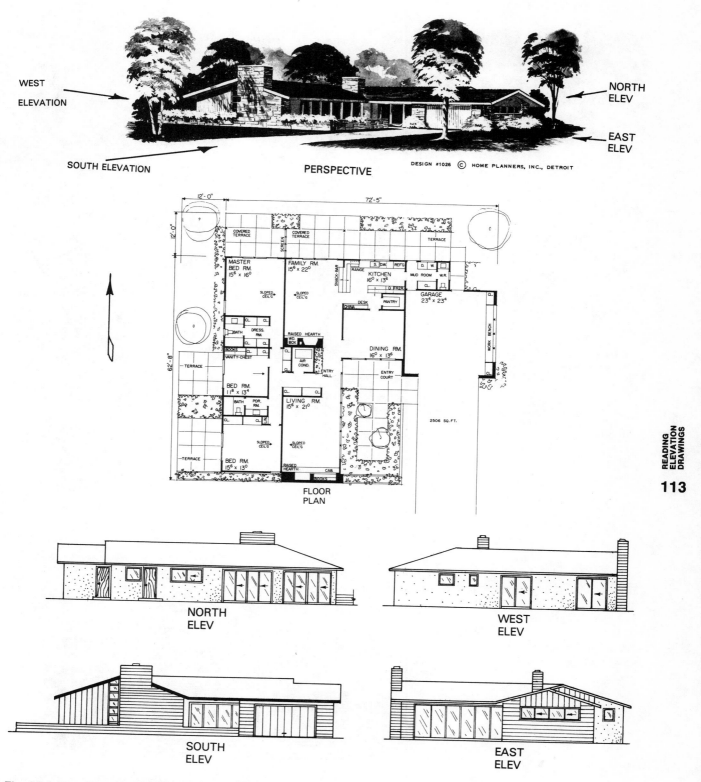

WEST ELEVATION

SOUTH ELEVATION

NORTH ELEV

EAST ELEV

PERSPECTIVE

DESIGN #1026 © HOME PLANNERS, INC., DETROIT

FLOOR PLAN

NORTH ELEV

WEST ELEV

SOUTH ELEV

EAST ELEV

Fig. 21-4. Use of compass direction to identify elevations. (Home Planners, Inc.)

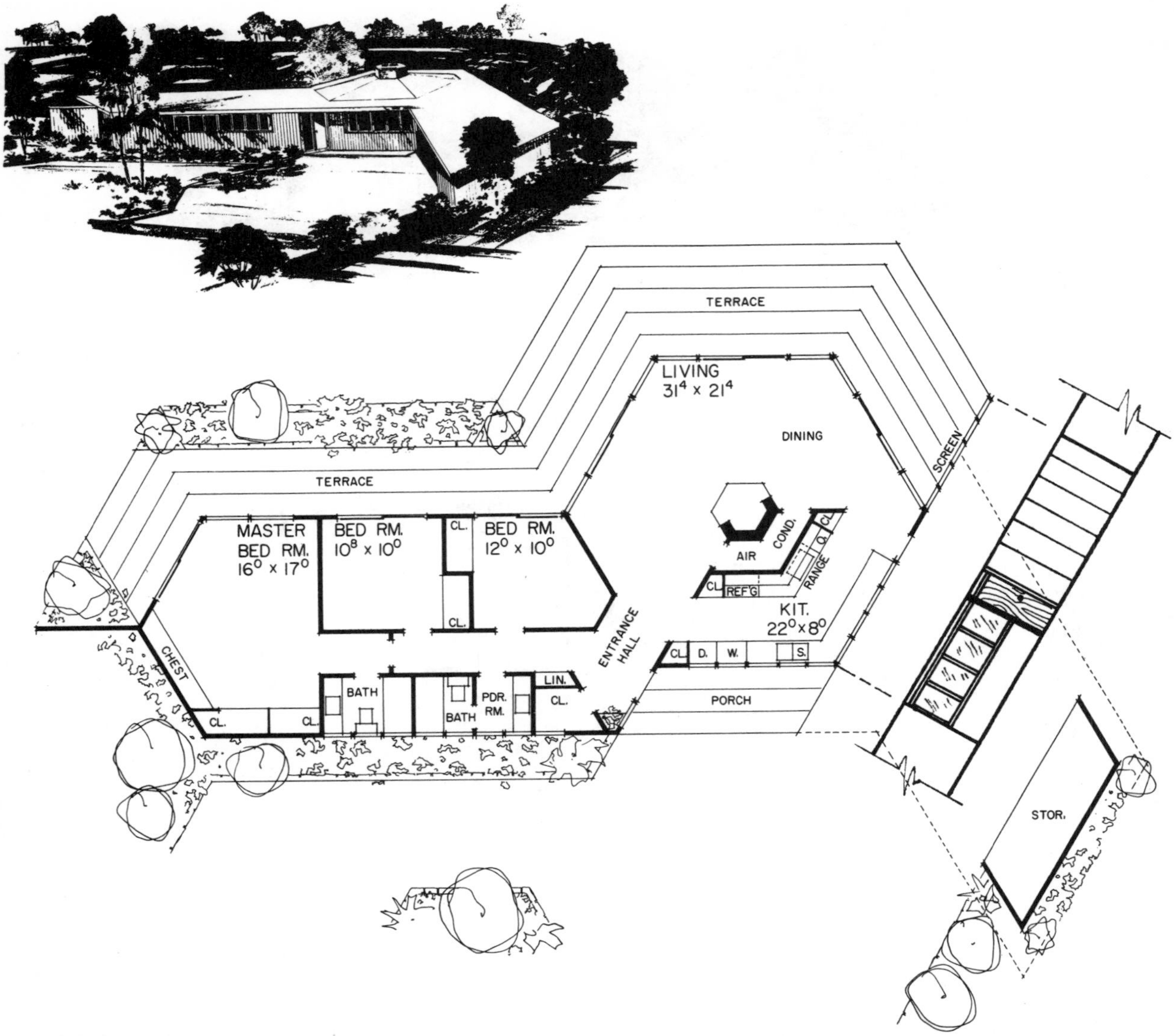
TERRACE
LIVING
31^{4} x 21^{4}
DINING
SCREEN
TERRACE
MASTER BED RM.
16^{0} x 17^{0}
BED RM.
10^{8} x 10^{0}
CL.
BED RM.
12^{0} x 10^{0}
AIR
COND.
CL.
REF'G
RANGE
KIT.
22^{0} x 8^{0}
CL.
ENTRANCE HALL
CL.
D. W.
S.
CHEST
CL.
CL.
BATH
BATH
PDR. RM.
LIN.
CL.
PORCH
STOR.

Fig. 21-5. Auxiliary elevation view. (Home Planners, Inc.)

Fig. 21-6. Effect of elevation design factors.

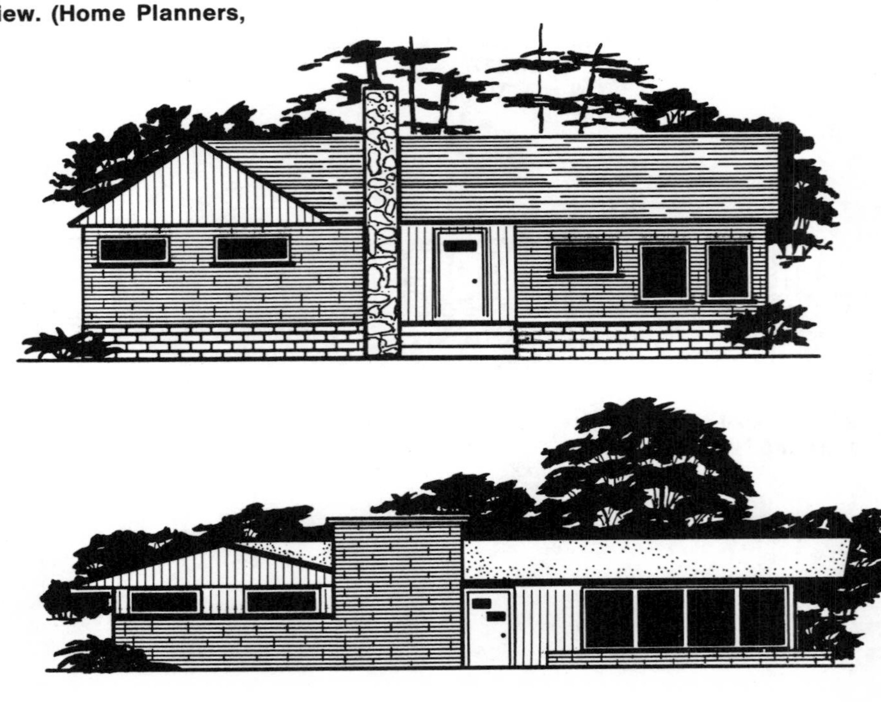

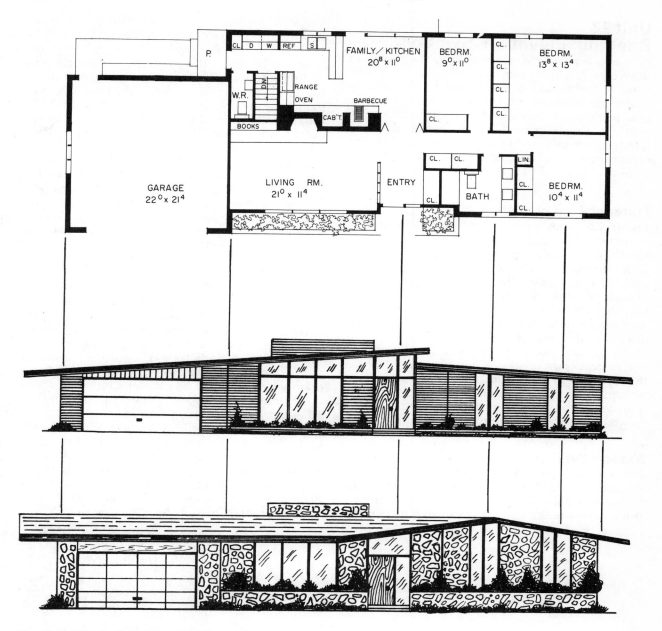

Fig. 21-7. Effect of roof on elevation appearance.

Unit 22
Reading Elevation Symbols

Symbols are used to clarify and simplify elevation drawings. Symbols help describe the basic features of the elevation. They show what building materials are used, and they describe the style and position of doors and windows. Elevation symbols also help make the elevation drawing look more realistic. Some of the most common elevation symbols are shown in the elevation drawings in Fig. 22-1.

Material Symbols

Figure 22-2 shows the relationship between elevation symbols and the actual construction materials they represent. Most elevation symbols look very similar to the actual material. However, in many cases the symbol does not show exactly the appearance of the material. For example, the symbol for brick as shown in Fig. 22-2 does not include all the lines shown in a pictorial drawing. Representing brick on the elevation drawing exactly as it appears is a long, laborious, and unnecessary process. Therefore, many elevation symbols are simplifications of the actual appearance of the material. The symbol often represents the appearance of the material as viewed from a great distance.

Window Symbols

The position and style of windows greatly affect the appearance of the elevation. Therefore, windows are usually drawn on the elevation with as much detail as the scale of the drawing permits.

Parts of windows shown on elevation drawings include the sills, sashes, mullions, and muntins, as shown in Figs. 22-3 and 22-4. Figure 22-5 shows methods of illustrating casement, awning, and sliding windows. It is also necessary to determine from a drawing the direction of the hinge for casement and awning windows. The direction of the hinge is shown by dotted lines. The point of the dotted lines shows the part of the window to which the hinge is attached.

Sometimes complete window details are not shown on each elevation drawing but are related to a large detail drawing. The detail drawing is indexed with a code to a window schedule and floor plan, as shown in Fig. 22-6. This is done to eliminate redrawing an identical window many times on elevation drawings.

Door Symbols

Doors are shown on elevation drawings by methods similar to those used for illustrating window style and position. They are either drawn completely, if the scale permits, or shown in abbreviated form. Sometimes the outline is indexed to a door schedule. The complete drawing of a door, whether shown on the elevation or on a separate detail, shows the division of panels, lights, sill, jamb, head and trim details.

The horizontal position of doors and windows is usually not dimensioned on elevation drawings, since their position and size is usually designated on the floor plan. However, heights of windows and doors cannot be dimensioned on floor plans; therefore, height (vertical) dimensions are found only on elevation drawings.

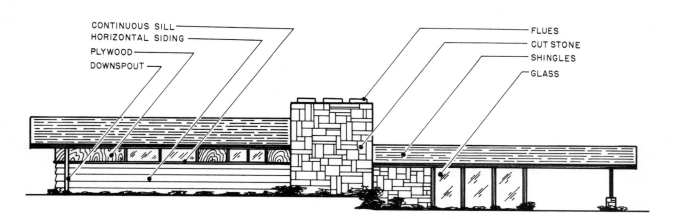

CONTINUOUS SILL
HORIZONTAL SIDING
PLYWOOD
DOWNSPOUT

FLUES
CUT STONE
SHINGLES
GLASS

Fig. 22-1. Application of common elevation symbols.

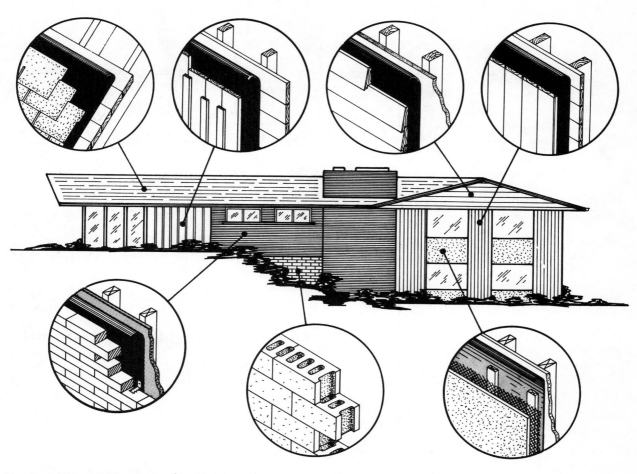

Fig. 22-2. The relationship between symbols and actual construction materials.

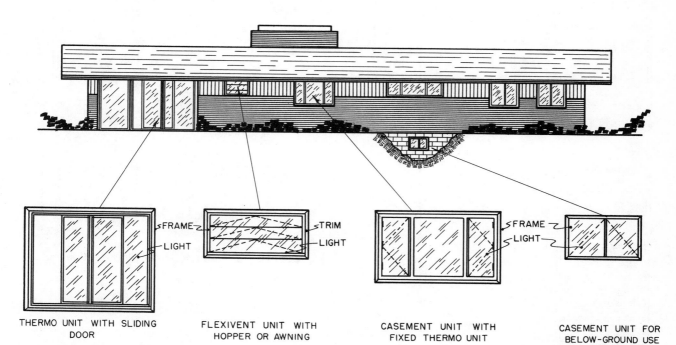

THERMO UNIT WITH SLIDING DOOR

FLEXIVENT UNIT WITH HOPPER OR AWNING

FRAME

TRIM

LIGHT

LIGHT

CASEMENT UNIT WITH FIXED THERMO UNIT

CASEMENT UNIT FOR BELOW-GROUND USE

FRAME

LIGHT

Fig. 22-3. Application of door and window elevation symbols.

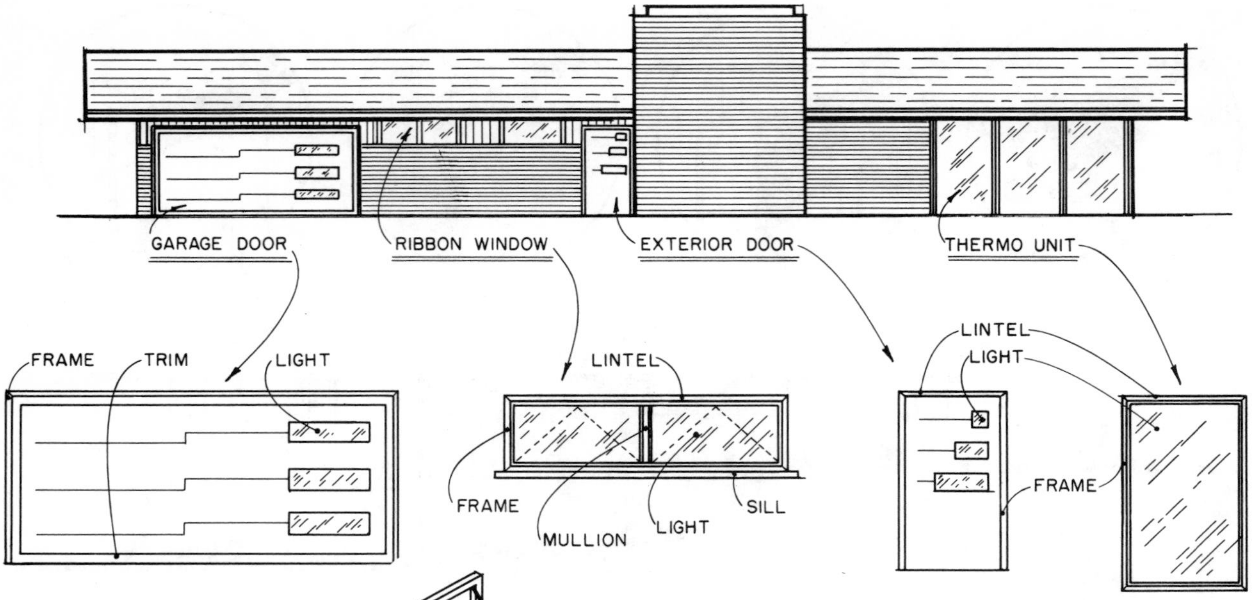

GARAGE DOOR — RIBBON WINDOW — EXTERIOR DOOR — THERMO UNIT

FRAME   TRIM   LIGHT

LINTEL

FRAME   MULLION   LIGHT   SILL

LINTEL
LIGHT
FRAME

Fig. 22-4. Application of door and window elevation symbols.

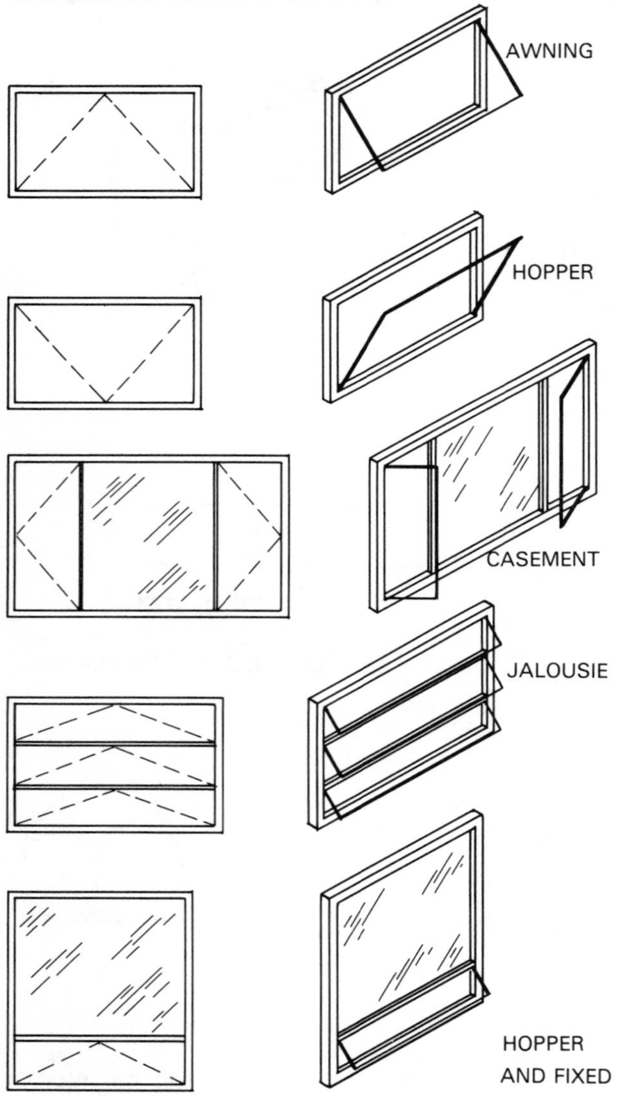

AWNING

HOPPER

CASEMENT

JALOUSIE

HOPPER
AND FIXED

Fig. 22-5. The placement of the hinge is shown by dotted lines.

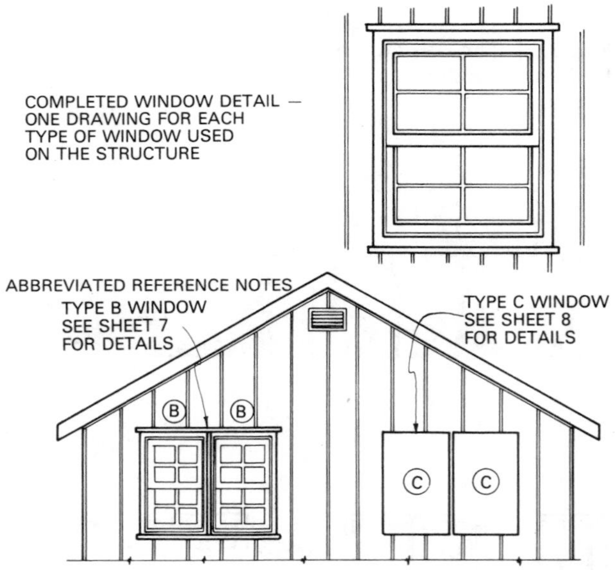

COMPLETED WINDOW DETAIL —
ONE DRAWING FOR EACH
TYPE OF WINDOW USED
ON THE STRUCTURE

ABBREVIATED REFERENCE NOTES
TYPE B WINDOW
SEE SHEET 7
FOR DETAILS

TYPE C WINDOW
SEE SHEET 8
FOR DETAILS

Ⓑ   Ⓑ   Ⓒ   Ⓒ

Fig. 22-6. Use of codes to show window style.

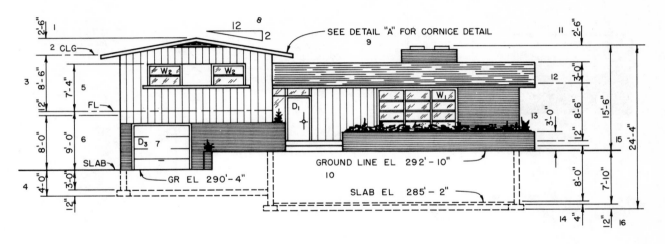

Fig. 23-1. Rules for reading elevation dimensions.

Unit 23
Reading Elevation Dimensions

Horizontal (width and length) dimensions of a building are found on the floor plans. Vertical (height) dimensions are found on elevation drawings. Elevation dimensions show the vertical distance from a datum line. A datum line is a horizontal plane that remains constant. Sea level is commonly used as the datum for many drawings although other consistent distances other than sea level or street markers are also used.

Dimensions on elevation drawings show the heights above the datum of the ground line. They also show the distance from the ground line to the floor, ceiling, ridge, and eave lines and to the tops of chimneys, doors, and windows. Distances below the ground line are shown by dotted lines but are dimensioned by solid dimension and extension lines.

Elevation dimensions are drawn to basic standards to ensure consistency of interpretation. The numbered arrows on the elevation drawing shown in Fig. 23-1 show the application of the following guides for reading elevation dimensions.

1. Vertical elevation dimensions are read from the right of the drawing.
2. Levels that are dimensioned are labeled with a note, term, or abbreviation.
3. Room heights are shown by dimensions extending from the floor line to the ceiling line.
4. The depth of footers (footings) is dimensioned from the ground line.
5. Heights of windows and doors are dimensioned from the floor line to the tops of windows or doors (Fig. 23-2).
6. Elevation dimensions show only vertical distances. Horizontal distances are found on the floor plan.
7. Windows and doors may be indexed to the door or window schedule, or the style of window or door may be shown directly on the elevation drawing.
8. The roof pitch is shown by indicating the rise over the run. Figure 23-3 shows common roof pitch terminology shown on elevation drawings. Figure 23-4 shows common roof pitches.
9. Dimensions for small, complex, or obscured areas are usually found indexed to a separate detail.

Fig. 23-2. Common vertical dimensions found on elevation drawings.

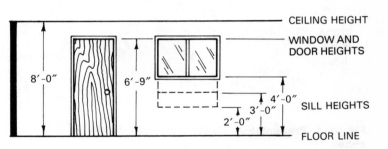

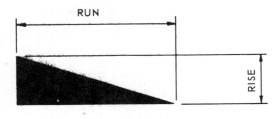

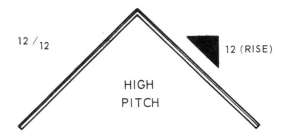

HIGH PITCH

12 / 12

12 (RISE)

2 / 12

12 (RUN)

2 (RISE)

LOW PITCH

Fig. 23-3. Roof pitch terminology.

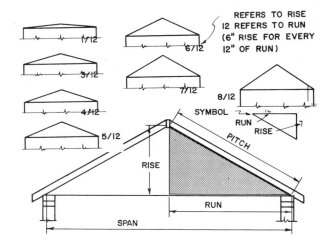

Fig. 23-4. Common roof pitches.

REFERS TO RISE
12 REFERS TO RUN
(6" RISE FOR EVERY
12" OF RUN)

1/12

6/12

3/12

7/12

8/12

4/12

SYMBOL

RUN

RISE

5/12

PITCH

RISE

RUN

SPAN

10. Ground line position is expressed as height above the datum.
11. Heights of chimneys above the ridge line are dimensioned from the top of the highest ridge.
12. Floor and ceiling lines are shown by thinner lines that function as extension lines.

13. Heights of planters and walls are dimensioned from the ground line.
14. Thicknesses of slabs are dimensioned with either a note or a dimension line.
15. Overall height dimensions are placed on the outside of subdimensions.
16. Thicknesses of footers (footings) are dimensioned with either a note or dimension line.

Unit 24
Presentation Elevations
Dimensions and hidden lines are omitted and landscape features are added to create presentation elevation drawings. Drawings of this type are prepared

Fig. 24-1. Presentation elevations. (John Seals, Architect.)

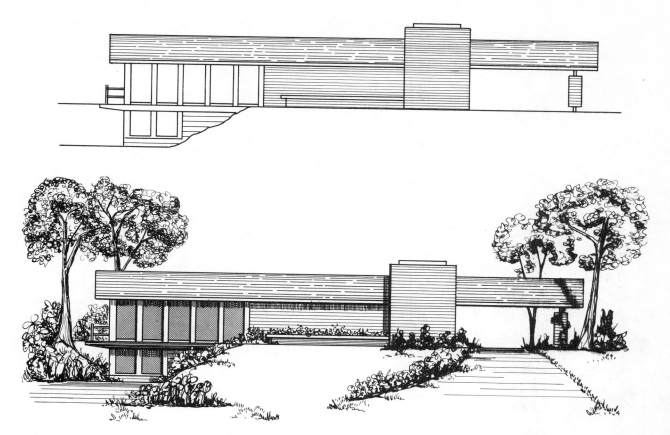

Fig. 24-2. Comparison of dimensioned elevation drawing and presentation elevation drawing.

only to reveal the final appearance of the structure and are not used for construction purposes. Figure 24-1 shows a typical presentation elevation drawing with dimensions and hidden lines omitted and with landscape features and shadows added for realism and effect. Figure 24-2 shows the differences between a working elevation and a presentation elevation drawing. The working elevation drawing is complete with appropriate symbols, dimensions, and hidden lines. The presentation elevation drawing shows landscape features and interpretive material symbols without dimensions or hidden foundation lines.

Unit 25
Interior Elevations
Floor plans show horizontal arrangements of partitions, fixtures, and appliances. Exterior elevations show the vertical design of the exterior walls. But neither of these shows the design of interior walls. Interior elevations are necessary to show the design of interior vertical planes. Because of the need to show

cabinet heights and soffit depths, interior wall elevations are often prepared for kitchen and bathroom walls. An interior wall elevation shows the appearance of a wall as viewed from the center of the room. Figure 25-1 shows the four wall elevations of a kitchen. Imagine yourself in the center of the room looking north. You will see the north wall of the kitchen, which is drawn in the top of Fig. 25-2. Now rotate the page so that you are facing east. You can now view the vertical height of the east wall and read the height dimensions for the base cabinets, wall cabinets, electrical outlets, and soffit line. Rotate the page upside down and you can read the south elevation as it appears from the center of the room. The same is true if you rotate the page so you are looking directly at the west wall.

Interior elevation drawings, like exterior elevations, are not prepared in the original position as projected from the floor plan. Interior elevations are positioned with the floor line on the bottom as normally viewed. For this reason, a symbol is used to identify the position of the elevation on the floor plan. A typical indexing system for this purpose is shown in Fig. 25-2.

NORTH

N

KITCHEN
8'-0" × 10'-0"

WEST

EAST

SOUTH

SOFFIT

HOOD/FAN

RANGE

REF

FORMICA TOP & SPLASH

8'-0"

Fig. 25-1. Relationship of floor plan to interior wall elevations.

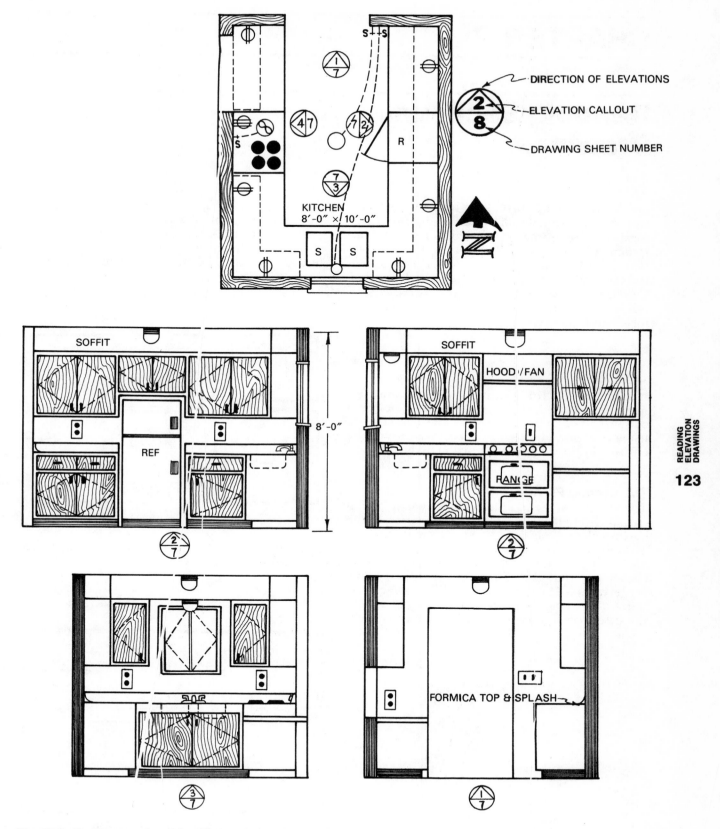

- DIRECTION OF ELEVATIONS
- ELEVATION CALLOUT
- DRAWING SHEET NUMBER

KITCHEN
8'-0" × 10'-0"

SOFFIT

REF

SOFFIT

HOOD/FAN

RANGE

8'-0"

FORMICA TOP & SPLASH

Fig. 25-2. Coding used to identify interior wall elevations.

CHAPTER 7

Reading Sectional Drawings

Sectional drawings reveal the internal construction of an object. Architectural sectional drawings are sometimes prepared for the entire structure. These are called full sections. Or sectional drawings are prepared for specific parts of the building. These are called detailed sections. The size and complexity of the part usually determine the type and number of sections prepared for a building.

Unit 26
Full Sections

Drafting technicians frequently prepare drawings which show a building cut in half. Figure 26-1 shows the principle of using a cutting-plane line, which cuts through the entire building, to reveal a portion of the interior.

The Cutting Plane

The cutting plane is an imaginary plane which passes through the building. The position of the cutting plane is shown by the cutting-plane line. The cutting-plane is a long, heavy line with two dashes and is found on the floor plan. Figure 26-2 shows a cutting-plane line and the cutting plane it represents. The cutting-plane line is placed on the sectioned part, and the arrows at its end show the direction from which the section is viewed.

For example, in Fig. 26-2, section BB is viewed from the right and section AA from the left.

Since the cutting-plane line often interferes with many dimensions, notes, and details, an alternative method of drawing cutting-plane lines

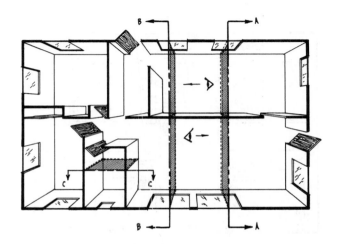

Fig. 26-2. Cutting-plane arrows show position from which sections are viewed.

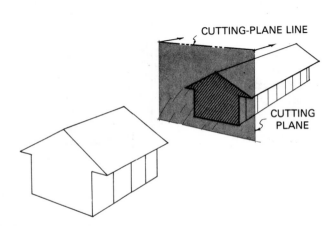

Fig. 26-1. The cutting plane and cutting-plane line.

CUTTING-PLANE LINE

CUTTING PLANE

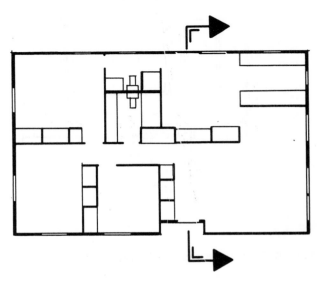

Fig. 26-3. Separated cutting-plane line.

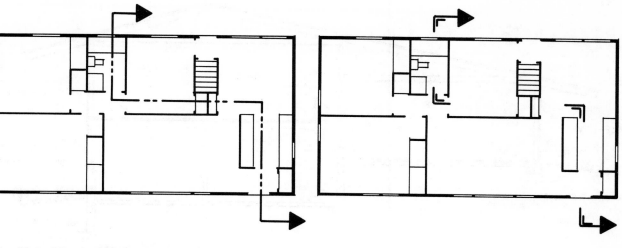

Fig. 26-4. Offset cutting-plane lines.

is used to overcome this interference. This alternative method, which includes only the extremes of the cutting-plane line, is shown in Fig. 26-3. The cutting-plane line is then assumed to be in a straight line between these extremes. When a cutting-plane is offset to show a different area, the corners are illustrated as shown in Fig. 26-4. The offset cutting-plane line is often used to show different wall sections on one sectional view. In reading cutting-plane lines, it is important to determine whether the cutting-plane line extends directly through the structure on a straight line or whether it contains an offset.

There are two kinds of full sections prepared for most architectural plans: the longitudinal section (Fig. 26-5), and the transverse section (Fig. 26-6). Longitudinal means lengthwise. A longitudinal section is one showing a lengthwise cut through a building. A transverse section is one showing a cut across the narrow width of the building. Figure 26-7 shows the use of transverse and longitudinal cutting-plane lines. Figures 26-8

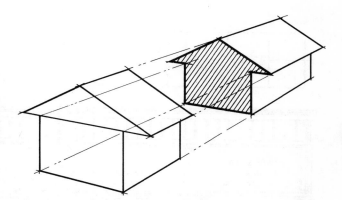

Fig. 26-6. A transverse section.

and 26-9 show the effect of these cutting-plane lines and how they produce a transverse or longitudinal section. A longitudinal section through a structure reveals construction methods as viewed parallel to the roof ridge. The transverse

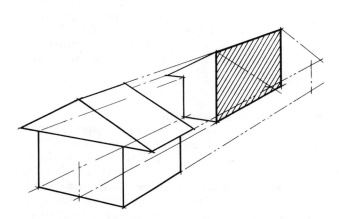

Fig. 26-5. A longitudinal section.

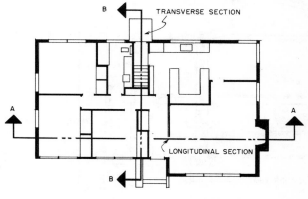

Fig. 26-7. Transverse and longitudinal cutting-plane lines.

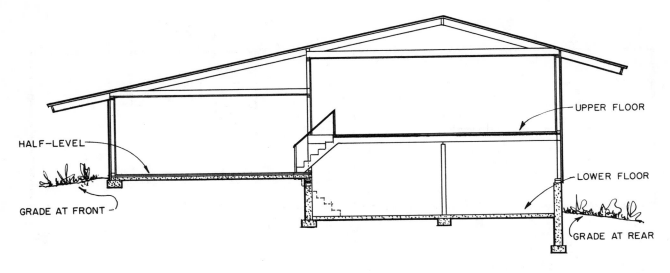

Fig. 26-8. A transverse section through a split-level building.

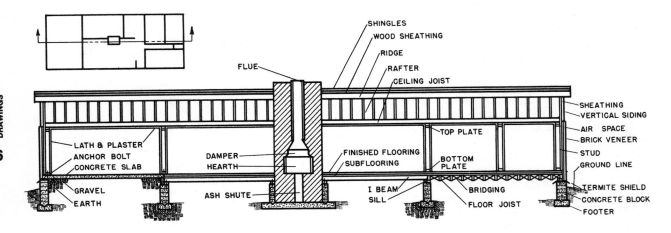

Fig. 26-9. Longitudinal section through a one-story building.

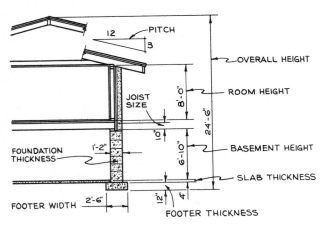

Fig. 26-10. Use of break lines used to enlarge details by eliminating unnecessary parts of a section.

section shows a section perpendicular to the roof ridge. This is the most popular type of full section since it shows truss and roof overhang details more clearly. Longitudinal sections are especially effective and necessary for showing the various methods of constructing multilevel buildings, since footers, grade lines, slabs, and floor lines are included in one section.

Sectional Dimensioning

Full sections expose the size and shape of building materials and components not revealed on floor plans or elevations. The full section is, therefore, an excellent place to locate many detailed dimensions. Look for many vertical dimensions on full sections. They are located in the same manner as elevation dimensions are located on elevation drawings (Fig. 26-10).

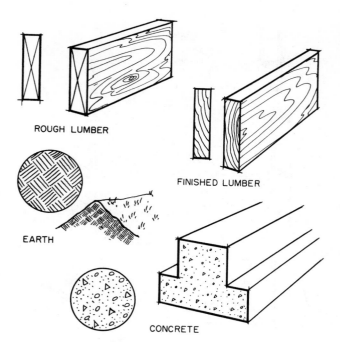

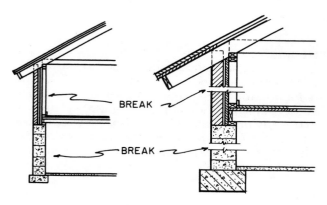

Fig. 27-1. Detail section symbols.

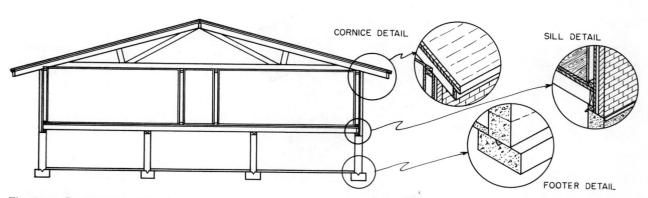

Unit 27
Detail Sections
Because full sections are usually drawn to a small scale, many parts are difficult to interpret and dimension. In order to reveal the exact position and size of many small members, larger (enlarged) detail sections are prepared.

Symbols
Section-lining symbols represent the way building materials look when they are sectioned or cut in half. A floor-plan drawing is actually a horizontal section. Many of the symbols used in floor plans also apply to other sections. Section-lining symbols as shown in Unit 4 are used for detail sections (Fig. 27-1).

Breaks
One method of showing sections larger than possible in full section drawings is through the use of break lines, which reduce long distances. Using break lines as shown in Fig. 27-2 allows the area to be drawn much larger than when the entire length is included in the drawing. Break lines are placed in areas where the material does not change over a long distance.

Removed Sections
When a very large section is needed for interpretation or dimensioning purposes, it is sometimes impossible to draw the entire wall section even with the use of break lines. In this case, a removed section as in Fig. 27-3 is shown. Removed sections are frequently drawn for ridge, cornice, sill, footing, and beam areas.

Footing Sections
A footing detail section is needed to show the width and length of the footing, the type of material used, and the position of the foundation

Fig. 27-2. Use of break lines to enlarge and dimension sections.

Fig. 27-3. Removed sections.

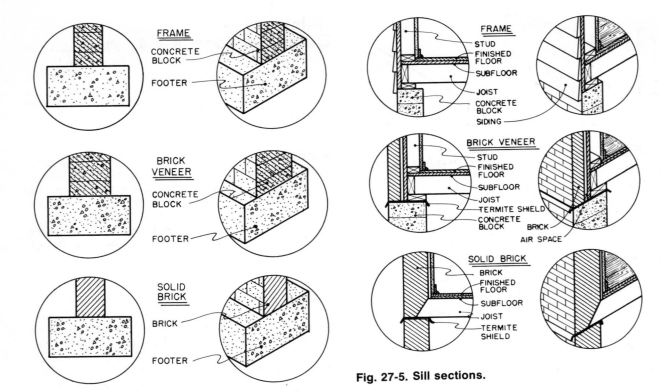

Fig. 27-4. Footing sections.

Fig. 27-5. Sill sections.

wall on the footing. Figure 27-4 shows several footing details and the pictorial interpretation of each.

Sill Sections

Sill sections as shown in Fig. 27-5 explain graphically how the foundation supports and intersects with the floor system and the outside wall.

Beam Details

Beam detail sections are necessary to show how joints are supported by beams and how columns or foundation walls support beams. As in all other sections, the position of the cutting-plane lines is extremely important. Figure 27-6 shows two possible positions of the cutting plane. If the cutting-plane line is placed parallel to the beam, you see a cross section of the joint as shown in (A). If the

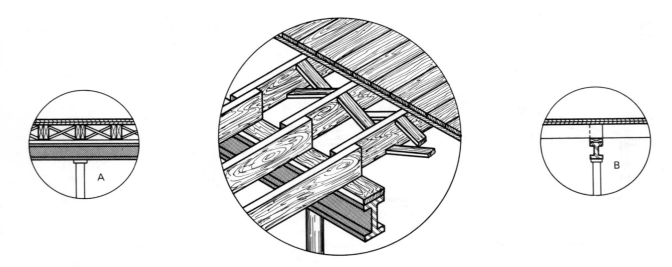

Fig. 27-6. Beam section from two different views.

Fig. 27-7. Full-wall section.

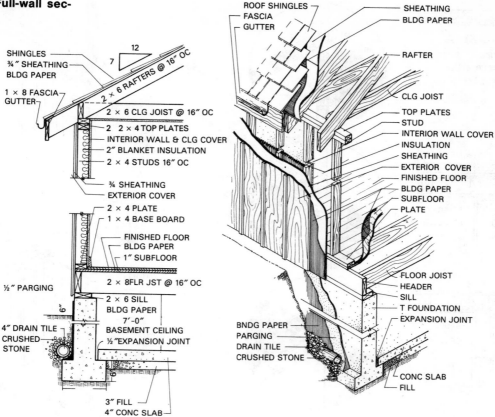

SHINGLES
¾" SHEATHING
BLDG PAPER

1 × 8 FASCIA
GUTTER

2 × 6 RAFTERS @ 16" OC

12
7

2 × 6 CLG JOIST @ 16" OC

2 2 × 4 TOP PLATES
INTERIOR WALL & CLG COVER
2" BLANKET INSULATION
2 × 4 STUDS 16" OC

¾ SHEATHING
EXTERIOR COVER

2 × 4 PLATE
1 × 4 BASE BOARD
FINISHED FLOOR
BLDG PAPER
1" SUBFLOOR

2 × 8 FLR JST @ 16" OC

½" PARGING

2 × 6 SILL
BLDG PAPER
7'-0"
BASEMENT CEILING
½"EXPANSION JOINT

6"

4" DRAIN TILE
CRUSHED
STONE

3" FILL
4" CONC SLAB

ROOF SHINGLES
FASCIA
GUTTER

SHEATHING
BLDG PAPER

RAFTER

CLG JOIST
TOP PLATES
STUD
INTERIOR WALL COVER
INSULATION
SHEATHING
EXTERIOR COVER
FINISHED FLOOR
BLDG PAPER
SUBFLOOR
PLATE

FLOOR JOIST
HEADER
SILL
T FOUNDATION
EXPANSION JOINT

BNDG PAPER
PARGING
DRAIN TILE
CRUSHED STONE

CONC SLAB
FILL

cutting-plane line is placed perpendicular to the beam, you see a cross section of the beam as shown in (B). Sections are also frequently shown through the vertical plane of a stair section. This reveals the dimensions of the treads, risers, stringers, total rise, total run, and stairwell vertical distances.

Wall Sections

The most typical wall section is the type shown in Fig. 27-7. This shows a full section through an exterior wall from the footing through the roof cornice area. Through one wall section of this type, the size of all exterior wall materials can be shown plus the exact vertical location of foundation walls, floors, ceilings, and roof pitch. Another type of exterior wall section is shown by removing the exterior sheeting and siding, which exposes the wall construction, as shown in Fig. (34-24).

A floor plan is a horizontal section. However, many details are omitted from floor plans because of the small scale used. Very few construction details are necessary to adequately interpret

the floor plan. If the floor plan is drawn exactly as a true horizontal section, it will appear similar to the section shown in Fig. 27-8. Floor plans omit the stud sections.

Window Sections

Because much of the actual construction of most windows is hidden, sections are necessary for the correct interpretation for most window construction methods. Figure 27-9 shows the areas of windows that are commonly sectioned. These include sections of the head, jamb, and sill construction.

Vertical Sections

Sill sections and head sections are vertical sections and are sometimes included on the same sectional drawing. Figure 27-10 shows the relationship between the cutting-plane line and the sill and head sections. The circled areas in Fig. 27-11 show the areas that are removed when a separate head and sill section is shown.

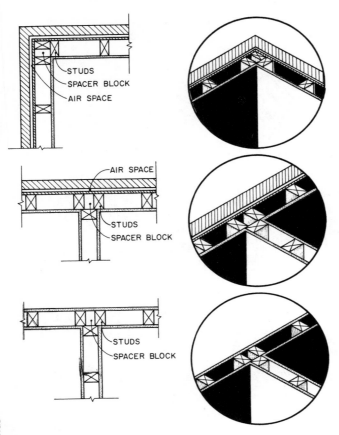

Fig. 27-8. Plan section through wall.

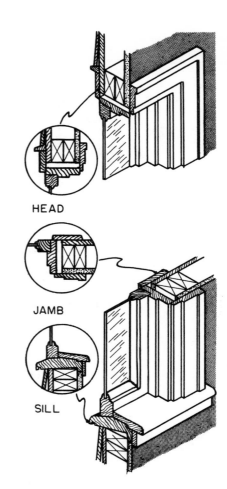

HEAD

JAMB

SILL

Fig. 27-9. Sectional details showing the relationship of head, sill, and jamb for a window.

Horizontal Sections

When a cutting-plane line is extended horizontally across the entire window, the resulting sections are known as the jamb sections. Figure 27-12 shows the derivation of jamb sections from a horizontal cutting-plane line. Figure 27-13 shows the method of projecting a jamb detail from the window elevation drawing. Since the construction of both jambs is the same, the right jamb drawing is the reverse of the left. Only one jamb detail is normally shown for this reason. The builder interprets the right jamb or left jamb construction detail as the reverse of the other.

Door Sections

A small horizontal section of all doors is actually shown on the floor plan. However, this single-line section is almost completely symbolic and lacks detail. It is, therefore, not used for construction purposes. An enlarged jamb, head, and sill section, as shown in Fig. 27-14, is necessary to completely describe door construction methods. When the cutting-plane line is extended vertically through the head and sill, a section similar to the one shown in Fig. 27-15 is revealed. However, these sections are often too small to show

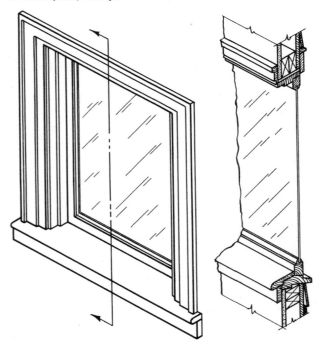

Fig. 27-10. Window head and sill sections are in the same plane.

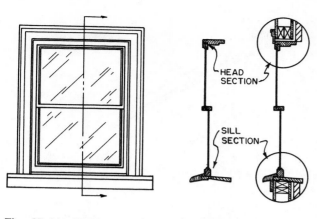

Fig. 27-11. Alignment of window head and sill sections.

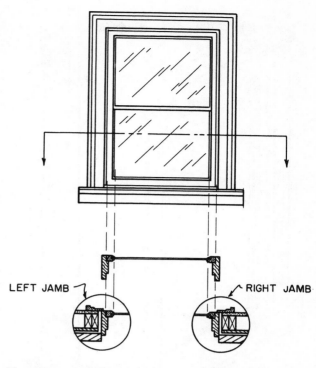

Fig. 27-13. Alignment of left and right window jamb sections.

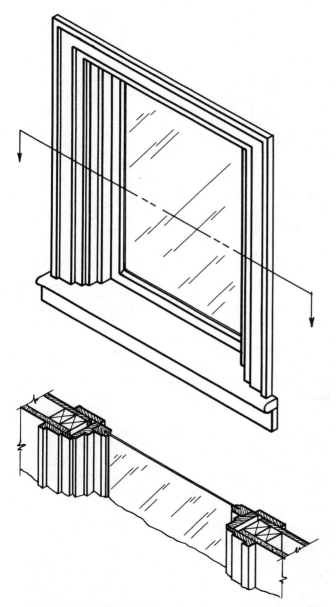

Fig. 27-12. Window right and left jamb are in the same plane.

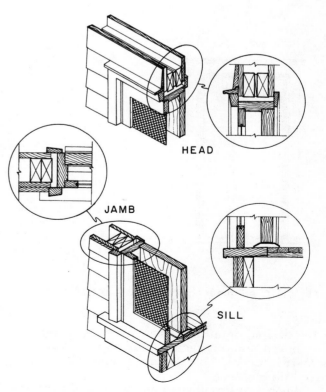

Fig. 27-14. The relationship of a door, head, sill, and jamb.

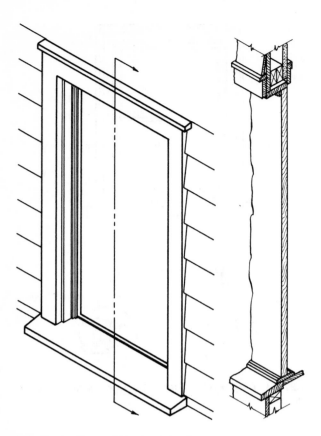

Fig. 27-15. Door, head, and sill sections are in the same plane.

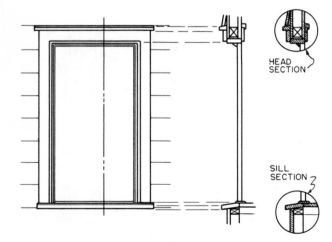

Fig. 27-16. Alignment of door, head, and sill sections.

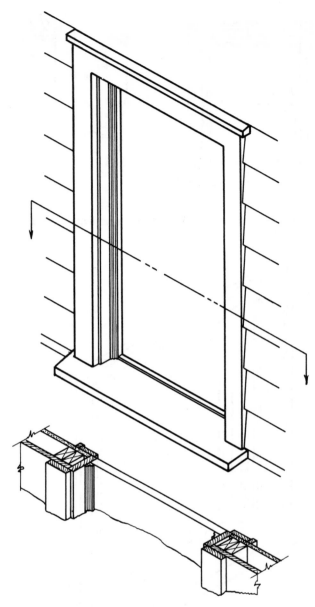

Fig. 27-17. Right and left door jambs are in the same plane.

Just as window jamb sections are often removed and enlarged, door jambs, either the right or left, are also often removed or enlarged for construction reading purposes. Just as with windows, the left door jamb is identical to the right when reversed.

Roof Sections

Detailed sections of roof designs usually show cornice sections as in Fig. 27-18, or sections through the roof ridge, as shown in Fig. 27-19. Both of these illustrations show the section through the wall related to a pictorial interpretation of the sectioned area.

the desired detail necessary for construction. A removed and enlarged section as shown in Fig. 27-16 is used to show the construction of the head and sill areas.

When a cutting-plane line is extended horizontally through a door assembly, both right and left door jambs are revealed as shown in Fig. 27-17.

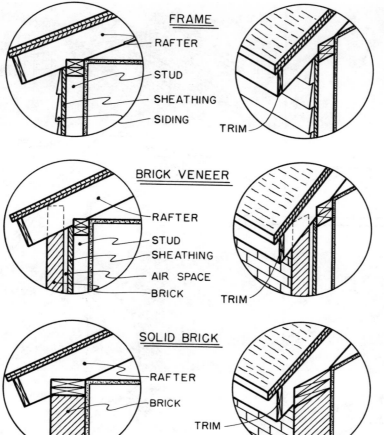

FRAME

— RAFTER
— STUD
— SHEATHING
— SIDING

TRIM

BRICK VENEER

— RAFTER
— STUD
— SHEATHING
— AIR SPACE
— BRICK

TRIM

SOLID BRICK

— RAFTER
— BRICK

TRIM

Fig. 27-18. Section through roof (eave) cornice.

Special Sections

In addition to foundation, wall, and roof sections, special sections are usually prepared to show unique construction methods, materials, or designs, such as the sunken roof design shown in Figure 27-20.

Elevation sections are also often prepared for the installation of built-in furniture, equipment, fixtures, or appliances, as shown in Fig. 27-21.

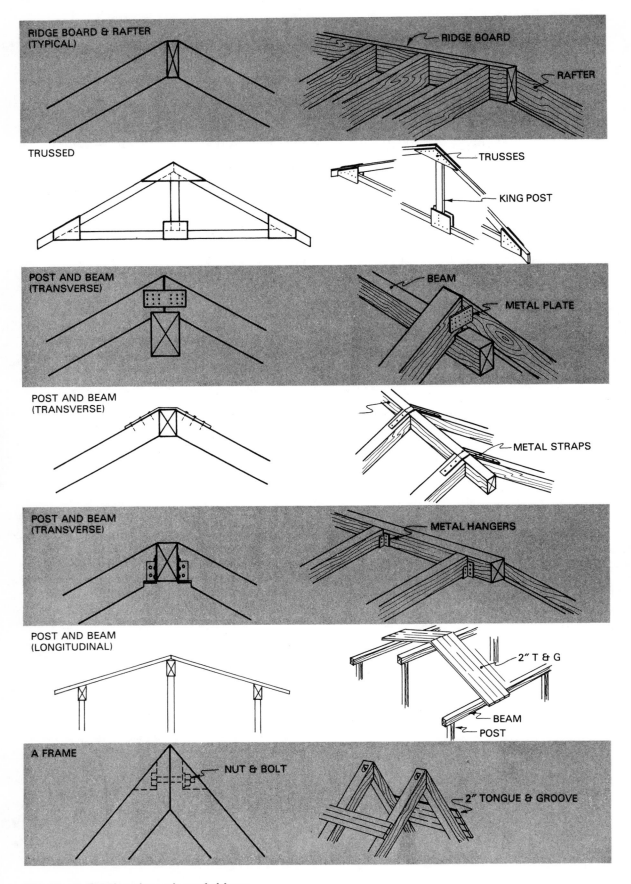

RIDGE BOARD & RAFTER (TYPICAL)

RIDGE BOARD

RAFTER

TRUSSED

TRUSSES

KING POST

POST AND BEAM (TRANSVERSE)

BEAM

METAL PLATE

POST AND BEAM (TRANSVERSE)

METAL STRAPS

POST AND BEAM (TRANSVERSE)

METAL HANGERS

POST AND BEAM (LONGITUDINAL)

2" T & G

BEAM

POST

A FRAME

NUT & BOLT

2" TONGUE & GROOVE

Fig. 27-19. Section through roof ridges.

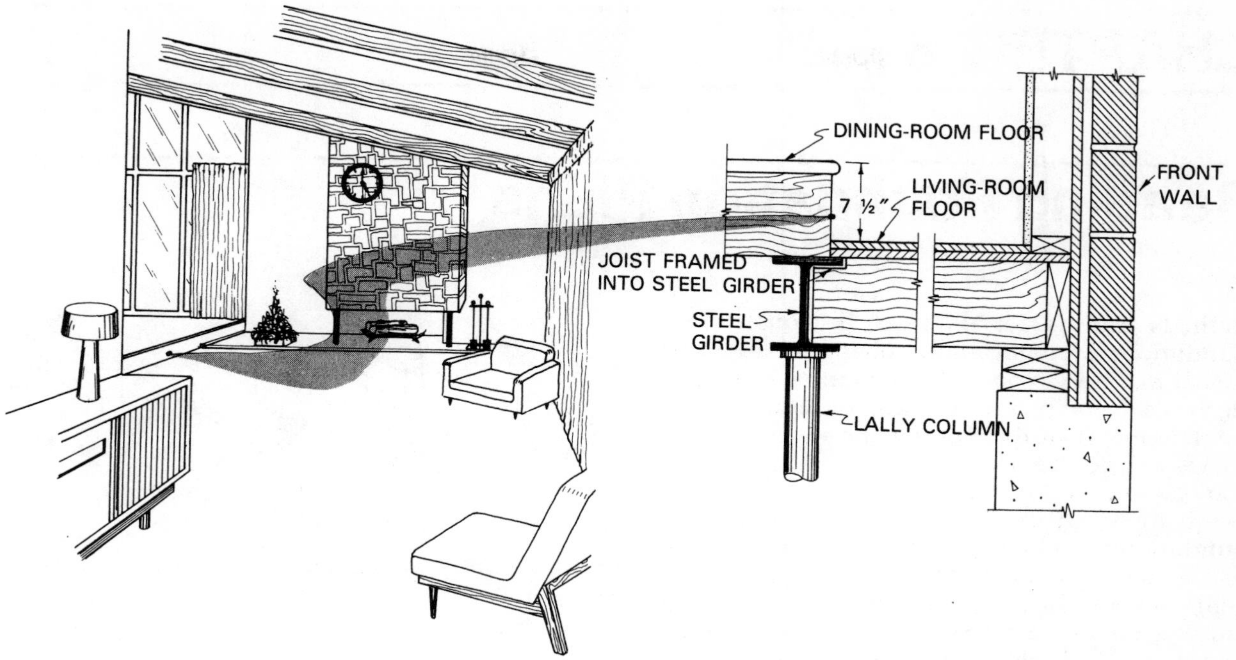

Fig. 27-20. Special framing detailed section for sunken room.

CHAPTER 8

Reading Foundation Plans

Methods and materials used in constructing foundations are continually changing and vary greatly in different parts of the country. However, the basic principles of foundation construction are the same regardless of specific application.

Every structure needs a foundation. The function of a foundation is to provide a level and uniformly distributed support for the structure. The foundation must be strong enough to support and distribute the load of the structure and sufficiently level to prevent the walls from cracking and the door and windows from sticking. The foundation also helps to prevent cold or warm air and dampness from entering the structure from beneath. The foundation waterproofs and forms the supporting walls of a basement.

Unit 28
Foundation Members The number, placement, design, and size of various foundation structural members varies according to the design and function of each member. Before learning to read foundation plans, you must, therefore, become familiar with the function and symbols used for the following foundation members.

Footing
The footing (footer) as shown in Fig. 28-1 distributes the weight of the structure over a large area.

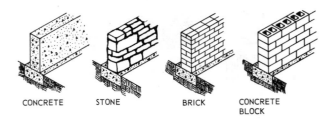

Fig. 28-2. Alternate materials for foundation walls.

Concrete is commonly used for footings because it can be poured to maintain a firm contact with the support soil. Concrete is also effective because it can withstand heavy weights and is relatively decay-proof. Steel reinforcement is sometimes added to the concrete footing to keep the concrete from cracking and to provide additional support. The footing must be positioned to support the weight of the building effectively and evenly. In cold climates the footing must be placed below the frost line. In soft soil areas it must extend to solid load supporting surfaces.

Foundation Walls
The function of foundation walls is to support the load of the building above the ground line and to transmit the weight of the building to the footing. Foundation walls are normally made of concrete, stone, brick, or concrete block, as shown in Fig. 28-2. When a complete excavation is made for a basement, foundation sides also provide walls for the basement (Fig. 28-3a).

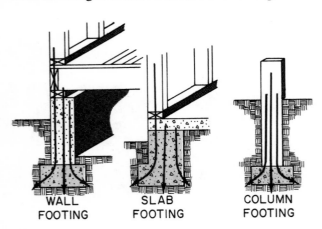

Fig. 28-1. Footings distribute the weight of a building over a wide area.

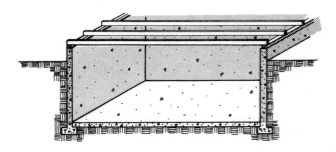

Fig. 28-3a. Foundation walls can become basement walls.

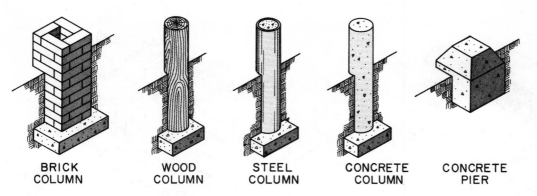

BRICK COLUMN WOOD COLUMN STEEL COLUMN CONCRETE COLUMN CONCRETE PIER

Fig. 28-3b. Piers and columns.

Piers and Columns

Piers and columns are vertical members, usually made of concrete, brick, steel, or wood, which are used to support floor systems (Fig. 28-3b). Piers and columns may be used as the sole support of a structure, or they may be used in conjunction with the foundation wall and provide only the intermediate support between girders and beams. Figure 28-4 shows the application of piers and columns to foundation systems.

Sills

Sills are members, usually wood, that are fastened with anchor bolts to the foundation (Fig. 28-5). Sills provide the base for attaching the exterior walls to the foundation. Most codes require a treated sill or a galvanized iron termite shield to be placed between the concrete and the wood, since the sill is positioned as the lowest wood member closest to the ground line in a frame structure.

Fig. 28-4. Pier and column construction.

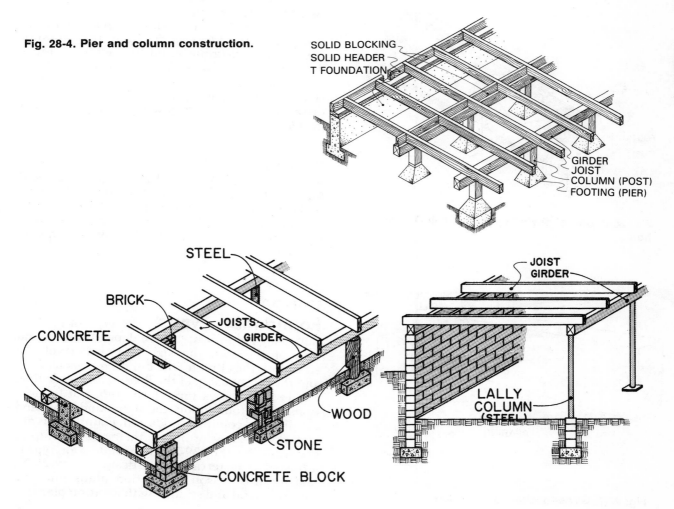

SOLID BLOCKING
SOLID HEADER
T FOUNDATION

GIRDER
JOIST
COLUMN (POST)
FOOTING (PIER)

STEEL

BRICK

CONCRETE

JOISTS
GIRDER

WOOD

STONE

CONCRETE BLOCK

JOIST
GIRDER

LALLY COLUMN (STEEL)

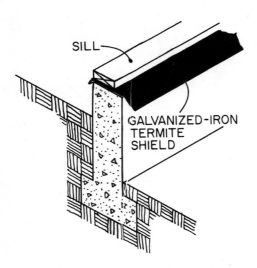

Fig. 28-5. Sill and termite shield placement.

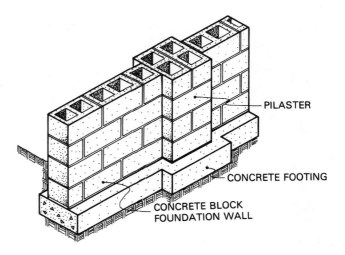

Fig. 28-8. Pilaster construction.

Posts

Posts are vertical wood members that support the weight of girders or beams and transmit the weight to a footing, as shown in Fig. 28-6.

Girders and Beams

Girders are major horizontal support members upon which the floor system is laid. They are supported by posts or piers and secured to the foundation wall (See Fig. 28-7). Steel beams perform the same function as wood girders. However, steel beams of an equivalent size can span larger distances and support greater loads than wood girders.

Pilasters

A pilaster is a column set within or against a wall for the purpose of strengthening the wall. Pilasters such as the one shown in Fig. 28-8 are used to reinforce long foundation walls. Pilasters are also used under girders to provide a wider support surface.

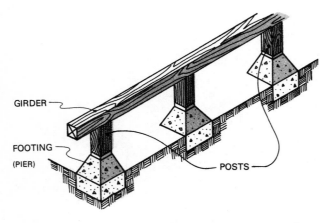

Fig. 28-6. Use of posts in pier and column construction.

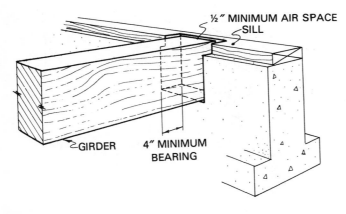

Fig. 28-7. Girder pocket construction.

Unit 29

Foundation Types The type of foundation designed for a structure depends on the nature of the soil, the size and weight of the structure, the climate, the building laws (codes), and the relationship of the floor to the grade line. Figure 29-1 shows three basic types of foundations: the T foundation, the slab foundation, and the pier and column foundation. Foundations are also classified according to their relationship to the grade line (Fig. 29-2). Each foundation type is adjusted to the terrain by the designer (Fig. 29-3), and for this reason foundation plans must be carefully read and studied with location plans.

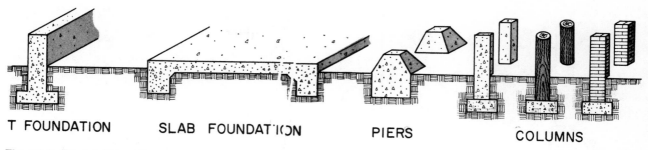

Fig. 29-1. Foundation components.

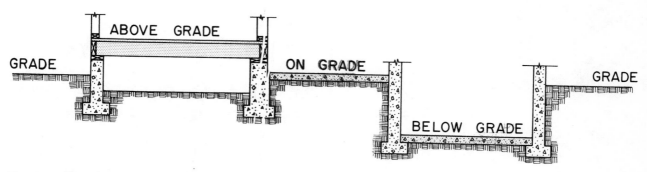

Fig. 29-2. Foundation relationship to grade.

Slab Foundations

A slab foundation is a poured solid mass of concrete. The slab is poured directly on the ground with footing placed where excess support is needed. Figure 29-4 shows common details relating to slab foundation plans. Each of these details is shown as a section as it usually appears on a detail drawing. A pictorial drawing related to each portion of the foundation plan is also shown. Figure 29-5 shows a cross section of a slab foundation as applied to various wall support needs. Figure 29-6 shows the use of external slabs, commonly used for porches, patios, and other light load areas.

T Foundations

The T foundation consists of a trench footer (footing) upon which is placed a solid concrete or a concrete block wall (Fig. 29-7). The combination of the footing and wall forms an inverted T. The T foundation is popular with structures with basements or when the bottom of the first floor must be accessible. Figure 29-8 shows a T foundation with several intermediate support systems consisting of steel lally column, I beam, wood post, and wood beam systems. The details of construction relating to the T foundation and methods used are shown in Fig. 29-9. Learn to recognize and associate these details with the symbols shown on foundation plans.

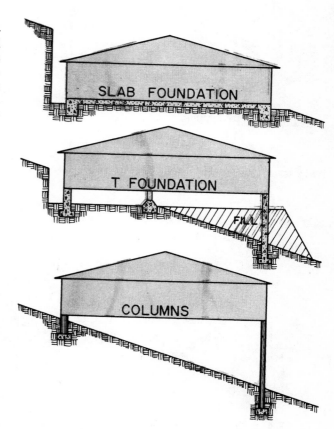

Fig. 29-3. Foundation application to terrain.

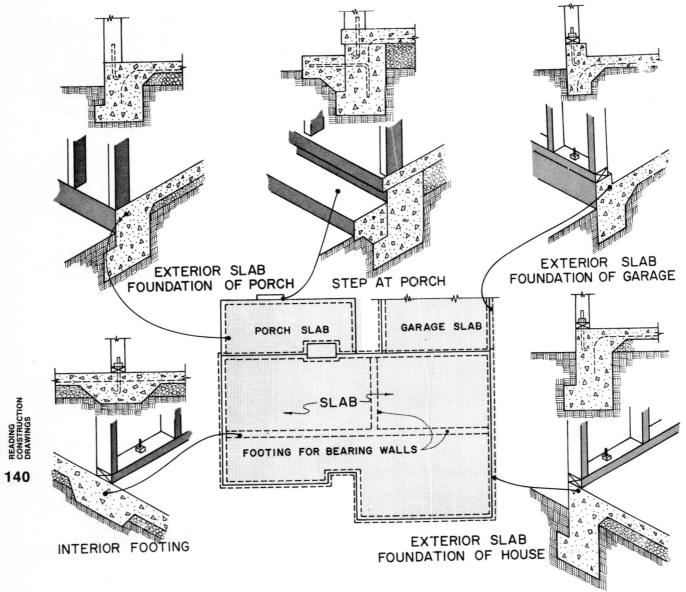

EXTERIOR SLAB
FOUNDATION OF PORCH

STEP AT PORCH

EXTERIOR SLAB
FOUNDATION OF GARAGE

PORCH SLAB

GARAGE SLAB

SLAB

FOOTING FOR BEARING WALLS

INTERIOR FOOTING

EXTERIOR SLAB
FOUNDATION OF HOUSE

Fig. 29-4. Slab foundation details.

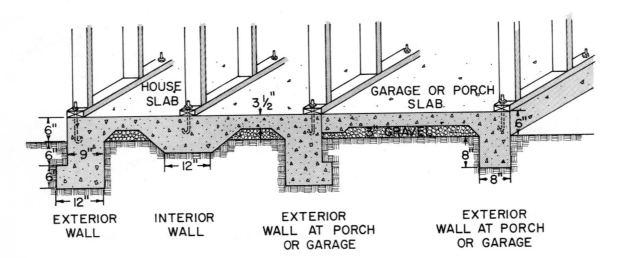

HOUSE
SLAB

3 ½"

GARAGE OR PORCH
SLAB

3" GRAVEL

6"

6"

9"

6"

12"

12"

8"

8"

6"

EXTERIOR
WALL

INTERIOR
WALL

EXTERIOR
WALL AT PORCH
OR GARAGE

EXTERIOR
WALL AT PORCH
OR GARAGE

Fig. 29-5. Slab foundation support method.

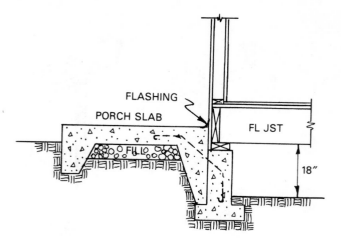

FLASHING

PORCH SLAB

FL JST

FILL

18"

Fig. 29-6. Exterior slab design.

Pier and Column Foundations

A drawing showing the intersection between exterior foundation walls and the floor system is known as a sill detail. The sill is the link between the framework of the structure and the foundation. Sill details are sections through the foundation extending through the floor line. Figure 29-10 shows a sill detail as it would appear for a brick-veneer structure, and Fig. 29-11 shows a sill detail for solid masonry and cavity masonry walls. Notice that in solid masonry or cavity masonry walls a fire cut is included on all wood members that extend into the masonry. This fire cut is made to eliminate destruction of the masonry wall if the beam collapses, as shown in Fig. 29-12.

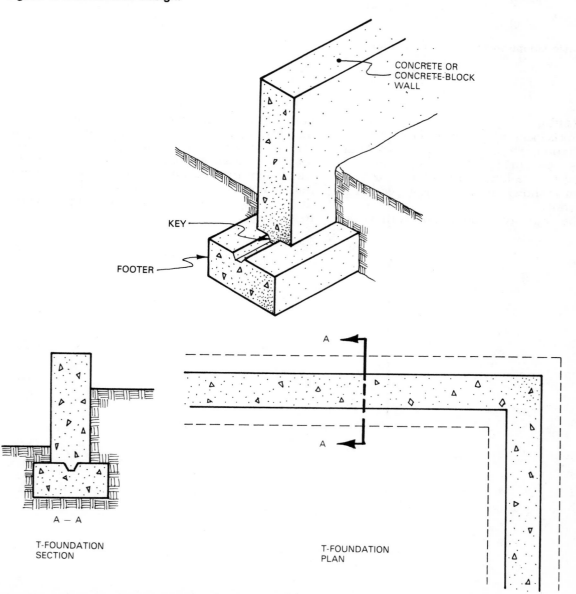

CONCRETE OR
CONCRETE-BLOCK
WALL

KEY

FOOTER

A

A

A — A

T-FOUNDATION
SECTION

T-FOUNDATION
PLAN

Fig. 29-7. T-foundation construction shown in sectional view.

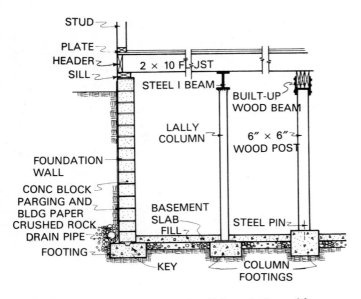

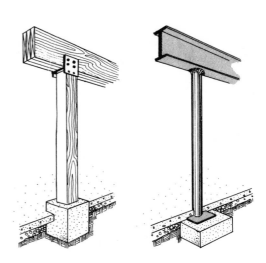

STUD

PLATE
HEADER
SILL

2 × 10 FL JST

STEEL I BEAM

BUILT-UP
WOOD BEAM

6″ × 6″
WOOD POST

LALLY
COLUMN

FOUNDATION
WALL

CONC BLOCK
PARGING AND
BLDG PAPER

CRUSHED ROCK
DRAIN PIPE

FOOTING

BASEMENT
SLAB
FILL

STEEL PIN

KEY

COLUMN
FOOTINGS

Fig. 29-8. Basic components of a T foundation with basement.

Level Variation

If the sill line is located a distance above the level of the foundation, cripples are used to extend that distance, as shown in Fig. 29-13. Cripples are normally noted on the foundation plan, and their exact height is shown in an elevation section or removed detail.

Where the depth of a foundation wall varies because of changes in grade line, the footing and foundation wall is stepped down at right-angle increments and the foundation elevation changes are shown in a drawing, as seen in Fig. 29-14. The exact depth of the foundation at each of these points, however, is dimensioned on the elevation drawing because it shows the area perpendicular to the steps.

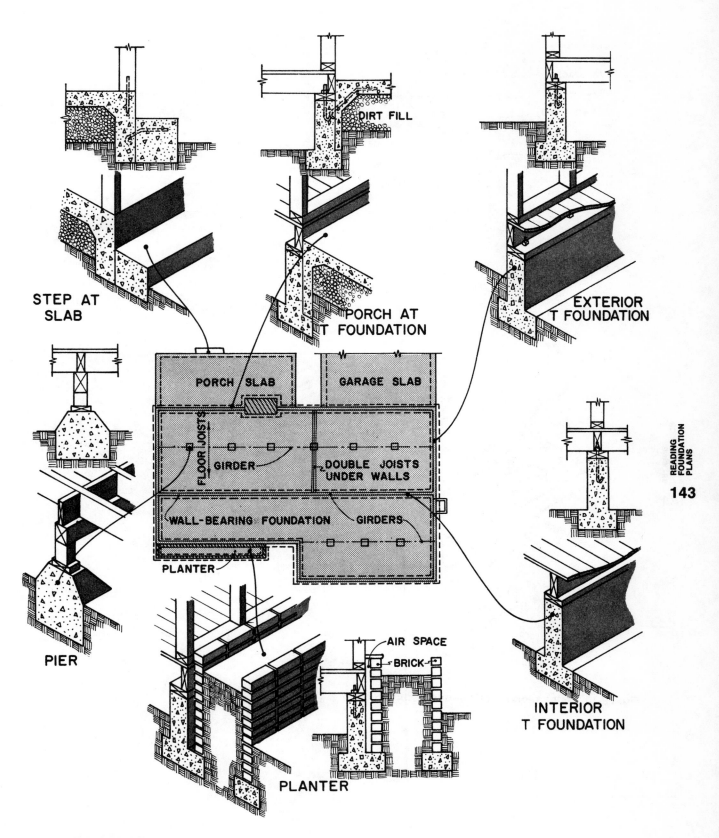

STEP AT
SLAB

PORCH AT
T FOUNDATION

DIRT FILL

EXTERIOR
T FOUNDATION

PORCH SLAB

GARAGE SLAB

FLOOR JOISTS

GIRDER

DOUBLE JOISTS
UNDER WALLS

WALL-BEARING FOUNDATION

GIRDERS

PLANTER

PIER

PLANTER

AIR SPACE

BRICK

INTERIOR
T FOUNDATION

Fig. 29-9. T-foundation details.

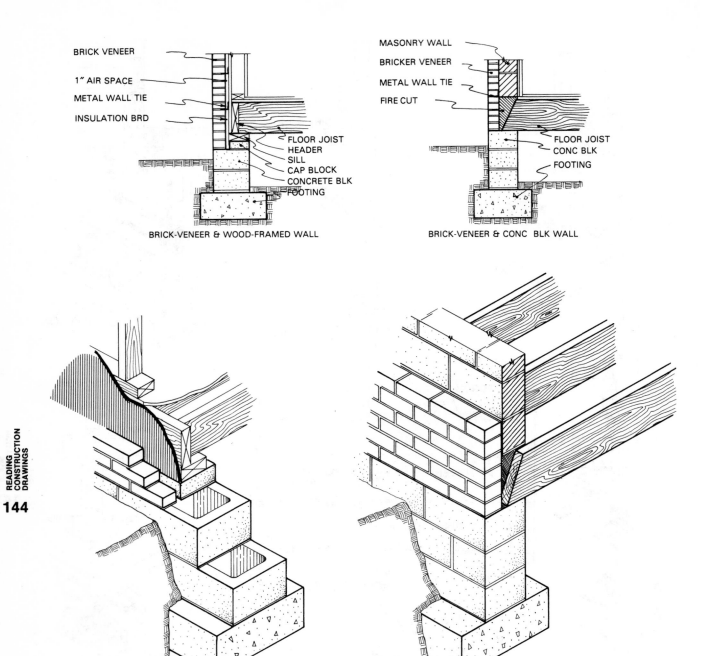

BRICK VENEER
1" AIR SPACE
METAL WALL TIE
INSULATION BRD

FLOOR JOIST
HEADER
SILL
CAP BLOCK
CONCRETE BLK
FOOTING

BRICK-VENEER & WOOD-FRAMED WALL

MASONRY WALL
BRICKER VENEER
METAL WALL TIE
FIRE CUT

FLOOR JOIST
CONC BLK
FOOTING

BRICK-VENEER & CONC BLK WALL

Fig. 29-10. Foundation for brick-veneer construction.

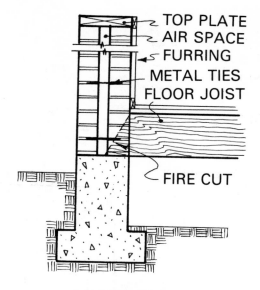

CAVITY BRICK WALL

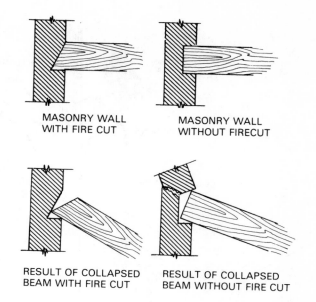

Fig. 29-12. Principle of fire-cutting shown in sectional view.

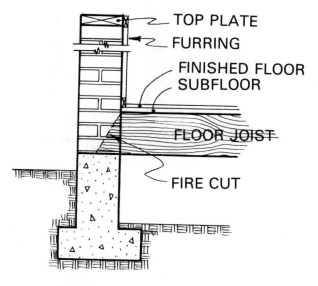

SOLID BRICK WALL

Fig. 29-11. Foundation for solid and cavity masonry walls.

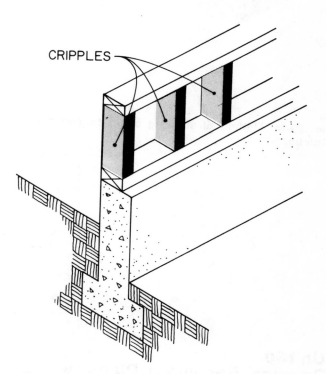

Fig. 29-13. Method of raising floor height to create basement.

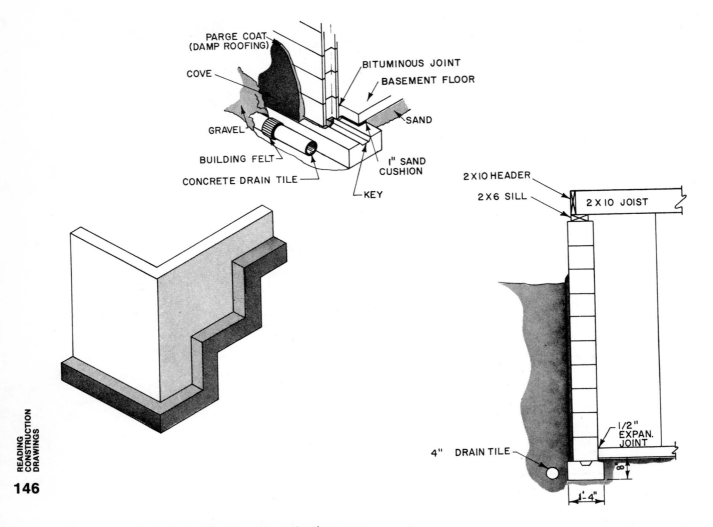

Fig. 29-14. Method of showing offset footing construction.

Unit 30
Reading Basement Plans
When a complete basement is planned for a structure, any facilities to be included, such as laundry and lavatory fixtures, shop equipment, fireplaces, and position of the floor drains, are found on the basement plan. A complete basement plan (Fig. 30-1) shows the position of footing, posts, and girders. It also shows the direction of the floor joists above. For example, the joists in the recreation room extend from left to right and are supported in the center with a beam. The joists in the shop and lavatory and laundry area span the area from the outside wall to the center wall of the basement area. Figure 30-2 shows a basement which is part finished, part crawl space, and part completely excavated. These areas are always labeled as shown.

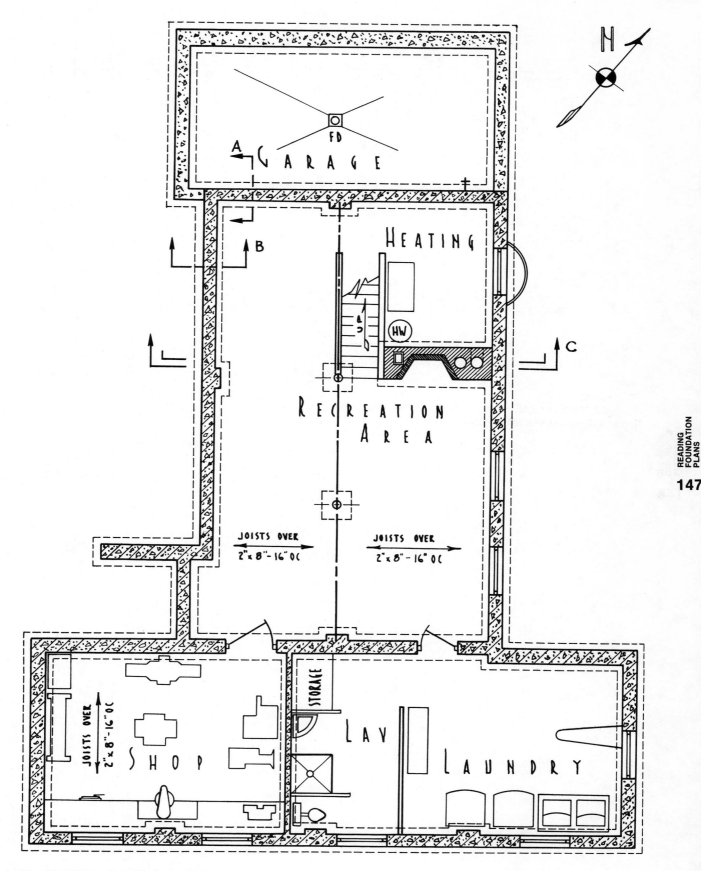

Fig. 30-1. Complete basement plan.

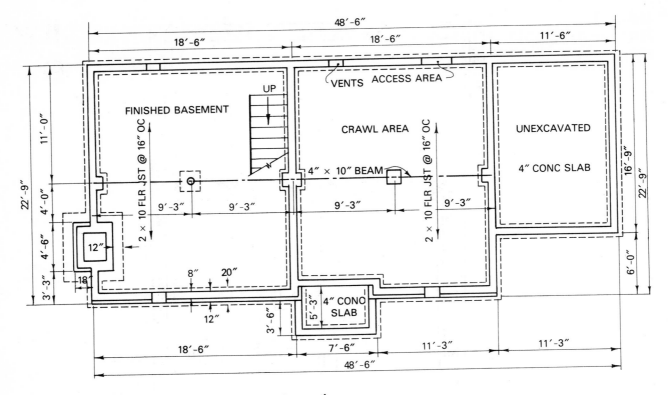

Fig. 30-2. Foundation with partial basement, crawl space, and unexcavated area.

Unit 31
Reading Fireplace Plans

Provisions are made in each foundation plan for the support and weight of fireplaces and other heavy structures. A solid, reinforced concrete footing is used in most plans. In residential work this footing is usually a minimum of 12 in. (300 mm) thick and extends at least 12 in. past the perimeter of the fireplace base.

Fireplace plans are of three types: those constructed totally on the site, those using a manufactured firebox and flow system, which are covered with masonry on the site, and those that are free-standing units.

The main component of a fireplace system is the firebox. The firebox reflects heat and draws smoke up through the chimney. Major parts of a firebox include: sides, back, smoke chamber, flue, throat, and damper (see Fig. 31-1). Most fireboxes are factory-constructed. A mason then places the firebox in the proper location in the chimney construction, according to detail plans, and lines it with firebrick. The type of masonry usually specified for fireplaces or chimneys is brick, stone, or concrete, although firebrick is specified as the lining of the firebox. The hearth is also constructed of fire-resistant material such as brick, tile, marble, or stone. Fireplace details are usually multiview plans with the side and plan views shown in sections. This reveals the positions of the firebox and the size of materials used in the footing, hearth, face, and flue. Since the front view (elevation) of most fireplaces is symmetrical, quite often one-half of this view and the full section is shown for the side view. Separate sections, such as the chimney intersection with ceiling joist, are also included in many chimney details.

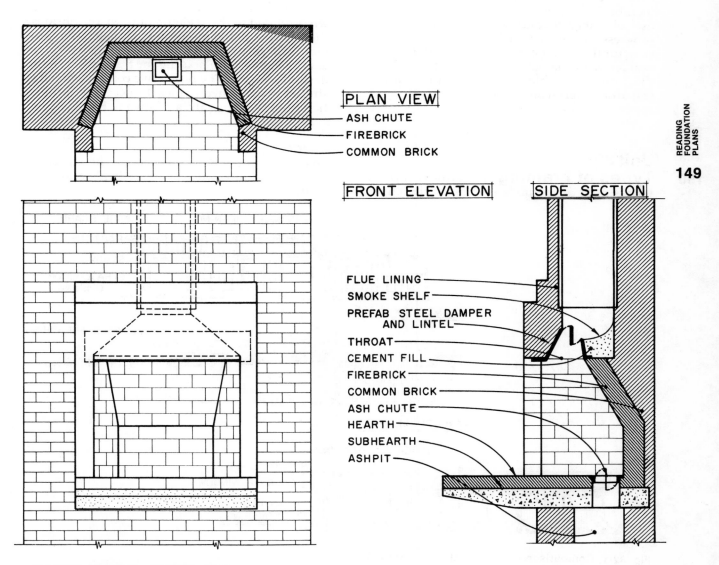

PLAN VIEW
- ASH CHUTE
- FIREBRICK
- COMMON BRICK

FRONT ELEVATION

SIDE SECTION

FLUE LINING
SMOKE SHELF
PREFAB STEEL DAMPER AND LINTEL
THROAT
CEMENT FILL
FIREBRICK
COMMON BRICK
ASH CHUTE
HEARTH
SUBHEARTH
ASHPIT

Fig. 31-1. Fireplace multiview plans.

CHAPTER 9

Framing Plans

The basic engineering principles upon which modern framing methods are based have been known for centuries. However, it has not been until recently that the development of materials and construction methods has allowed the full use of these principles. Today's designer chooses from among many basic materials in the design of the basic structural framework of a building. New and improved methods of erecting structural steel, new developments in laminating members, developments and refinements in the use of concrete and masonry products, and progress in the standardization of the design of structural components provide the modern builder with the flexibility to erect buildings at the lowest possible cost and with a minimal waste of material and time.

principles upon which structural design is based remain constant. In most structures, the roof is supported by the wall framework and interior (bearing) partitions or columns. Each exterior wall and bearing partition is supported by the foundation, which in turn is supported by a footing.

Today most buildings are constructed using a basic skeleton frame. A structural tie such as sheathing or diagonal bracing is covered with a protective siding. The structural system is somewhat related to the skeleton of most vertebrates: the framework functions like the skeleton in providing the basic rigid frame and the structural tie acts like muscles in holding the framework in the desired position. The protective coating, which is similar to the skin, provides the necessary protection from the elements, as shown in Fig. 32-1.

Unit 32
Types of Framing
Regardless of the materials used or methods employed, the physical

Loads
Loads that must be supported by the structure are divided into two types: live loads and dead loads.

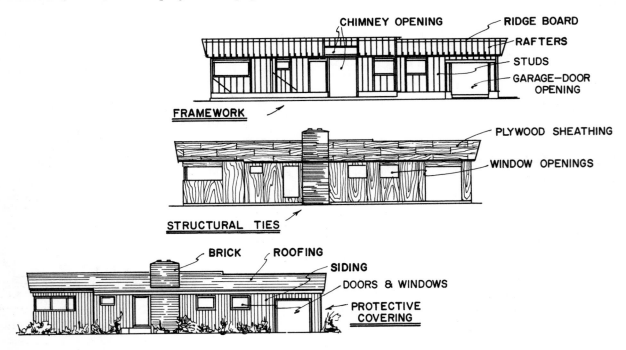

Fig. 32-1. Comparison of biological and construction skeleton frame.

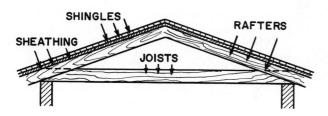

Fig. 32-2. Dead loads.

Fig. 32-3. Live loads.

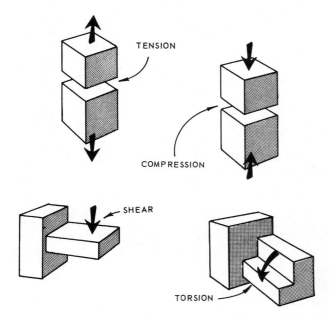

Fig. 32-4. Building stability depends on tension, compression, sheer, and torsion control.

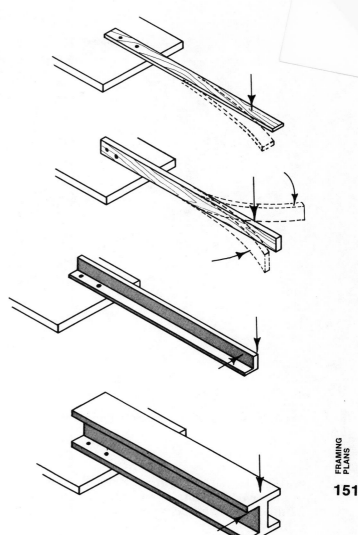

Fig. 32-5. Structural shape related to support capability.

clude the weight of snow and impact of wind and rain, as shown in Fig. 32-3.

Strength of Materials

The stability of a building depends on the strength of the materials used and the connection of members to overcome tension, compression, sheer, and torsion, as shown in Fig. 32-4. The strength of building materials is irrelevant if the building is structurally unstable. The strength of building materials is significant only when related to the structure. Most lumber and even steel is relatively flexible until tied into the structure. Figure 32-5 shows that turning a member on its side will reduce the vertical deflection (bending) but that horizontal deflection will be unaffected. Combining the horizontal and vertical components of the member to make a channel reduces

Dead loads are those loads caused by the weight of construction materials. Dead loads, such as the roof in Fig. 32-2, must be supported by the walls or bearing partitions. The dead loads (weight) of the wall must in turn be supported by the foundation. Every piece of lumber, glass, metal, and every nail and brick add to the total dead load of a structure.

Live loads are those loads that may vary from structure to structure. Live loads on a floor include furniture and people. Live loads also in-

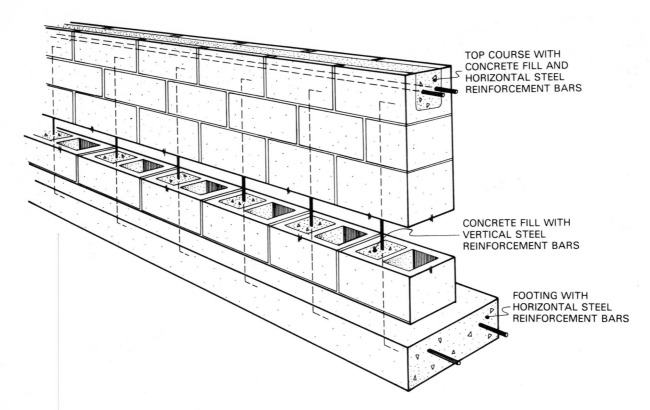

Fig. 32-6. Solid concrete-block bearing wall.

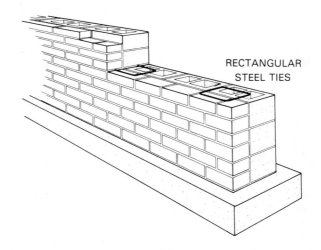

RECTANGULAR STEEL TIES

Fig. 32-7. Double solid bearing wall—brick and concrete-block.

TOP COURSE WITH CONCRETE FILL AND HORIZONTAL STEEL REINFORCEMENT BARS

CONCRETE FILL WITH VERTICAL STEEL REINFORCEMENT BARS

FOOTING WITH HORIZONTAL STEEL REINFORCEMENT BARS

both the vertical and horizontal deflection. Combining two horizontal members with one vertical member as in an I beam provides even more stability, as shown in Fig. 32-5.

Structural Framing Types

Architectural structures are divided into two basic structural types: bearing wall structures and skeleton frame structures.

BEARING WALL The bearing wall has walls that are solid and support themselves and the roof. Figure 32-6 shows a solid concrete block bearing wall. Bearing walls are often faced with other materials, as shown in Fig. 32-7, which either add to the support of the wall or do not function as part of the bearing wall. The brick facing on the wall in Fig. 32-7 is part of the bearing structure; however, if wood paneling were attached to the concrete, it would not add additional support. Some bearing walls are solid and others have hollow cores or cavities. These are known as cavity walls. Most bearing walls are constructed of concrete block, solid concrete, or brick. Stone is sometimes used for bearing wall construction, but its use is usually decorative—unless the stone is cut on blocks with flat, even, horizontal surfaces to provide stability.

SKELETON FRAME Since the passing of the log cabin, which was a bearing wall design, skeleton frame construction has been the most popular building type. In skeleton frame design the framework is self-supporting and the exterior walls may provide only partial support or no protection. The outer skin (wall) in this type of construction is known as a curtain wall. Skeleton

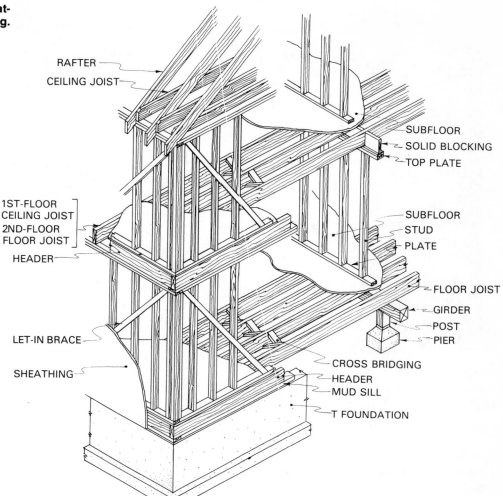

Fig. 32-8. Two-story platform (western) framing.

RAFTER

CEILING JOIST

SUBFLOOR

SOLID BLOCKING

TOP PLATE

1ST-FLOOR
CEILING JOIST
2ND-FLOOR
FLOOR JOIST

HEADER

SUBFLOOR

STUD

PLATE

FLOOR JOIST

GIRDER

POST

PIER

LET-IN BRACE

CROSS BRIDGING

HEADER

MUD SILL

SHEATHING

T FOUNDATION

frame design was made possible through the development of machinery capable of mass-producing sized and seasoned lumber that could be used interchangeably for framing purposes. However, limits on the size of the members that could be processed effectively on the job led to the design and use of relatively small structural members placed at close intervals. This type of construction is known as conventional framing.

When two or more floors are constructed using wood skeleton frame, either the platform (western) framing method, as shown in Fig. 32-8, or the balloon (eastern) framing method is used. The basic difference between western framing and eastern framing is that in western framing the second floor rests directly on the first-floor exterior walls; in eastern framing, the exterior vertical structural members (studs) extend continuously from the first floor through the second floor. Notice that in the platform (western) framing the first-floor joist rests on a sill plate, the second floor joists bear on ribbon (leger) strips set into studs, the studs are continuous for the full

height of the building, and floor joists lap over the sides of studs.

In studying the balloon (eastern) framing details in Fig. 32-9, notice that studs are continuous and the subfloor extends to the outer edge of the frame. This system is most adaptable to off-site prefabrication. In balloon framing first-floor joists bear directly on the foundation sill and second-floor joists bear on the double top plate.

POST AND BEAM CONSTRUCTION Post and beam construction is a type of skeleton frame based on the use of larger members spaced at greater intervals than in conventional framing. The use of post and beam construction methods has been increased by the practicality of manufacturing large heat-resistant windows and window walls. The accessibility, manufacture, and transportation of larger wood members have also popularized post and beam construction. More expansive spacing of members accommodates the placement of large windows and sliding doors that unite the indoors and outdoors in a single

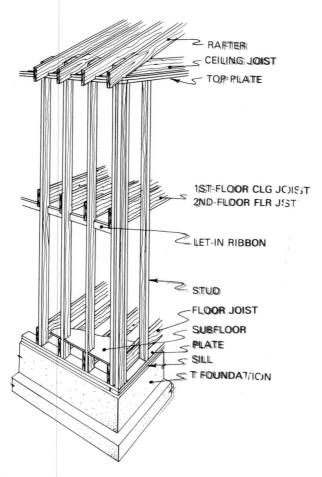

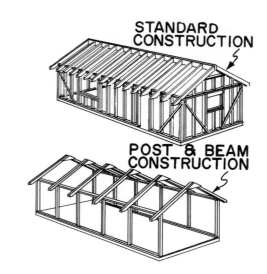

Fig. 32-10. Comparison of conventional and post and beam construction.

RAFTER
CEILING JOIST
TOP PLATE
1ST-FLOOR CLG JOIST
2ND-FLOOR FLR JIST
LET-IN RIBBON
STUD
FLOOR JOIST
SUBFLOOR
PLATE
SILL
FOUNDATION

Fig. 32-9. Two-story balloon (eastern) framing.

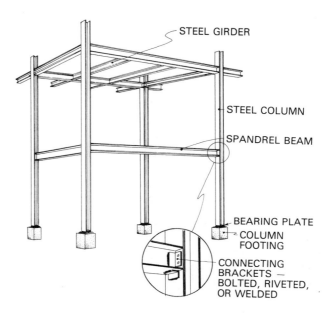

STEEL GIRDER
STEEL COLUMN
SPANDREL BEAM
BEARING PLATE
COLUMN FOOTING
CONNECTING BRACKETS — BOLTED, RIVETED, OR WELDED

Fig. 32-11. Structural-steel skeleton framing.

living space. Figure 32-10 shows a comparison of the post and beam method of construction and conventional methods. In post and beam construction, floor and roof systems are supported by beams which in turn are supported by posts which transfer the load to the foundation. Posts and beams are often left exposed to form a visible framework within which wall panels, doors, and windows are placed. Members placed between bearing post and beam can be of nonbearing quality since they serve only to fill in space and are not needed for bearing purposes.

Secure joints such as those shown in Fig. 32-11 are extremely important in post and beam and structural steel construction since there is no structurally sound intermediate support. The areas between beams or posts may be only glass or thin siding or may be completely open.

STEEL FRAMING Steel framing is similar in principle to post and beam framing. Figure 32-11 shows the skeleton frame of a structural steel building. Steel columns perform the same function as wooden posts in post and beam construction by providing vertical support. Steel beams

like wood girders support the floor and roof systems. Steel framing, however, can support and span larger distances because of the rigidity of structural steel members. For this reason, steel framing is used to erect tall, multiple-story buildings. The greatest utilization of structural steel framing has been for large commercial and industrial buildings such as schools, churches, and office buildings. Types of steel framing are shown in Fig. 32-12. Because of the rigid attachment of steel members to each other and to footing and walls, a minimum of cross-bracing is needed. Structural steel framework is designed to span long distances without intermediate support.

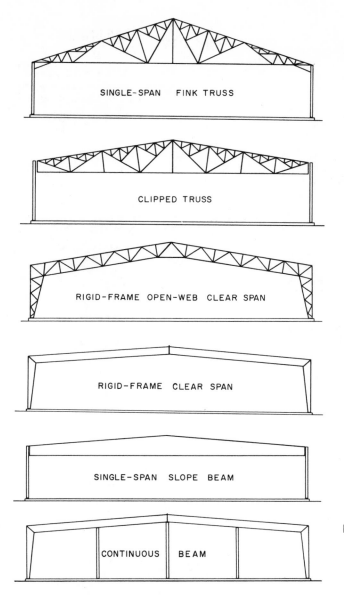

SINGLE-SPAN FINK TRUSS

CLIPPED TRUSS

RIGID-FRAME OPEN-WEB CLEAR SPAN

RIGID-FRAME CLEAR SPAN

SINGLE-SPAN SLOPE BEAM

CONTINUOUS BEAM

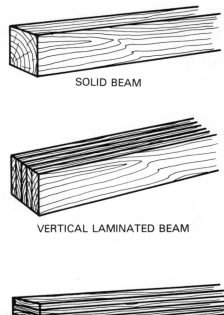

SOLID BEAM

VERTICAL LAMINATED BEAM

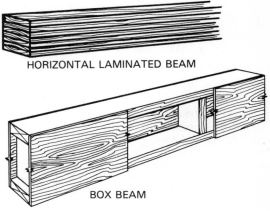

HORIZONTAL LAMINATED BEAM

BOX BEAM

Fig. 32-14. Types of structural wood members.

Fig. 32-12. Types of structural-steel framing.

Fig. 32-13. Reinforced-concrete structural frame.

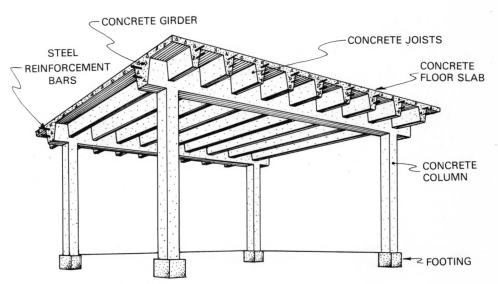

CONCRETE GIRDER

CONCRETE JOISTS

STEEL REINFORCEMENT BARS

CONCRETE FLOOR SLAB

CONCRETE COLUMN

FOOTING

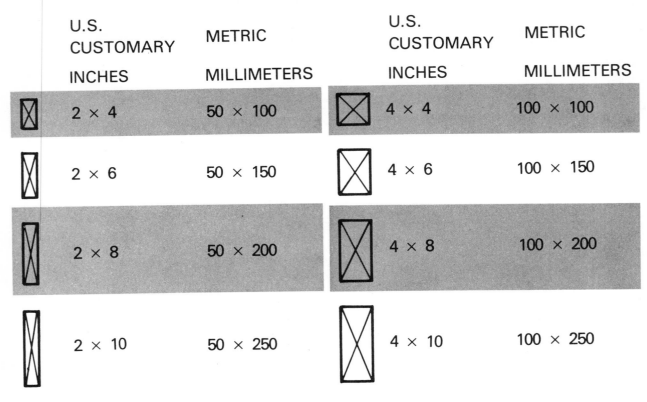

U.S. CUSTOMARY INCHES		METRIC MILLIMETERS	U.S. CUSTOMARY INCHES		METRIC MILLIMETERS
	2 × 4	50 × 100		4 × 4	100 × 100
	2 × 6	50 × 150		4 × 6	100 × 150
	2 × 8	50 × 200		4 × 8	100 × 200
	2 × 10	50 × 250		4 × 10	100 × 250

Fig. 32-15. Comparison of typical cross-sectional sizes of structural wood members in customary and metric sizes.

Longer spans create large unobstructed areas by eliminating the need for columns or bearing partitions.

REINFORCED CONCRETE STRUCTURES Reinforced concrete structural frames such as the ones shown in Fig. 32-13 are constructed of concrete with steel reinforcing rods imbedded in the concrete. Reinforced concrete structures are made of round or square reinforced columns supporting concrete slabs. Slab ribs extend either in one direction, as shown in Fig. 32-13, or in both directions to create a waffle effect.

STRUCTURAL MEMBERS Structural members, whether of wood, steel, or concrete, are available in a variety of shapes and sizes. The size of the most commonly used block is $7^5/8'' \times 7^5/8'' \times 15^5/8''$. When $3/8$ in. is added for the mortar joint, the total space covered is $8'' \times 8'' \times 16''$, which fits into a modular grid. Figures 32-14 and 32-15 show some typical types of structural wood members, and Figs. 32-16 and 32-17 show the sizes of structural concrete members. Figure 32-18 shows structural steel member symbols.

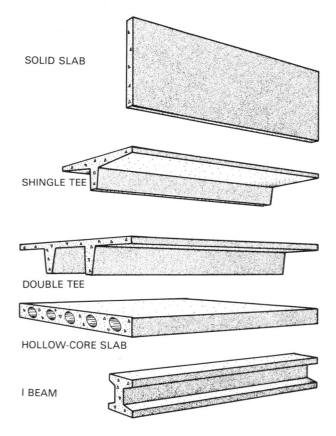

SOLID SLAB

SHINGLE TEE

DOUBLE TEE

HOLLOW-CORE SLAB

I BEAM

Fig. 32-16. Forms of precast structural concrete members.

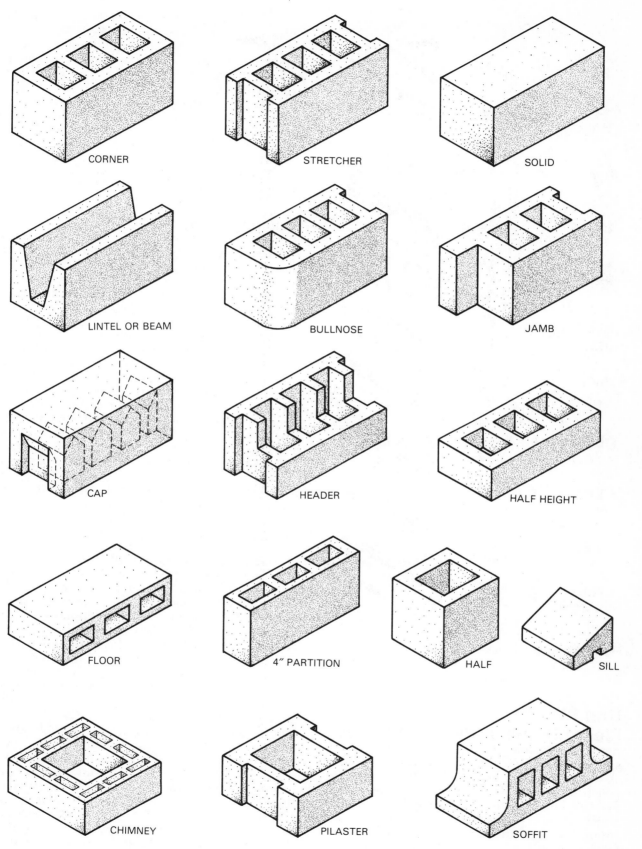

CORNER

STRETCHER

SOLID

LINTEL OR BEAM

BULLNOSE

JAMB

CAP

HEADER

HALF HEIGHT

FLOOR

4″ PARTITION

HALF

SILL

CHIMNEY

PILASTER

SOFFIT

Fig. 32-17. Commonly used concrete blocks.

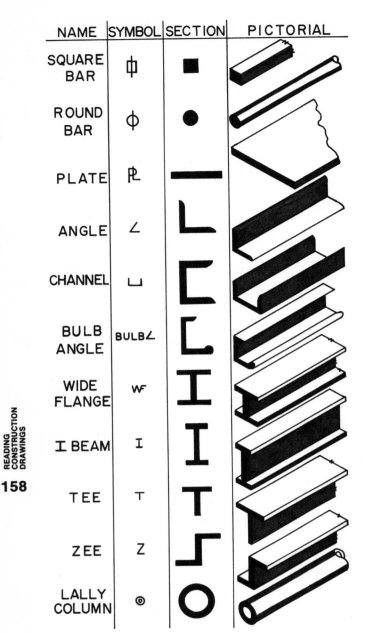

NAME	SYMBOL	SECTION	PICTORIAL
SQUARE BAR	⏀		
ROUND BAR	φ		
PLATE	℞		
ANGLE	∠		
CHANNEL	⊔		
BULB ANGLE	BULB∠		
WIDE FLANGE	W		
I BEAM	I		
TEE	T		
ZEE	Z		
LALLY COLUMN	◎		

Fig. 32-18. Structural-steel members.

Unit 33
Floor Framing Plans Floor framing plans range from those plans which show the structural support for the floor platform, as shown in Fig. 33-1, to drawings which show the construction details of the intersections of the floor system with foundation walls, fireplaces, stairwells,,and so forth. Figure 33-2 shows a floor support system attached to the foundation and footings. Girders in this type of floor system reduce the span between column supports. On top of the girders are joists, as shown in Fig. 33-2. The flooring surface is supported by the joists and the joists in turn are supported by the girders.

Girders are the major horizontal support members upon which the floor is laid. Steel beams perform the same function as wood girders. However, steel beams of comparable size can span larger distances than wood girders. Joists are the part of the floor system that is placed directly on the girders. Joists span either from girder to girder or from the girder to a foundation wall. The ends of the joists butt against a header and extend to the end of the sill with blocking placed between them, as shown in Fig. 33-2.

Platform Floor Systems
Platform floor systems are divided into three types: the conventional floor framing system, plank and beam floor systems, and panelized floor systems. These types are shown in Fig. 33-3. The conventionally framed platform system provides the most flexible method of floor framing for a wide variety of design conditions. Floor joists are spaced at 16-in. (400-mm) intervals and are supported by the sidewalls of the foundation or by beams. The plank and beam (post and beam) method of floor framing utilizes individual framing members that are much larger but less numerous than conventional framing members. Because of the size and rigidity of the members in this system, the necessity for additional bridging and reinforcement between joists is eliminated.

Panelized floor systems are composed of preassembled sandwich panels using a variety of skin and core materials. Core panel systems are used for long, clear spans over basement construction and for shorter spans in nonbasement structures.

General Floor Framing Plans
If a floor framing plan is not prepared to accompany basic architectural plans, the design of the framing for the floor system is delegated entirely to the builder. Some architectural plans include only floor framing plan notes in which the direction of the joists and the location of beams or girders are shown on the floor plan. Figure 33-4 shows a plan of this type. Figures 33-5 and 33-6 show other methods used in presenting floor framing plans. All of these plans are related to the same basic floor plan shown in Fig. 32-4. The plan shown in Fig. 32-5 clearly shows exactly how the framing is to be placed on the floor support system. However, the drawing shown in Fig. 32-6 is a simplified version of the same plan. In this plan only single lines are used to show the position of floor framing joists, girders, posts, and bridging. Figure 33-5 is the most complete and most acceptable method used to show floor framing plan design since the thickness of each structural member is exactly represented. The abbreviated plan shown in Fig. 33-7 is a shortcut method of

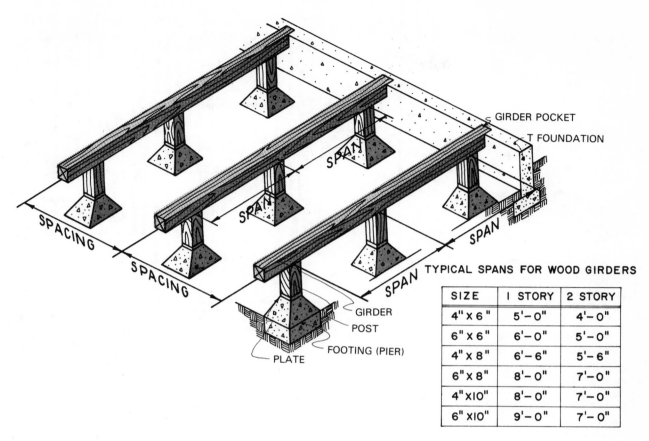

TYPICAL SPANS FOR WOOD GIRDERS

SIZE	I STORY	2 STORY
4" x 6 "	5'– 0"	4'– 0"
6" x 6"	6'– 0"	5'– 0"
4" x 8"	6'– 6"	5'– 6"
6" x 8"	8'– 0"	7'– 0"
4" x 10"	8'– 0"	7'– 0"
6" x 10"	9'– 0"	7'– 0"

Fig. 33-1. Girders reduce the joist span between column supports.

JOIST SPAN FOR STANDARD GRADE WOOD

JOIST SIZE	JOIST SPACING	JOIST SPAN
2 × 6	12″	10′
	16″	9′
	24″	7′–6″
2 × 8	12″	13′
	16″	12′
	24″	10′–6″
2 × 10	12″	16′
	16″	15′
	24″	12′
2 × 12	12″	20′
	16″	18′
	24″	15′
2 × 14	12″	23′
	16″	21′
	24″	17′

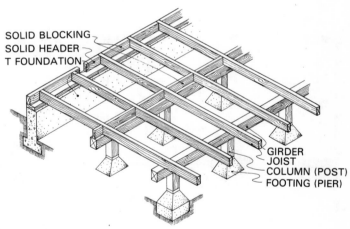

GIRDER SPAN FOR STANDARD GRADE WOOD

GIRDER SIZE	SUPPORTING WALLS	NO WALL SUPPORT
4 × 4	3′–6″	4′–0″
	3′–0″	3′–6″
4 × 6	5′–6″	6′–6″
	4′–6″	5′–6″
4 × 8	7′–0″	8′–6″
	6′–0″	7′–6″

Fig. 33-2. The floor is supported by joists, and joists are supported by girders.

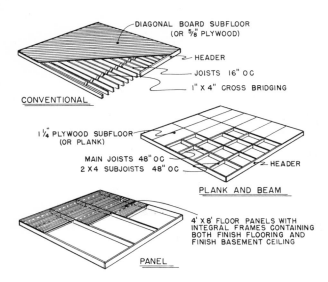

Fig. 33-3. Types of floor systems.

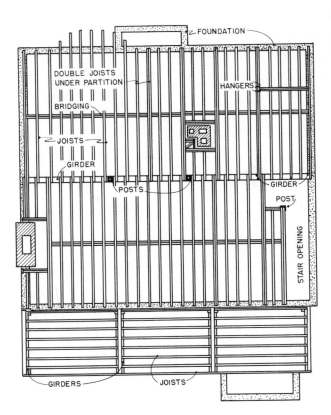

Fig. 33-5. Complete floor framing plan showing material thicknesses.

drawing floor framing plans. A single line is used to designate each structural member. Chimney and stair opening areas are shown by diagonal lines. Only the outline of the foundation and post locations are shown on the drawing. The abbreviated floor plan in Fig. 33-7 uses a technique similar to the one used on floor plans to show the entire area where uniformly spaced joists are

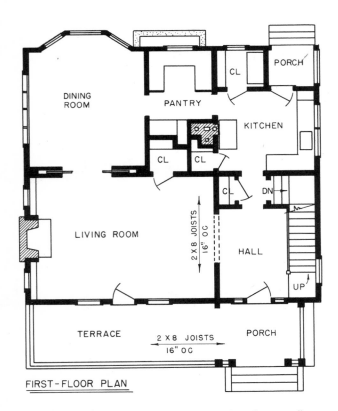

Fig. 33-4. Method of showing joist direction on floor plan.

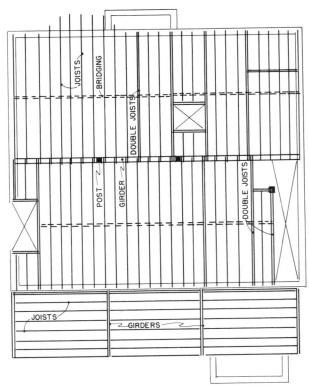

Fig. 33-6. Simplified method of showing floor framing using single lines for materials.

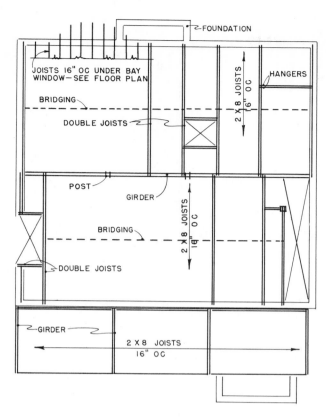

Fig. 33-7. Abbreviated method of showing floor framing design.

placed. The direction of joist is shown by an arrow. The size and spacing of joists are shown by notes placed on the arrow. This type of framing plan is usually accompanied by numerous drawings of intersection details.

The method of cutting and fitting subfloor and finished floor panels is usually determined by the builder. However, where off-site or mass-produced floor systems are built, a plan similar to the one shown in Fig. 33-8 is often prepared to ensure a maximum utilization of materials with a minimal amount of waste.

Floor Framing Detail Drawings

Although many floor framing plans are easily interpreted by experienced builders, others may require that a detail of some segment of the plan be prepared separately to explain more clearly the construction methods required. Details are drawn to eliminate the possibility of error in interpreting a plan or to explain more thoroughly some unique condition of the plan. Details may be merely enlargements of what is already on the floor framing plan. They may be prepared for dimensioning purposes, or they may show a view from a different angle to reveal the underside or elevation view for better interpretation. Details are indexed to a general floor framing plan either through a coding system, as shown in Fig. 33-9, or by identifying the area with a detail number, as shown in Fig. 33-10*a*, *b*, and *c*. The coding system shown in Fig. 33-9 is used when details are placed on separate sheets. This coding system identifies the number of the detail drawing, the sheet where it is located, how many sections are related to the base drawing, and the cutting-plane line if the detail is also a section.

A more simplified reference system, as used in Fig. 33-10, is also used when the details are shown on the same sheet as the base drawing. Study the

ALL SUBFLOOR PANELS
4'-0" × 8'-0" & 4'-0" × 8'

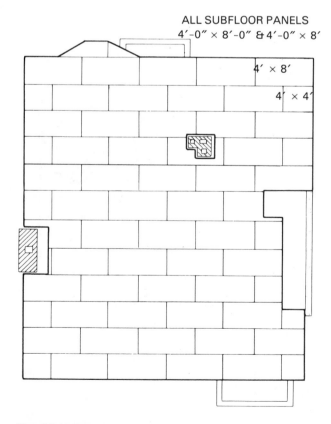

Fig. 33-8. Subfloor panel layout plan.

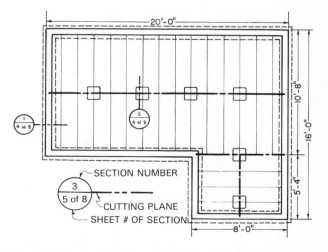

Fig. 33-9. Indexing details to floor framing plan with call-outs.

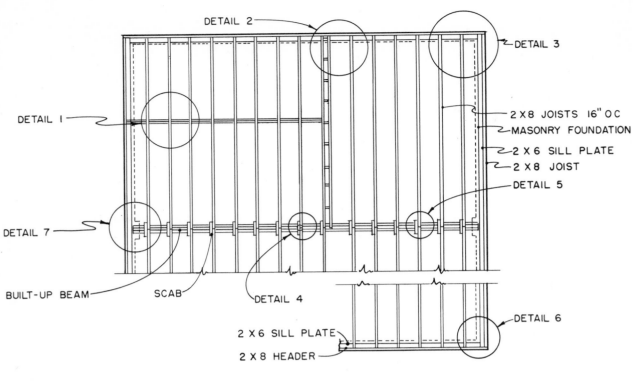

Fig. 33-10a. Floor framing details.

162

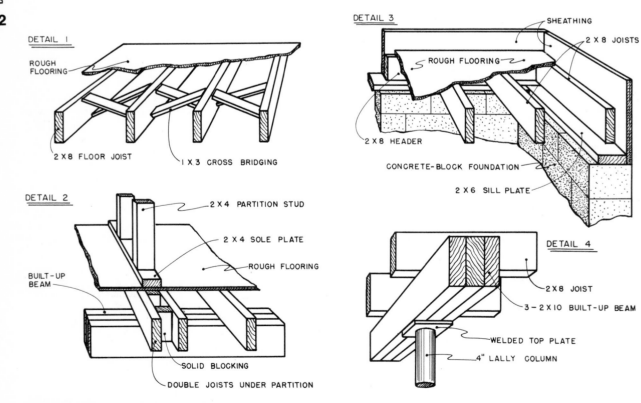

Fig. 33-10b. Floor framing details.

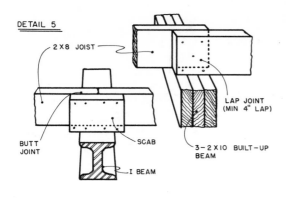

DETAIL 5

2 X 8 JOIST

LAP JOINT
(MIN 4" LAP)

BUTT
JOINT

SCAB

3 – 2 X 10 BUILT – UP
BEAM

I BEAM

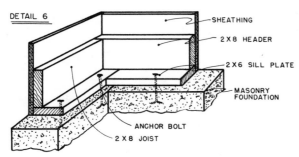

DETAIL 6

SHEATHING

2 X 8 HEADER

2 X 6 SILL PLATE

MASONRY
FOUNDATION

ANCHOR BOLT

2 X 8 JOIST

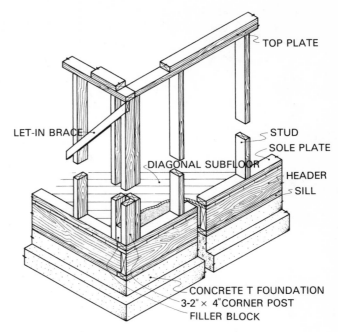

TOP PLATE

LET-IN BRACE

DIAGONAL SUBFLOOR

STUD

SOLE PLATE

HEADER

SILL

CONCRETE T FOUNDATION

3-2" × 4" CORNER POST

FILLER BLOCK

Fig. 33-11. Exterior view of sill corner.

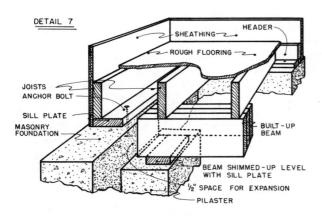

DETAIL 7

SHEATHING

HEADER

ROUGH FLOORING

JOISTS

ANCHOR BOLT

SILL PLATE

MASONRY
FOUNDATION

BUILT – UP
BEAM

BEAM SHIMMED – UP LEVEL
WITH SILL PLATE

½" SPACE FOR EXPANSION

PILASTER

Fig. 33-10c. Floor framing details.

details shown in Fig. 33-10*a*, *b* and *c*, and be sure you understand and visualize what the detail represents on the floor framing plan. Detail 1 shows the position of cross-bridging; detail 2 shows the relationship of the built-up beam and the double joist under the partition in the solid bridging; detail 3 shows the sill construction in relation to the floor joist and subflooring to the foundation; detail 4 shows the method of supporting the built-up beam by the lally (steel) column and the joist position on the beam; detail 5 shows several alternative methods of supporting the joist over a built-up beam (or I beam), thus giving the builder a material option; detail 6 shows the attachment of a typical box sill to the masonry foundation; detail 7 shows the method of supporting the built-up beam with a pilaster and the tie-in with the box sill and joist.

SILL SUPPORT DETAILS Detail drawings showing sill construction details reveal not only the construction of the sill but also the method of attaching the sill to the foundation. This type of exterior wall framing is revealed in a sill detail. Since the sill is the transition between the foundation and the exterior walls of a structure, a sill detail is usually included in most sets of architectural plans.

Some sill details are shown in pictorial form, as in Fig. 33-11 and 33-12. However, if pictorial drawings are used, two drawings must be used to show the exterior and interior view. Figure 33-11 shows an exterior pictorial view of a sill corner, and Fig. 33-12 shows an interior pictorial view of a sill corner. Pictorial drawings are easy to interpret but are more difficult and time-consuming to draw; therefore, most sill details are prepared in sectional form, as shown in Fig. 33-13. The floor area in a sill detail sectional drawing is usually extended to show at least one joist which indicates the direction of the joist and its size and placement in relation to the placement of the subfloor and finished floor. This is also sometimes shown on a floor framing plan as it is in Fig. 33-13. However, the attachment of the floor system and sill to the foundation and the intersection of the exterior walls with the floor and sill are not revealed in this type of drawing. Figure 33-13 also shows a floor framing plan and elevation section compared to a pictorial drawing of a platform framed sill. Figure 33-14 shows the same three drawings related to a balloon framing sill, and Fig. 33-15 shows the relationship between a post

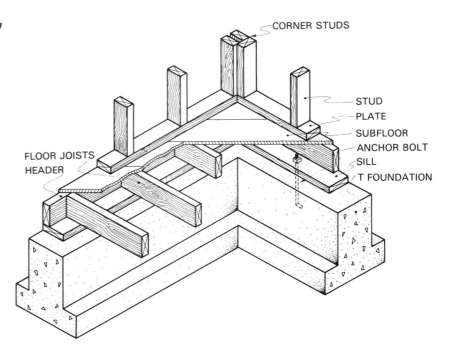

Fig. 33-12. Interior view of T-foundation sill.

CORNER STUDS

STUD
PLATE
SUBFLOOR
ANCHOR BOLT
SILL
T FOUNDATION

FLOOR JOISTS
HEADER

and beam floor framing plan, a pictorial drawing, and an elevation section drawing. The floor framing plan in this illustration shows the spacing of girders, blocking, and peers more clearly. However, the elevation section shows the intersections between the floor system foundation sill and exterior wall more clearly. For this reason, sometimes both drawings are used to fully describe the type of construction required.

Sill details of this type are also required to show the relationship and joining of materials, such as masonry, wood, precast concrete, and structural steel. Figure 33-16 shows a typical masonry sill detail. This section is necessary to show the fire-cut portion of the joist on the foundation because this fire-cut detail does not show on the floor framing plan. Likewise, a sill detail shows the position of brick veneer in relation to the foundation framing wall, floor, and sill members, as shown on the sectional elevation and on a floor framing plan in Fig. 33-17.

Precast concrete sill and floor details and structural steel sill and floor details are also used to show the size and spacing of structural members. They are also included in many sets of drawings to show the fastening methods and devices used to anchor concrete and masonry to wood, structural concrete, and steel members.

INTERMEDIATE SUPPORT DETAILS

These show the position and method of attachment of intermediate support members, such as girders and beams. These details are often shown by either a pictorial drawing, a floor framing plan detail, or an elevation section, as shown in Fig. 33-18a, b, and c. In these three illustrations notice

how the elevation section through the sill and floor plan detail is used to show the difference between construction methods used with a standard girder. Figure 33-18b shows a girder supported with a box sill. Figure 33-18c shows a fire-cut girder and Fig. 33-18a shows a girder supported in a pocket in a masonry foundation wall. Study the relationship between the pictorial drawing in each of these illustrations and the plan and elevation detail. Since pictorial drawings are seldom used for construction, a builder must learn to read and interpret the plan and elevation details and visualize what is shown in the pictorial drawing.

Since there are so many different methods of attaching intermediate supports to foundation walls, a support detail is always necessary to complete a full set of plans. Figure 33-19 shows several alternative methods of attaching steel joists to masonry walls. Sometimes the builder is provided with options or alternatives of this type through the use of detail drawings.

If a beam or girder cannot span the distance between foundation walls, an intermediate vertical support such as a wood or steel column, as shown in Fig. 33-20, is used. This illustration also shows methods of attaching the column to a girder or beam. The plan and elevation view of the support intersections is shown to the right of the pictorial drawing. Study each drawing and be sure you can visualize the construction details as shown in the pictorial drawing.

Figure 33-21 shows several methods of connecting intersecting horizontal members such as beams and girders to a post. The method used on plan and elevation drawings to show these inter-

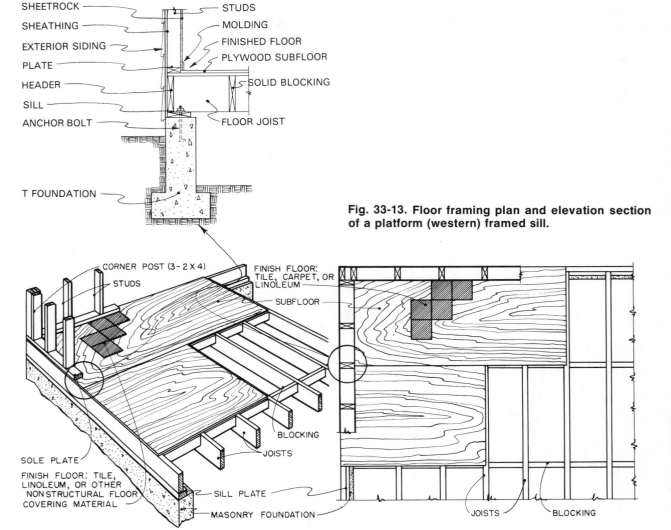

SHEETROCK

SHEATHING

EXTERIOR SIDING

PLATE

HEADER

SILL

ANCHOR BOLT

T FOUNDATION

STUDS

MOLDING

FINISHED FLOOR

PLYWOOD SUBFLOOR

SOLID BLOCKING

FLOOR JOIST

Fig. 33-13. Floor framing plan and elevation section of a platform (western) framed sill.

CORNER POST (3 - 2 X 4)

STUDS

FINISH FLOOR: TILE, CARPET, OR LINOLEUM

SUBFLOOR

BLOCKING

JOISTS

SOLE PLATE

FINISH FLOOR: TILE, LINOLEUM, OR OTHER NON STRUCTURAL FLOOR COVERING MATERIAL

SILL PLATE

MASONRY FOUNDATION

JOISTS

BLOCKING

sections is shown at the right of each pictorial drawing. Other intersections between perpendicular beams and floor joists are shown in Fig. 33-22. Here, the plan and elevation views are shown to the right of each pictorial drawing. From these drawings the builder determines the type of beam and girder and the method of intersecting the joist with the beam. The joist in the left column rests directly on the beam or girder, while the joist in the right column intersects the girder, thus allowing the top of the girder and the top of the joist to be at or near the same level. Figure 33-23 shows the intersection of wood joists with steel beams. The methods of showing these intersections on plan and elevation views are shown.

Bridging or blocking is used between joists to keep them from swaying (twisting) in the center of the span. Bridging is either of the cross bridging type or solid blocking, as shown in Fig. 33-24.

Where complete floor framing plans are not used, bridging is usually specified through the use of a note on the foundation plan.

Another type of intermediate support is the use of double joists under partitions. Double floor joists are used under bearing partitions, as shown in Fig. 33-25, when the partition wall is parallel to the floor joist. When space is provided for pipes or wires, a spacer is used between double joists under bearing partitions, as shown in Fig. 33-26. When the joists are to be level with the top of a girder, as shown in Fig. 33-27, shorter joists are used which rest on a ledger board attached to the girder. The plan and elevation drawing used for each of these double-joist arrangements is shown to the right of the pictorial drawing. You should be able to recognize the construction method used by referring to either the plan or the elevation drawing without reference to the pictorial

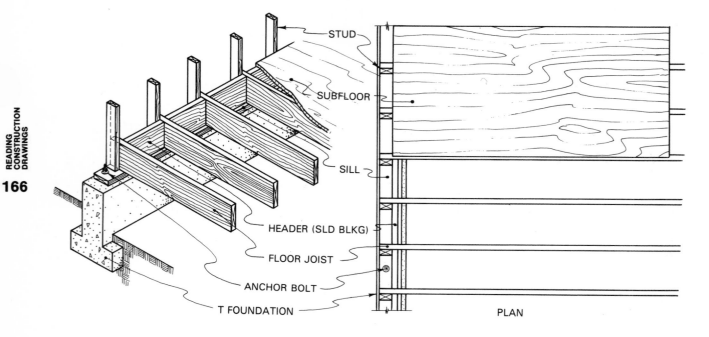

SHEATHING
EXTERIOR SIDING
SHEETROCK
STUD
MOLDING
FINISHED FLOOR
PLYWOOD SUBFLOOR
HEADER
FLOOR JOIST
ANCHOR BOLT
SILL

T FOUNDATION — ELEVATION

Fig. 33-14. Floor framing plan and elevation section of a balloon (eastern) framed sill.

STUD
SUBFLOOR
SILL
HEADER (SLD BLKG)
FLOOR JOIST
ANCHOR BOLT
T FOUNDATION

PLAN

drawing. Support blocking is used to provide additional support under partitions when the partition is a nonbearing wall (Fig. 33-28).

When an interior partition is perpendicular to the floor joist then solid blocking is placed directly under the partition. This provides additional support for the partition and saves materials, since blocking would be used between joists in anchor location. Figure 33-29 shows the drawings used to illustrate this support system.

The best method of providing intermediate support for a floor system is the use of another foundation wall. Where interior foundation walls are needed, this additional foundation wall replaces a girder or beam but, of course, does not

provide the clear span that the use of a girder would allow (Fig. 33-30).

Intermediate support for steel joist framing systems, as shown in Fig. 33-31, also utilizes I beams. In this case, the steel joists are welded or bolted to the I beam and continuous horizontal bridging is welded to the top and bottom cords to prevent excessive lateral movement of the joists.

FLOOR DECKING Decking is the top surface of a floor system. The floor deck, in addition to bridging, provides additional lateral support for the joists. Decking usually consists of a subfloor and a finished floor, although in some systems they are combined. Subfloor decking materials

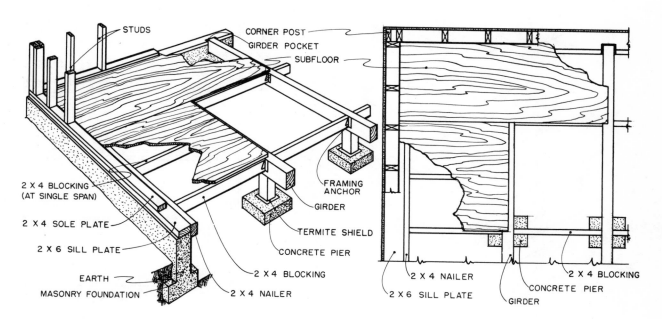

STUD
PLATE
SILL
FRAMING ANCHOR
BLOCKING
T FOUNDATION

STUDS
CORNER POST
GIRDER POCKET
SUBFLOOR
2 X 4 BLOCKING (AT SINGLE SPAN)
FRAMING ANCHOR
2 X 4 SOLE PLATE
GIRDER
2 X 6 SILL PLATE
TERMITE SHIELD
EARTH
CONCRETE PIER
MASONRY FOUNDATION
2 X 4 BLOCKING
2 X 4 NAILER
2 X 4 NAILER
2 X 6 SILL PLATE
2 X 4 BLOCKING
CONCRETE PIER
GIRDER

Fig. 33-15. Post and beam floor framing plan and elevation section.

consist of diagonal wood members, plywood sheets, prefabricated panels, plank boards, concrete slabs, or corrugated steel sheets. Plank flooring, as shown in the example in Fig. 33-32, is laid directly over joists, and partitions rest directly on the plank flooring. Finished flooring is installed over the subfloor and butted against the partition framing members when a subfloor system is used. When steel floor decks are used they are usually constructed of corrugated sheet steel (Fig. 33-33). These subfloors act as platform surfaces during construction and also provide the necessary subfloor surface for a concrete slab floor. When precast concrete plank floor systems are used, one of the concrete slab members, as shown in Fig. 33-34, functions as subflooring.

SECOND-FLOOR FRAMING Second-floor framing details are usually shown with a full section through the exterior wall (Fig. 33-35). Notice how the intersection of the second-floor joist reveals that Fig. 33-35 is balloon (eastern) framing in which the studs are continuous from the foundation to the eave. Compare this to the section shown in Fig. 33-36, which shows a western (platform) frame second-floor construction. In this plan, the second-floor joist rests directly on a top plate, which rests directly on the first-floor studs. Recognizing these construction features through interpreting floor system sectional drawings is extremely critical to the builder. Study the relationship between the elevation section shown in Figs. 33-35 and 33-36. Be sure you can visualize

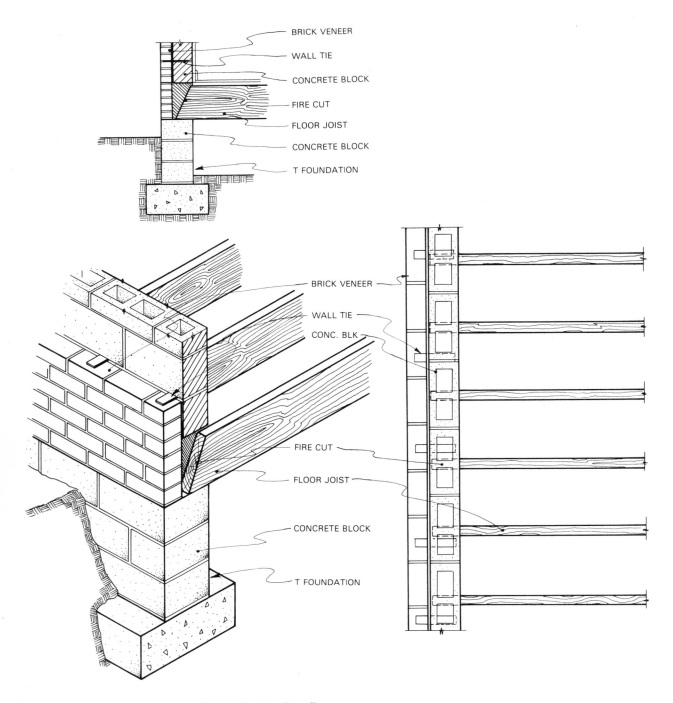

BRICK VENEER
WALL TIE
CONCRETE BLOCK
FIRE CUT
FLOOR JOIST
CONCRETE BLOCK
T FOUNDATION

BRICK VENEER
WALL TIE
CONC. BLK
FIRE CUT
FLOOR JOIST
CONCRETE BLOCK
T FOUNDATION

Fig. 33-16. Typical masonry sill detail showing fire cut.

what is shown in the pictorial drawing. You should see in balloon framing that studs are nailed directly to the sill and joist and that no header is used. Joists are supported by a ribbon board and nailed directly to the studs on the second floor, as shown in Fig. 33-35. When combinations of materials, such as brick veneer, are shown on sectional drawings (Fig. 33-37), the relationship between the floor system and the exterior wall system can be read easily. Notice that in Fig. 33-37 the brick-veneer wall extends only up to the level of the second floor. The second-floor header and floor joists rest directly on the top of the brick veneer. This is not revealed in the plan view, which is the reason for showing floor detail plans in elevation sections.

Floor extensions, such as the cantilevered (unsupported extension) second floor shown in Figs. 33-38 and 33-39, are usually shown in section drawings. Figure 33-38 shows the cantilevering of

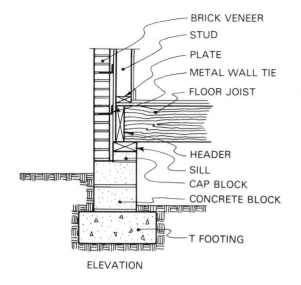

BRICK VENEER
STUD
PLATE
METAL WALL TIE
FLOOR JOIST

HEADER
SILL
CAP BLOCK
CONCRETE BLOCK

T FOOTING

ELEVATION

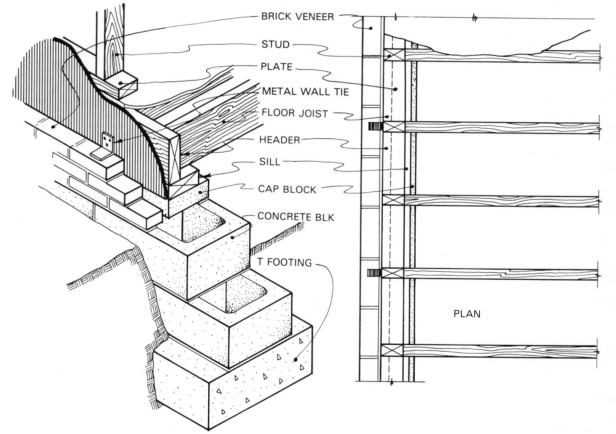

BRICK VENEER
STUD
PLATE
METAL WALL TIE
FLOOR JOIST
HEADER
SILL
CAP BLOCK
CONCRETE BLK
T FOOTING

PLAN

Fig. 33-17. Brick-veneer sill detail.

a second floor with joists parallel to an exterior wall. And Fig. 33-39 shows the cantilevering of a second floor with joists perpendicular to an exterior wall. The change in direction of the joists (Fig. 33-38) is thus shown on the plan view but can also be interpreted from the elevation section.

FIREPLACE DETAILS When it is necessary to cut floor joists to provide an opening for a fireplace, supplemental joists called headers are used. Headers are placed at right angles to the regular joists to support and tie together the ends of the joists that are cut. A header cannot be of greater depth than any other joist. Therefore,

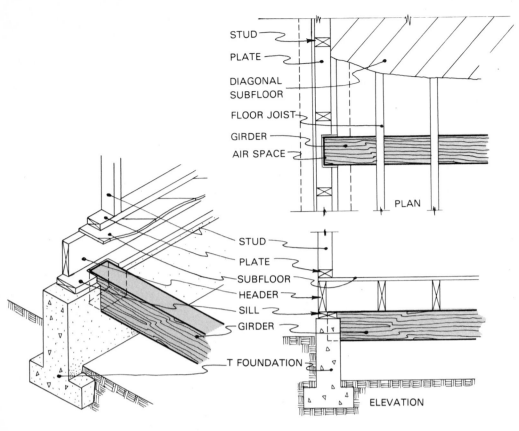

STUD
PLATE
DIAGONAL
SUBFLOOR
FLOOR JOIST
GIRDER
AIR SPACE

PLAN

STUD
PLATE
SUBFLOOR
HEADER
SILL
GIRDER

T FOUNDATION

ELEVATION

Fig. 33-18a. Standard girder (beam) pocket in exterior T-foundation wall.

Fig. 33-18b. Girder supported with a box sill on an exterior T-foundation wall.

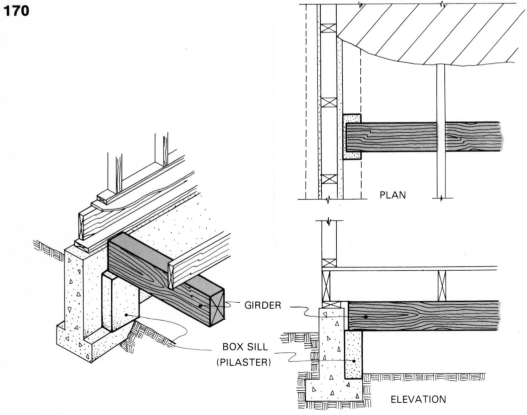

PLAN

GIRDER

BOX SILL
(PILASTER)

ELEVATION

Fig. 33-18c. Fire-cut girder supported in a pocket in a masonry foundation wall.

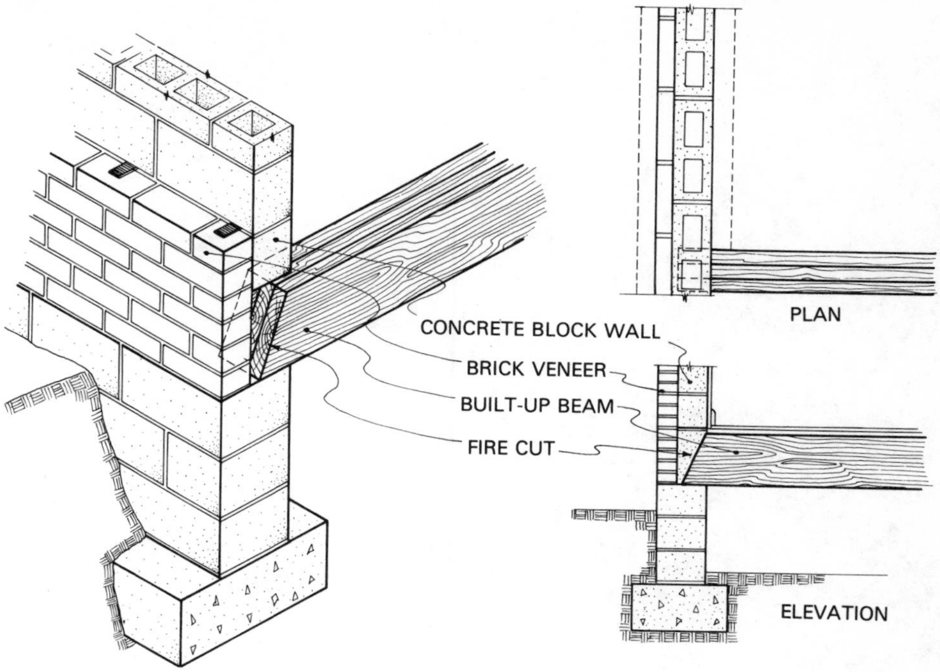

CONCRETE BLOCK WALL

BRICK VENEER

BUILT-UP BEAM

FIRE CUT

PLAN

ELEVATION

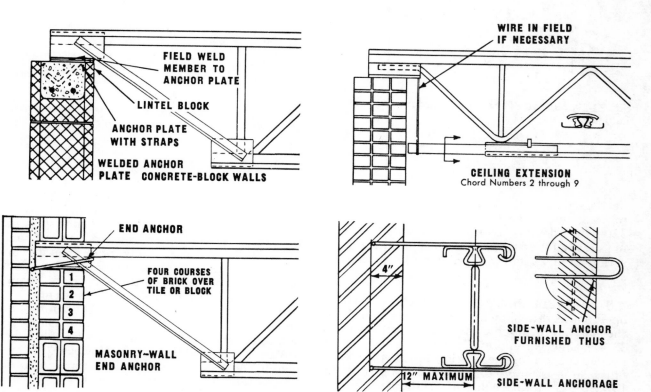

Fig. 33-19. Alternate methods of attaching steel joists to masonry walls.

FIELD WELD MEMBER TO ANCHOR PLATE

LINTEL BLOCK

ANCHOR PLATE WITH STRAPS

WELDED ANCHOR PLATE CONCRETE-BLOCK WALLS

WIRE IN FIELD IF NECESSARY

CEILING EXTENSION
Chord Numbers 2 through 9

END ANCHOR

FOUR COURSES OF BRICK OVER TILE OR BLOCK

MASONRY-WALL END ANCHOR

4"

12" MAXIMUM

SIDE-WALL ANCHOR FURNISHED THUS

SIDE-WALL ANCHORAGE

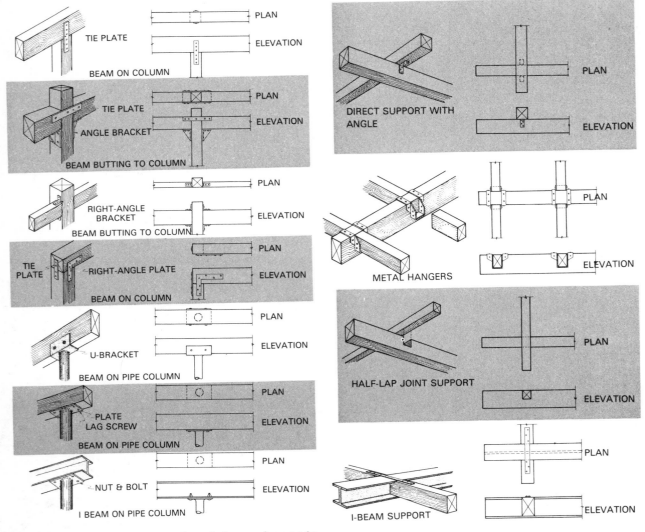

Fig. 33-20. Method of connecting girders or beams to columns.

Fig. 33-21. Method of connecting intersecting girders and beams to a vertical post.

headers are usually doubled (placed side by side) to compensate for the additional load. Figure 33-40 includes an elevation and a pictorial drawing showing the use of double headers as compensation for joists that were cut to provide space for a fireplace or chimney.

STAIRWELL FRAMING DETAILS Stairwell openings as drawn on a floor framing plan show the position of double joists and headers (Fig. 33-41). However, more information is frequently needed concerning the relationship of other parts of the stair assembly to the stairwell opening. In this case, an elevation drawing of the stair assembly, such as the one shown in Fig. 33-42, is included. Information concerning the size and position of the various parts of the stair assembly

is derived from such a detail drawing. Although the stairwell opening can be shown on a floor framing plan, an elevation is required to show the rise, head clearance, stringer design, and total run of the stairway assembly. Stair assembly details are also used, as shown in Fig. 33-43, to show the relationship between threads, risers, stringers, and landings on more complex systems.

When constructing the stairwell opening in a floor framing system, the only dimensions required are the width, length, and size of headers around the stairwell opening. However, to completely construct the stairwell system, an elevation section such as the one shown in Fig. 33-44 is usually supplied. This drawing shows all additional dimensions relating to the stair assembly. This includes all horizontal distances the length

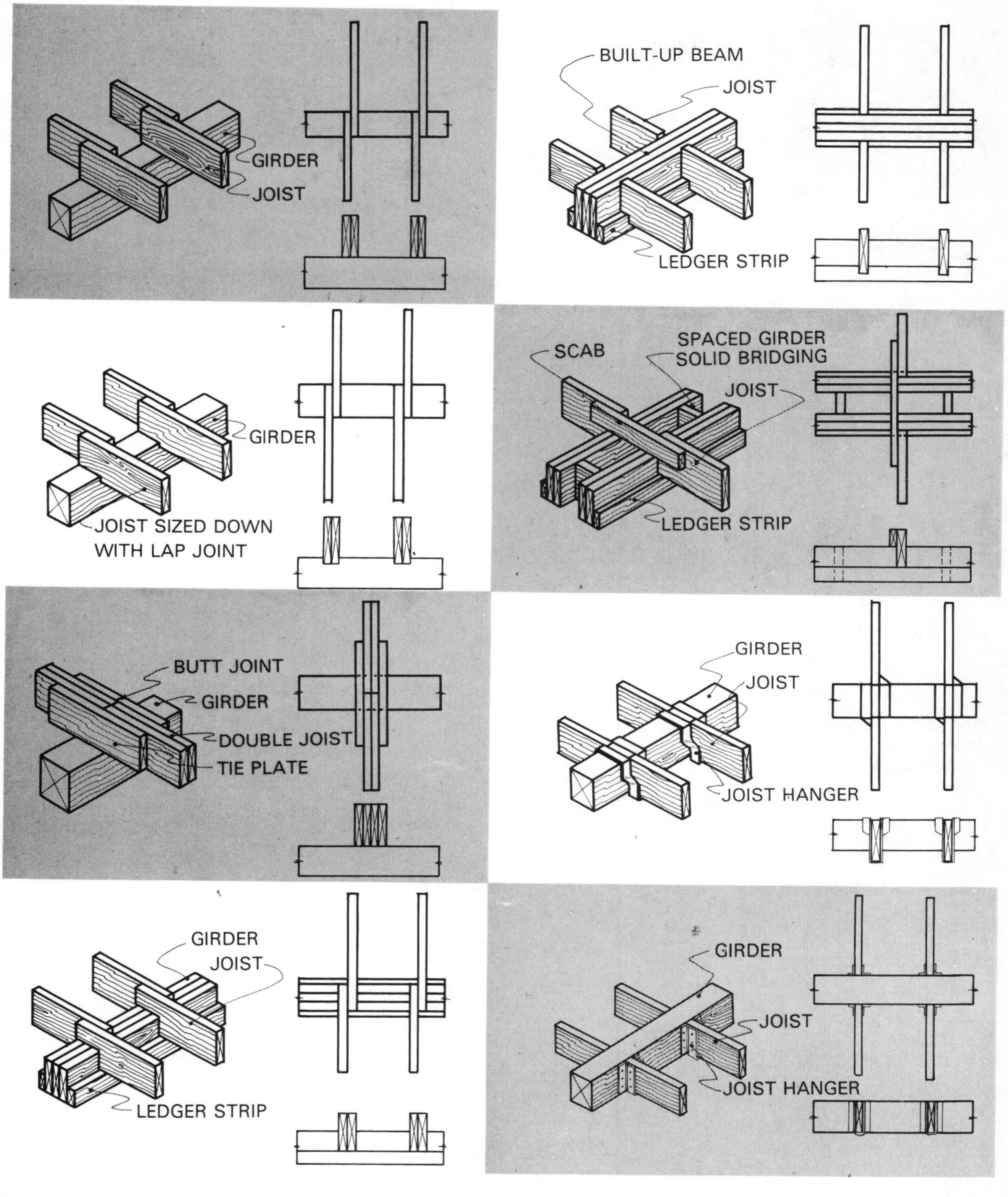

Fig. 33-22. Wood beam (girder) support system for joists.

of the run, and the length of each thread. It also includes all vertical dimensions such as the height of the risers, total rise from floor to floor, and the size of joists, subfloor, finished floor, headers, and stringers. This same kind of elevation section is also necessary for structural steel stairways and for the concrete stairs, as shown in Fig. 33-45.

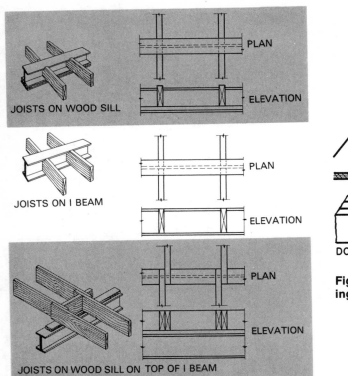

JOISTS ON WOOD SILL

PLAN

ELEVATION

JOISTS ON I BEAM

PLAN

ELEVATION

JOISTS ON WOOD SILL ON TOP OF I BEAM

PLAN

ELEVATION

Fig. 33-23. Joist support systems using steel beams.

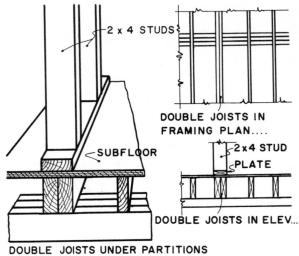

2 x 4 STUDS

SUBFLOOR

DOUBLE JOISTS IN FRAMING PLAN....

2 x 4 STUD
PLATE

DOUBLE JOISTS IN ELEV...

DOUBLE JOISTS UNDER PARTITIONS

Fig. 33-25. Double floor joist is placed under a bearing partition when the wall is parallel to floor joist.

Fig. 33-24. Alternate bridging as seen in plan and elevation view.

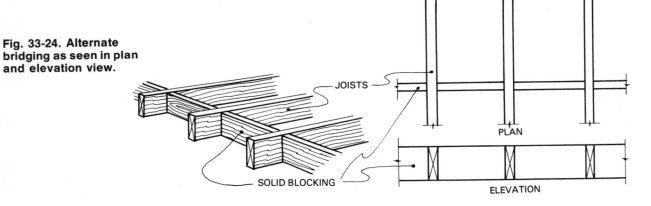

JOISTS

SOLID BLOCKING

PLAN

ELEVATION

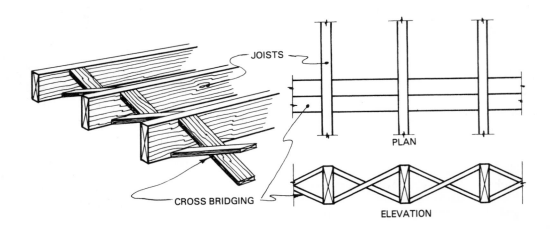

JOISTS

CROSS BRIDGING

PLAN

ELEVATION

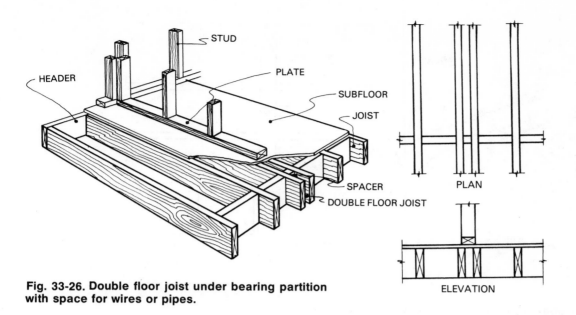

Fig. 33-26. Double floor joist under bearing partition with space for wires or pipes.

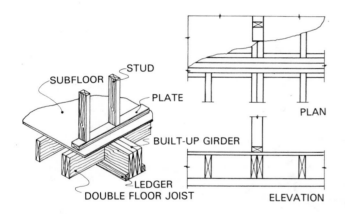

Fig. 33-27. Double floor joist level with top of girder.

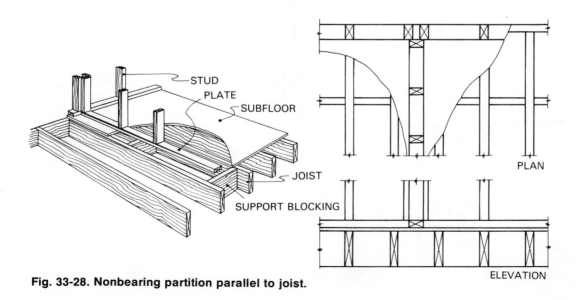

Fig. 33-28. Nonbearing partition parallel to joist.

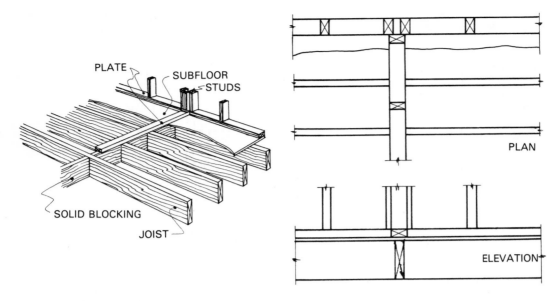

Fig. 33-29. Support of wall perpendicular to joist.

PLATE
SUBFLOOR
STUDS
SOLID BLOCKING
JOIST
PLAN
ELEVATION

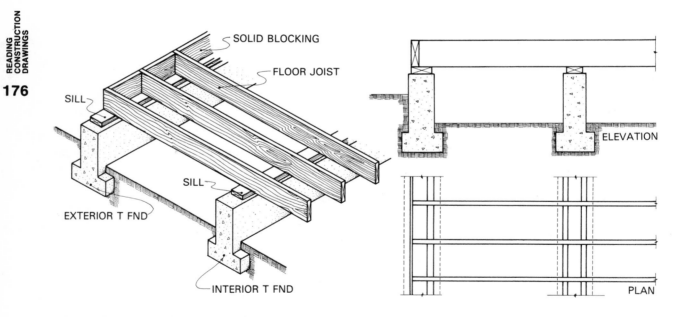

SOLID BLOCKING
FLOOR JOIST
SILL
SILL
EXTERIOR T FND
INTERIOR T FND
ELEVATION
PLAN

Fig. 33-30. Support joist on interior foundation wall.

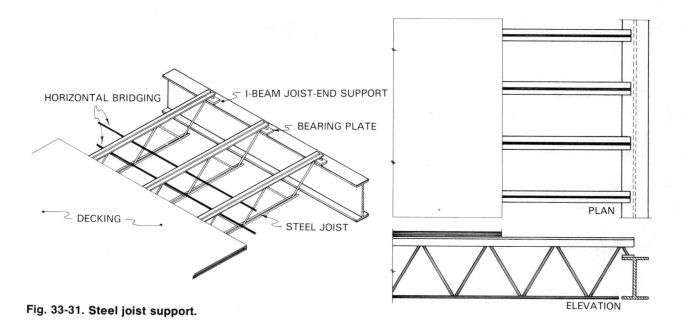

Fig. 33-31. Steel joist support.

HORIZONTAL BRIDGING

I-BEAM JOIST-END SUPPORT

BEARING PLATE

DECKING

STEEL JOIST

PLAN

ELEVATION

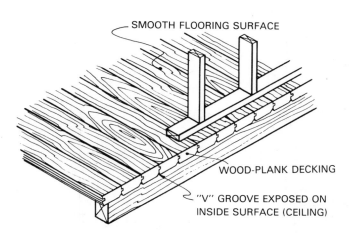

SMOOTH FLOORING SURFACE

WOOD-PLANK DECKING

"V" GROOVE EXPOSED ON INSIDE SURFACE (CEILING)

Fig. 33-32. Wood plank decking cover.

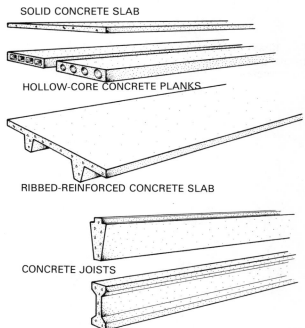

SOLID CONCRETE SLAB

HOLLOW-CORE CONCRETE PLANKS

RIBBED-REINFORCED CONCRETE SLAB

CONCRETE JOISTS

Fig. 33-34. Precast concrete floor systems and supports.

Fig. 33-33. Corrugated-steel floor decks.

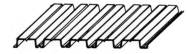
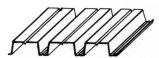

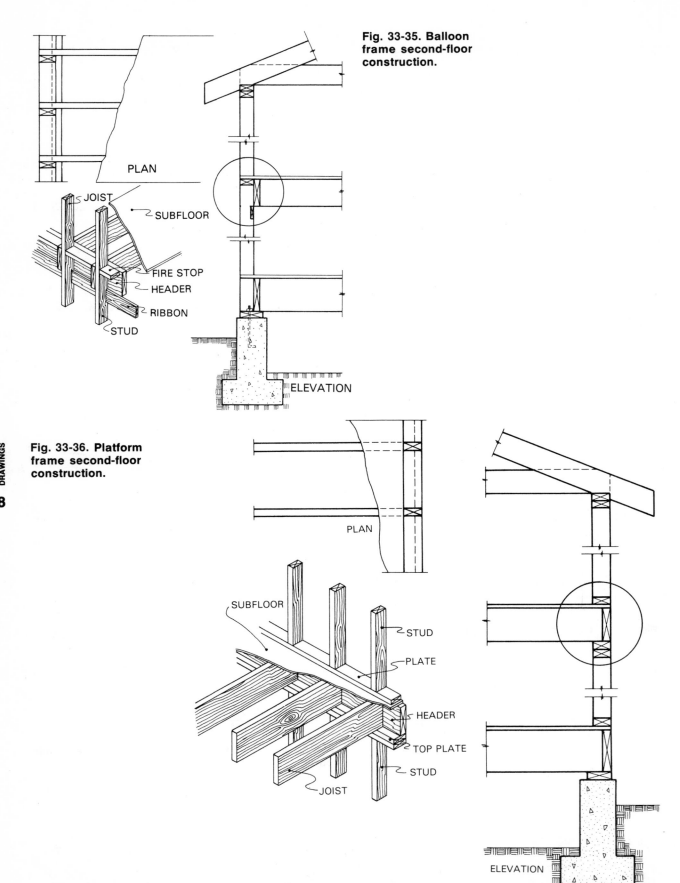

Fig. 33-35. Balloon frame second-floor construction.

PLAN

JOIST
SUBFLOOR
FIRE STOP
HEADER
RIBBON
STUD

ELEVATION

Fig. 33-36. Platform frame second-floor construction.

PLAN

SUBFLOOR
STUD
PLATE
HEADER
TOP PLATE
STUD
JOIST

ELEVATION

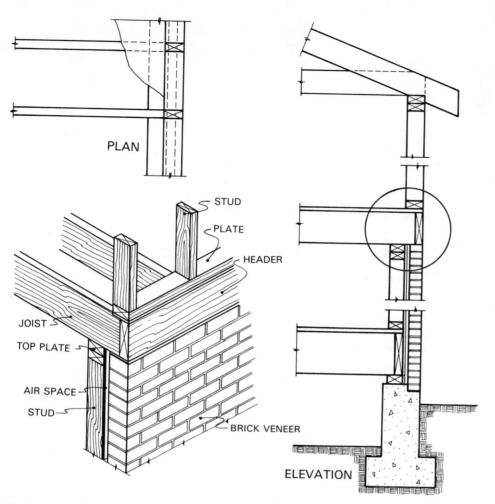

PLAN

STUD

PLATE

HEADER

JOIST

TOP PLATE

AIR SPACE

STUD

BRICK VENEER

ELEVATION

Fig. 33-37. First-floor brick veneer under second-floor frame construction.

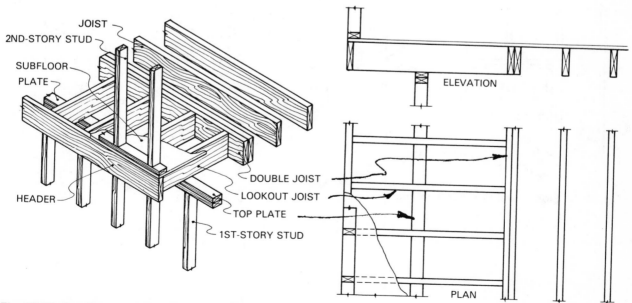

JOIST

2ND-STORY STUD

SUBFLOOR

PLATE

HEADER

ELEVATION

DOUBLE JOIST

LOOKOUT JOIST

TOP PLATE

1ST-STORY STUD

PLAN

Fig. 33-38. Cantilevered second floor with joist parallel to exterior wall.

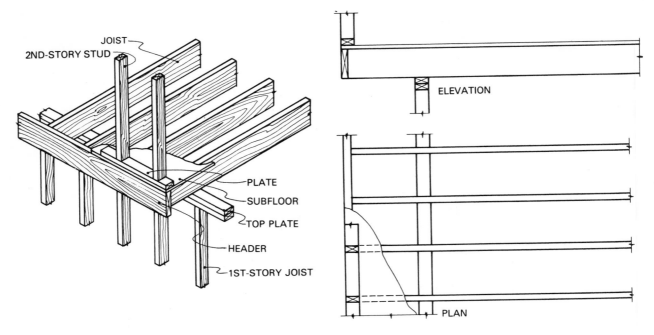

Fig. 33-39. Cantilevered second floor with joist perpendicular to exterior wall.

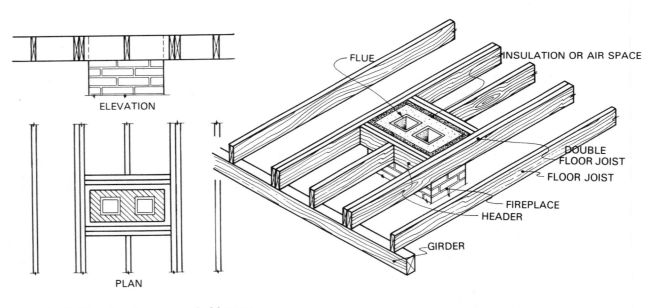

Fig. 33-40. Floor framing around chimney.

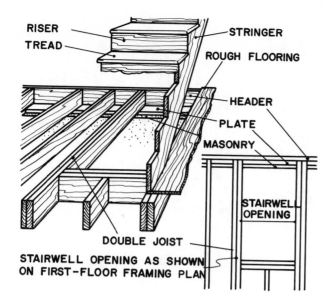

Fig. 33-41. Stairwell opening framing plan.

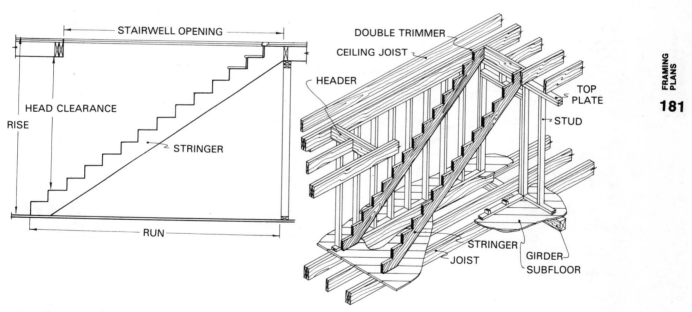

Fig. 33-42. Stair system elevation.

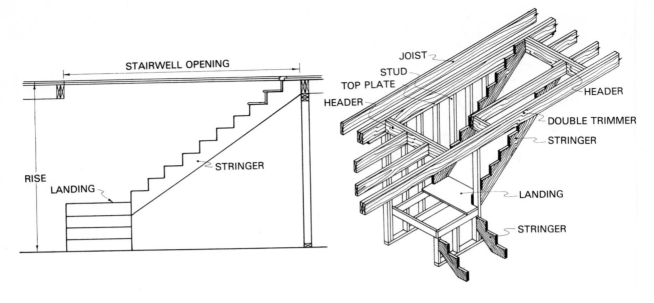

Fig. 33-43. Elevation of stairway system with landing.

182

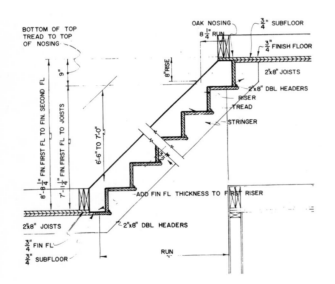

Fig. 33-44. Wood stairway details.

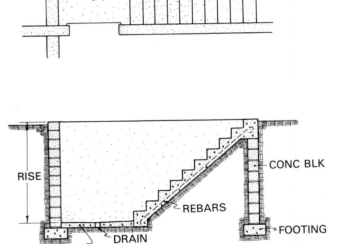

Fig. 33-45. Elevation and plan of exterior concrete construction stairway.

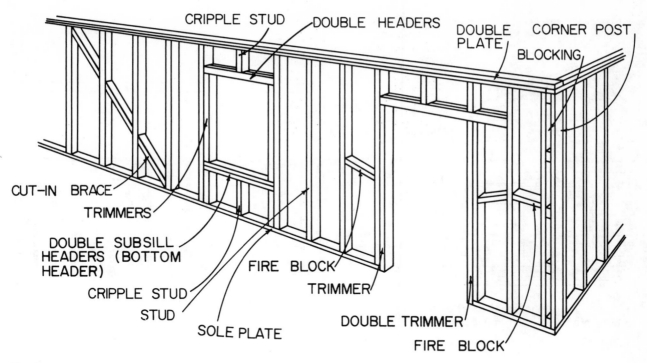

Fig. 34-1. Basic exterior wood framing members.

Unit 34
Wall Framing Plans

Wall framing drawings fall into several categories: exterior wall framing plans, interior wall framing plans, stud layouts, and construction details.

Exterior Wall Framing Drawings

Exterior wall framing panels such as the ones shown in Fig. 34-1 are best constructed by using a framing elevation drawing as a guide. The wall framing elevation drawing is the same as the north, south, east, or west elevations of a building with all of the wall covering materials removed. Figure 34-2 shows a framing elevation drawing compared with a pictorial drawing of the same wall. Notice that the framing elevation is an orthographic drawing and does not reveal a third (depth) dimension. To better understand the framing elevation, study Fig. 34-3. Notice how the framing elevation is projected from the floor plan and elevation drawing. Door and window openings as outlined on manufacturer's specifications, or on the door/window schedule, should agree with the framing openings for doors and windows shown on the framing elevation. If aligned correctly, the framing elevation will show the exact position of the framing members before doors and windows are installed.

COMPLETE SECTIONS Another method of illustrating framing methods used in exterior wall construction is shown in Fig. 27-7. In this drawing the elevation framing information is incorporated into a complete sectional drawing of the entire structure. The advantage of this drawing is that it shows the relationship of elevation panel framing to the construction of the foundation, floor, and roof systems. Since this is a sectional drawing, joists or other materials that intersect the cutting-plane line are shown by crossed diagonals.

BRACING One of the problems in correctly interpreting framing elevation drawings is to determine whether materials are placed on the out-

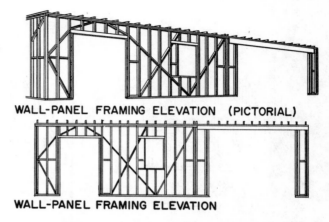

WALL-PANEL FRAMING ELEVATION (PICTORIAL)

WALL-PANEL FRAMING ELEVATION

Fig. 34-2. Wall framing elevation compared to wall framing pictorial drawing.

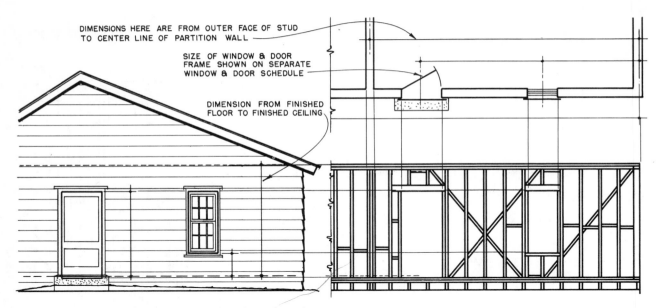

DIMENSIONS HERE ARE FROM OUTER FACE OF STUD
TO CENTER LINE OF PARTITION WALL

SIZE OF WINDOW & DOOR
FRAME SHOWN ON SEPARATE
WINDOW & DOOR SCHEDULE

DIMENSION FROM FINISHED
FLOOR TO FINISHED CEILING

Fig. 34-3. Relationship of exterior framing elevation to floor plan and elevation drawings.

side of a wall, on the inside of a wall, or between the studs. Figure 34-4a shows one method of illustrating let-in braces that are notched on the outside of the walls so that the outer surface of the brace is flush with the stud. Figure 34-4b shows a method of illustrating cut-in braces that are nailed between the studs, and Fig. 34-4c shows a method of illustrating diagonal braces that are placed on the inside faces of the studs.

Similar difficulties often occur in interpreting the true position of headers, cripple studs, plates, and trimmers. Figure 34-5 shows a method of illustrating the position of these members on the framing elevation drawing to eliminate confusion and to simplify proper interpretation.

PANEL ELEVATIONS Panel elevations show the attachment of sheathing to the frame. They are used to show the relationship between the

panel layout and the framing layout of an elevation by combining the two into one drawing. In this type of drawing the dotted diagonals indicate the position of panels (Fig. 34-6). When only panel layouts are shown, the outline of the panels are shown with diagonal solid lines. In this case, a separate framing plan would be prepared and correlated with the panel elevation.

Reading dimensions on panel elevation drawings involves interpreting widths, heights, and spacing of studs. Controlled dimensions for the heights of horizontal members and rough openings (framing openings) for windows are also included on framing drawings. If the spacing of studs does not automatically provide the rough opening necessary for the window or door, the rough opening widths of the window or door are dimensioned on the framing elevation as shown in Figs. 34-7 and 34-8.

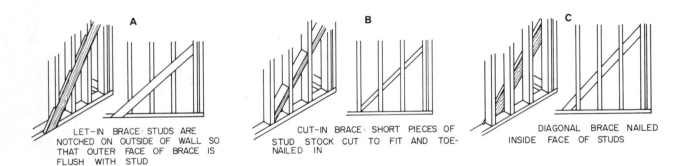

A
LET–IN BRACE: STUDS ARE NOTCHED ON OUTSIDE OF WALL SO THAT OUTER FACE OF BRACE IS FLUSH WITH STUD

B
CUT–IN BRACE: SHORT PIECES OF STUD STOCK CUT TO FIT AND TOE-NAILED IN

C
DIAGONAL BRACE NAILED INSIDE FACE OF STUDS

Fig. 34-4. Interpreting the position of braces.

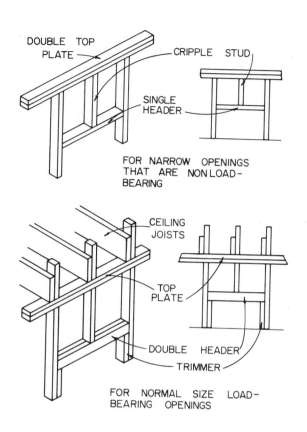

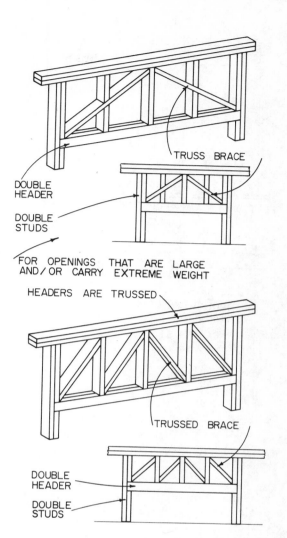

Fig. 34-5. Interpreting the position of headers and cripples.

Although structural steel exterior framing walls utilize different materials, the position and location of steel studs on exterior wall panels are shown (Fig. 34-9) in the same manner as wood skeleton frame walls. Masonry walls, however, are normally drawn on a plan view as shown in Fig. 34-10 if they are solid. If masonry is used only as a facing, then the exterior wall framing drawing is shown without the masonry. Study the relationship between the pictorial drawings in Fig. 34-10 and learn to identify each of the walls by observing only the plan or elevation view.

Interior Wall Framing Drawings
Interior wall framing drawings include plan, elevation, and pictorial drawings of partitions and wall coverings. Detail drawings of interior partitions are also prepared to show intersections between walls, ceilings, floors, windows, and doors.

PARTITION FRAMING PLANS Interior partition elevation drawings are most often used to show the construction of interior partitions. Interior partitions are projected from the partition on

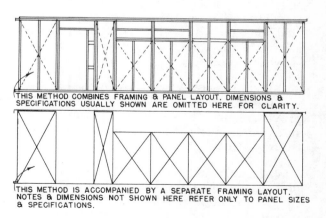

Fig. 34-6. Use of diagonal lines designates panel position.

Fig. 34-7. Door and window rough openings shown in framing elevation.

ROUGH OPENING WINDOW

CRIPPLE STUD

6'-10" 2090 mm

8'-1 1/2" 2500 mm

ROUGH OPENING DOOR

TOP PLATE

HEADER

TRIMMER STUD

STUD

PLATE

PLATFORM FRAMING (WESTERN) STUDS @ 16" (400 mm) OC

ROUGH OPENING WINDOW

CRIPPLE STUD

ROUGH OPENING DOOR

RIBBON

HEADER

STUD

DOUBLE TRIMMER

TRIMMER STUD

PLATE

BALLOON FRAMING (EASTERN) STUDS @ 16" (400 mm) OC

the floor plan in a manner similar to the projection of exterior partitions. To insure the correct interpretation of the partition elevation, each interior elevation drawing includes a label indicating the room and compass direction of the wall. For example, the elevation shown in Fig. 34-11 is labeled north wall-living room. If either the room name or the compass direction is omitted, the elevation may be misinterpreted and confused with a similar wall in another room. The elevation drawing is always viewed from the inside center of the room it represents. It is extremely important in interpreting interior elevations that the position of studs at corners be accurately interpreted from the drawing, as shown in Fig. 34-12. For this reason, a complete study of the floor plan, elevation, plumbing diagrams, and electrical plans should be made while reading interior wall framing drawings, since provisions must be made in the framing for soil stacks and other plumbing facilities and electrical equipment.

WALL COVERING DETAILS The basic types of wall covering materials used for finished interior walls include plaster, dry wall construction, paneling, tile, and masonry. The type of material is shown on either a wall detail pictorial drawing or on a plan or elevation section through the wall,

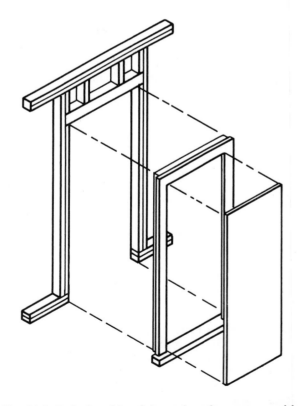

Fig. 34-8. Relationship of door, door frame assembly, and wall frame.

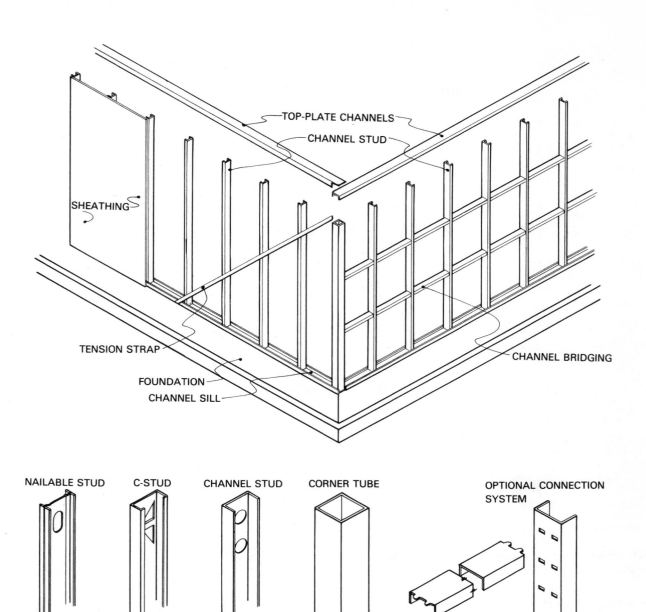

TOP-PLATE CHANNELS

CHANNEL STUD

SHEATHING

TENSION STRAP

CHANNEL BRIDGING

FOUNDATION

CHANNEL SILL

NAILABLE STUD

C-STUD

CHANNEL STUD

CORNER TUBE

OPTIONAL CONNECTION SYSTEM

TYPICAL FASTENING SYSTEM AND LIGHTENERS

Fig. 34-9. Exterior structural-steel paneling.

as shown in Fig. 34-13. In this drawing, the construction details relating to the positioning of dry-wall paneling and the intersection treatment of joints is graphically shown. Sometimes interior wall elevations are also prepared to show the design and surface treatment of selected interior partitions.

Stud Layouts

A stud layout is a plan similar to a floor plan but shows the position of each wall framing member

(stud). The stud layout is a section through each framed wall elevation, as shown in Fig. 34-14. The cutting-plane line, for purposes of projecting a stud layout, is placed approximately at the midpoint of the framed wall elevation. Figure 34-15 shows a stud layout which represents the framing plan of the wall. Stud layouts are of two types: the complete plan and the stud detail plan. The complete plan shows the position of all framing members on a floor plan, and the stud detail plan shows only the position and relationship of several studs, usually at framing intersections.

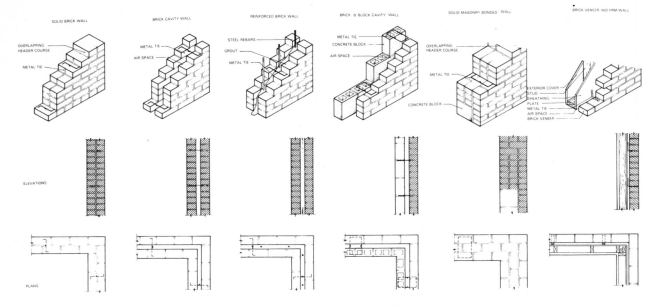

Fig. 34-10. Typical types of brick walls.

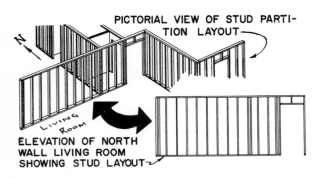

PICTORIAL VIEW OF STUD PARTI-
TION LAYOUT

ELEVATION OF NORTH
WALL LIVING ROOM
SHOWING STUD LAYOUT

Fig. 34-11. Panel elevation of interior wood frame wall.

STUD DETAILS The position of each stud in a corner post layout is frequently shown in a plan view as illustrated in Fig. 34-16. In some drawings, siding and inside wall covering materials are shown on this plan. Notice that the pictorial drawing on the left shows corner post construction without wall covering materials, and the stud layout on the right includes wall covering materials. Figure 34-17 shows alternate corner post stud positions without wall covering materials in place. When wall covering materials are shown on stud layouts, the nailing surfaces used to attach the covering to the studs is more observable, as shown in the right portion of Fig. 34-16.

Details of the exact position of each stud and blocking of an intersection are shown by a plan section (Fig. 34-18). If wall coverings are shown on the detail, the complete wall structure is often shown as in Fig. 34-19. Blocking is always labeled to prevent the possibility of identifying the blocking as a full-length stud because it does not extend completely from the sill to the top plate.

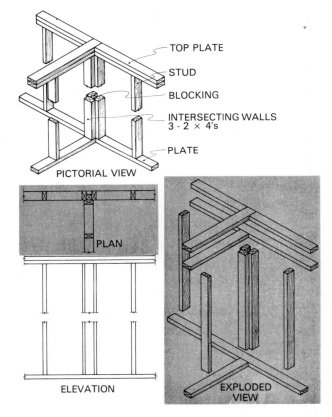

TOP PLATE
STUD
BLOCKING
INTERSECTING WALLS
3 - 2 × 4's
PLATE
PICTORIAL VIEW

PLAN

ELEVATION

EXPLODED
VIEW

Fig. 34-12. Methods of showing typical intersection of wood frame walls.

COMPLETE STUD PLANS The main purpose of a complete stud layout is to show how interior partitions fit together and how studs are spaced on the plan. Figure 34-20 shows parts of a stud layout. The outline of the plate and the exact posi-

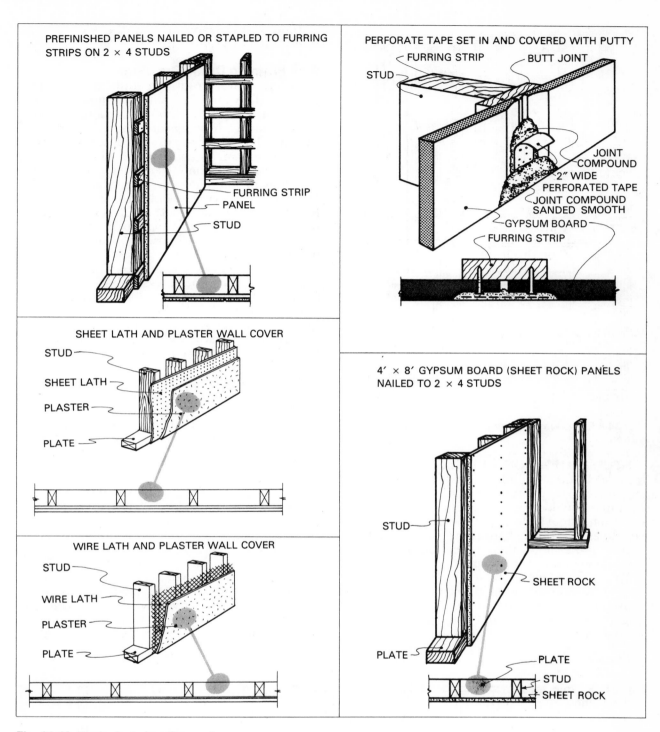

Fig. 34-13. Typical stud wall coverings.

tion of each stud that falls on an established center (16 in., 24 in. or 400 mm) are normally identified by diagonal lines. Studs, other than those that are placed on regular interval centers, are shown by different symbols. Studs that are short, blocking, or different in size are identified differently, as shown in Fig. 34-21. Using a coding system of this type eliminates the need for dimensioning the position of each stud if it is part of a regular partition pattern. The practice of coding studs and other members shown on the stud plan also eliminates the need for showing detailed dimensions of the spacing between each stud in a partition.

When laying-out the position of all studs, remember that the finished dimension of a 2 × 4 is actually 1½″ × 3½″. The exact dressed sizes for other rough stock is shown in Fig. 34-22.

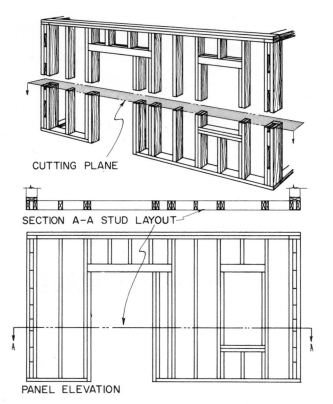

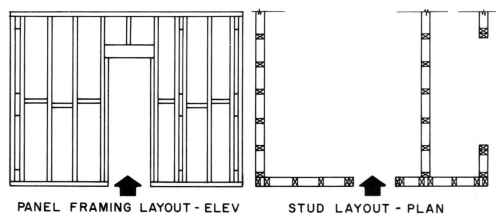

CUTTING PLANE

SECTION A-A STUD LAYOUT

PANEL ELEVATION

Fig. 34-14. Stud layout compared to wall framing elevation.

190

STUD DIMENSIONS Distances dimensioned on stud layouts include: inside framing dimensions of each room, framing widths of halls, rough openings for doors and arches, lengths of each partition, and widths of partitions where dimension lines pass through from one room to another. Figure 34-22 shows the application of these dimensional practices to a typical stud layout plan.

To conserve space where full partition widths are not important, as in closet walls, studs are sometimes turned so that the 2-in. dimension aligns with the partition. This rotation is reflected in the stud layout shown in Fig. 34-23.

Without a stud layout, this wall could easily be interpreted as a full-width partition.

Wall Framing Detail Drawings

Drawings of wall framing details are often prepared to show building materials in relationship to the basic framework of the structure. Sometimes cutaway drawings or removed sections, as shown in Fig. 34-24, are used to describe the wall construction methods and to define the use of wall framing materials. Removed sections of this type may be indexed to the floor plan, elevation, or pictorial drawing as indicated in Fig. 34-24a. In this example, Fig. 34-24b describes the framing method used on the wall and roof intersections by using break lines to expose the framing with a pictorial drawing. Figure 34-24c shows a wall section at the eave, and section C shows a sill section revealing the intersection between the foundation floor system and the exterior wall. These sections also show the inside wall treatment, insulation, sheathing, and exterior siding. Removed sections are effective in showing enlarged details of this type.

SIDING MATERIALS New siding materials are constantly being developed and new applications found for existing materials. Growing emphasis on the use of siding panel components designed to modular limits increases the necessity for carefully reading the relationship between the basic framing details and the exterior wall covering materials of a building. Figures 34-25 through 34-28 show three methods of showing siding material and its relation to the foundation, floor system, and exterior wall framing of a structure. Figure 34-25 shows an application for wood siding. Figure 34-26 shows the plan view, elevation section, and pictorial drawing used to describe a board and batter exterior siding design. Figure 34-27 shows the same type drawing used to describe brick-veneer walls, and Fig. 34-28 shows a stucco wall in a plan, elevation, and pictorial drawing. The most effective method

Fig. 34-15. Stud layout and elevation.

PANEL FRAMING LAYOUT - ELEV   STUD LAYOUT - PLAN

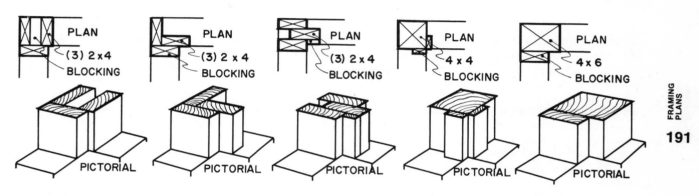

Fig. 34-16. Corner post stud details.

(3) 2 × 4'S CORNER POST

SUBFLOOR
PLATE
STUD

HEADER
SILL
JOIST
FOUNDATION

(3) 2 × 4'S

PLASTER
SHEATHING

CORNER POST DETAIL AS
SHOWN IN PICTORIAL

4 × 6   4 × 4

PLAN
(3) 2 x 4
BLOCKING
PICTORIAL

PLAN
(3) 2 x 4
BLOCKING
PICTORIAL

PLAN
(3) 2 x 4
BLOCKING
PICTORIAL

PLAN
4 x 4
BLOCKING
PICTORIAL

PLAN
4 x 6
BLOCKING
PICTORIAL

Fig. 34-17. Alternate corner post stud positions.

used to show the exact position of siding materials is the vertical or horizontal section. Compare the plan section and the elevation section in Figs. 34-25, 34-26, 34-27, and 34-28. You should be able to visualize the pictorial drawing by studying only the sectional drawings of either the plan or the elevation. Exterior wall elevation sections are the drawings most commonly used to describe wall construction details. Figure 34-29, for example, shows the details necessary for constructing a masonry wall, and Fig. 34-30 shows the details used for constructing a structural steel wall through the use of elevation sections.

SPECIAL WALL FRAMING When special framing requirements exist, such as framing for plumbing fixtures, a special (or partial) plan or elevation drawing may be prepared for the walls involved. For example, in Fig. 34-31 a special wall panel drawing is prepared to show the position of

framing for heating outlets. Figure 34-32 shows several methods used to detail plumbing wall construction.

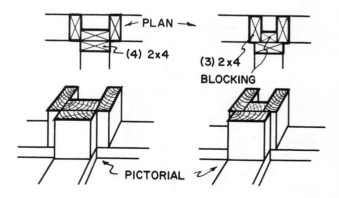

PLAN
(4) 2x4

(3) 2x4
BLOCKING

PICTORIAL

Fig. 34-18. Exterior partition intersection as shown on stud layout.

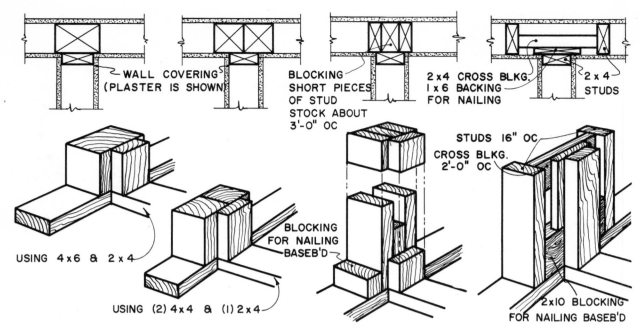

Fig. 34-19. Wall covering shown on stud layout.

Door Framing Details

The use of modular component door units is increasing throughout the building industry. Therefore, maintaining accurate rough opening dimensions for these units is most critical to their installation. Whether the door framing is conventional or of a component design, the exact position and dimension of the rough opening must be clearly understood from the framing drawing.

Head, sill, and jamb sections are used frequently and effectively in describing the framing construction. Since the door extends to the floor, the relationship of the floor framing system to the position of the door is critical. The method of intersecting the door and hinge with the wall framing is also important. The head section shows the intersection of the door and the header framing. The sill section shows how the door relates to the

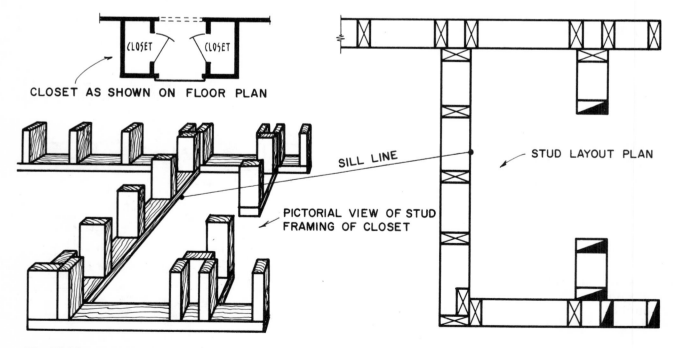

Fig. 34-20. Partition layout as shown on stud layout.

Fig. 34-21. Stud layout symbols.

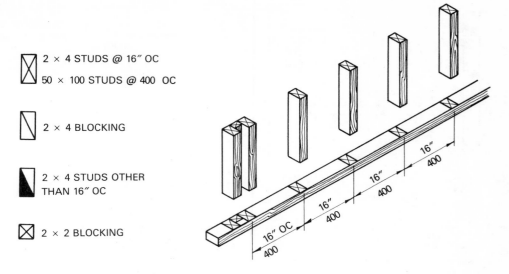

2 × 4 STUDS @ 16″ OC

50 × 100 STUDS @ 400 OC

2 × 4 BLOCKING

2 × 4 STUDS OTHER THAN 16″ OC

2 × 2 BLOCKING

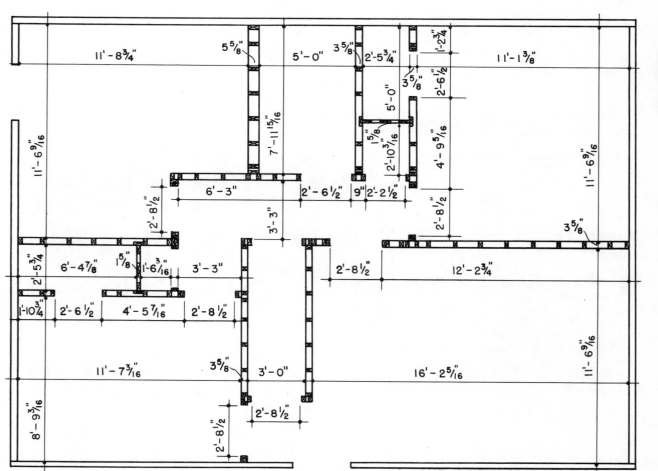

Fig. 34-22. Complete stud layout plan.

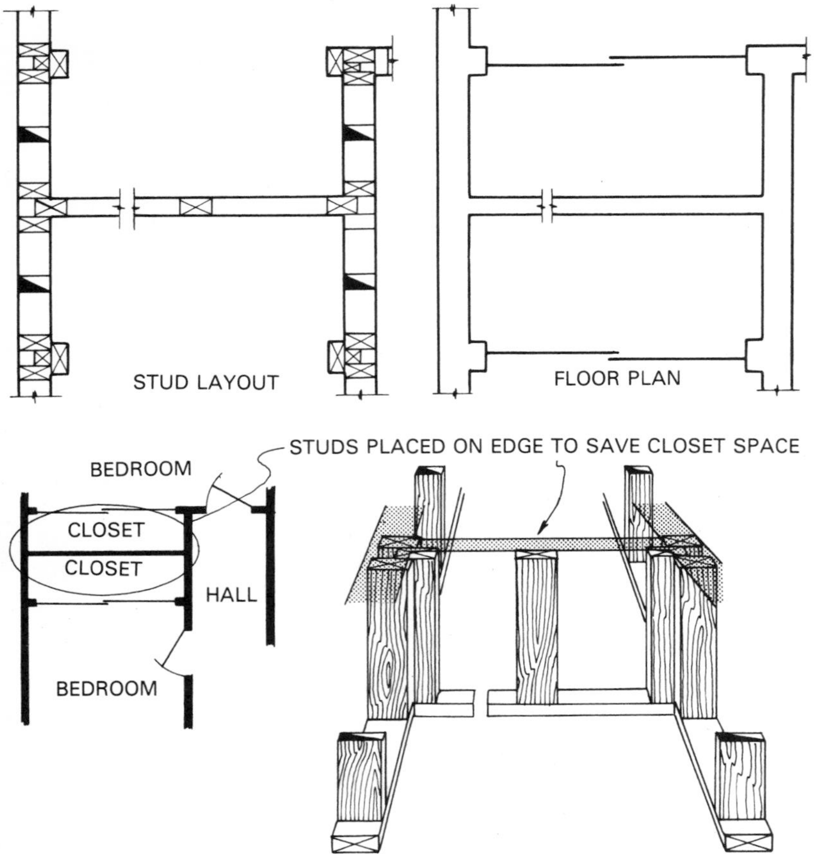
STUD LAYOUT
FLOOR PLAN
STUDS PLACED ON EDGE TO SAVE CLOSET SPACE
BEDROOM
CLOSET
CLOSET
HALL
BEDROOM

Fig. 34-23. Special stud positions shown on a stud layout floor plan.

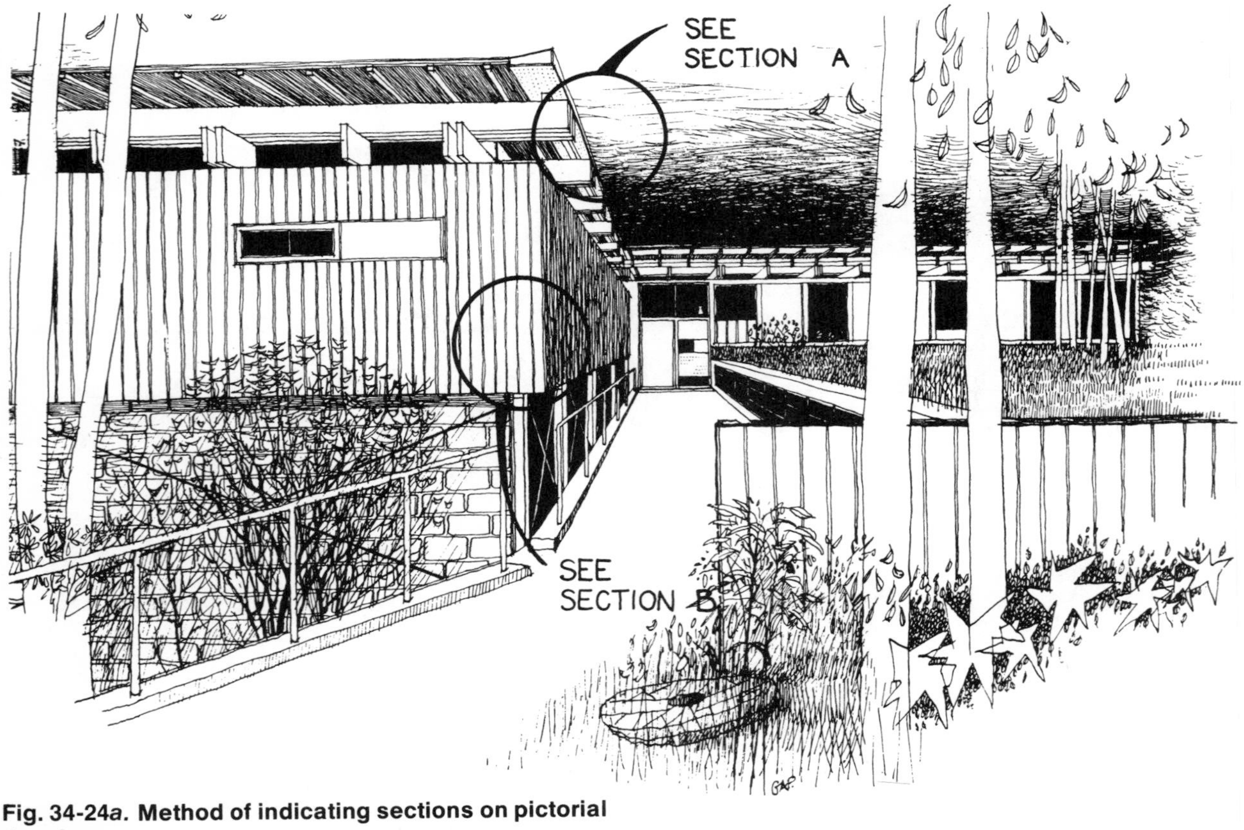
SEE
SECTION A
SEE
SECTION B

Fig. 34-24*a*. Method of indicating sections on pictorial drawings.

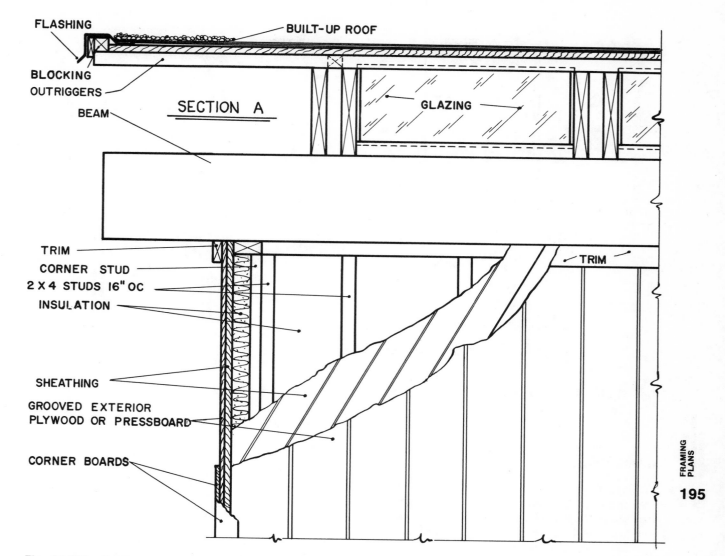

FLASHING
BUILT-UP ROOF
BLOCKING
OUTRIGGERS
BEAM
SECTION A
GLAZING
TRIM
CORNER STUD
2 X 4 STUDS 16" OC
INSULATION
SHEATHING
GROOVED EXTERIOR
PLYWOOD OR PRESSBOARD
CORNER BOARDS
TRIM

Fig. 34-24b. Cutaway drawing indexed to pictorial drawing.

floor framing, and the jamb section shows how the hinge and butt side is constructed in relationship to the walls.

Since there are many types of doors specified for a structure, be sure to locate the door framing detail that applies to each specific type of door. Figure 34-33 shows various types of interior doors and the accompanying jamb and head detail.

Although the actual door construction is usually not included in a set of architectural plans, nevertheless, shop drawings or sections of doors are included with the specifications to ensure that the doors specified meet the standards established by the designer. Figure 34-34 shows a drawing and section depicting door construction details used for this purpose.

Window Framing Drawings
One of the most frequently and effectively used

methods of showing window framing details is through the use of the head, jamb, and sill sections drawings, as shown in Fig. 34-35. Most windows are factory-made components ready for installation.

The most critical framing dimensions are those that describe the size of the rough framing opening. Window framing drawings include the dimensions of the rough framing openings in addition to the dimensions of the sash opening, as shown in Fig. 34-36. Figure 34-37 shows rough stud openings and sash openings for some of the most common window sizes.

When fixed windows or unusual window treatments are to be constructed, complete framing details are usually prepared as shown in Fig. 34-38. The more unusual the use of nonstandard sizes and components, the more complete is the detailed framing drawing that accompanies the design.

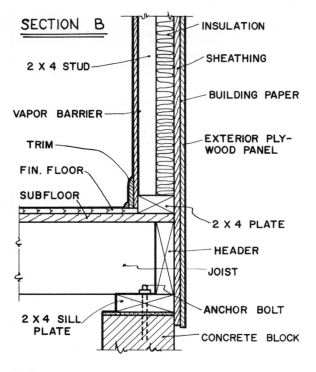

Fig. 34-24c. Sectional drawing indexed to pictorial drawing.

Labels on Section B:
- SECTION B
- 2 X 4 STUD
- VAPOR BARRIER
- TRIM
- FIN. FLOOR
- SUBFLOOR
- 2 X 4 SILL PLATE
- INSULATION
- SHEATHING
- BUILDING PAPER
- EXTERIOR PLY-WOOD PANEL
- 2 X 4 PLATE
- HEADER
- JOIST
- ANCHOR BOLT
- CONCRETE BLOCK

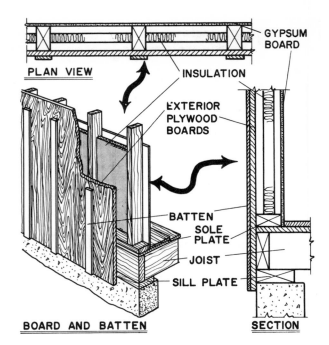

Fig. 34-25. Wood siding plan and elevation.

Labels:
- PLAN VIEW
- GYPSUM BOARD
- INSULATION
- EXTERIOR PLYWOOD BOARDS
- BATTEN
- SOLE PLATE
- JOIST
- SILL PLATE
- BOARD AND BATTEN
- SECTION

Intersections

Erecting the correct type of framing at wall, floor, and ceiling intersections is extremely critical; therefore, it is important to read construction drawings accurately and construct the framing precisely as described on the drawings.

BASE INTERSECTIONS The method of intersecting the finished wall materials with the floor is usually detailed. The details may be in a sectioned or a pictorial drawing, as shown in Fig. 34-39. The position of the wallboard (or plaster), baseboard, and molding is shown on this type of drawing.

CEILING INTERSECTIONS Details are also prepared to show the intersection between the ceiling and the wall. Details show the position of the top plate wallboard ceiling finish and the position of the molding used at the intersection, as shown in Fig. 34-40.

CORNER DETAILS Methods of framing inside corners (Fig. 34-41) and outside corners of partitions (Fig. 34-42) are also shown on detail construction drawings. Outside corners are often intersected by mitering or overlapping and exposing the paneling. Corner boards, metal strips, and molding are also used on these intersections. The same type of drawing (Fig. 34-43) is used to show wall covering joint construction.

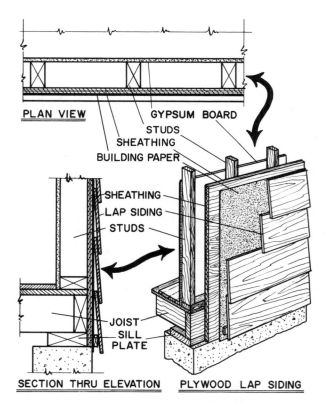

Labels:
- PLAN VIEW
- GYPSUM BOARD
- STUDS
- SHEATHING
- BUILDING PAPER
- SHEATHING
- LAP SIDING
- STUDS
- JOIST
- SILL PLATE
- SECTION THRU ELEVATION
- PLYWOOD LAP SIDING

Fig. 34-26. Board and batter plan and elevation.

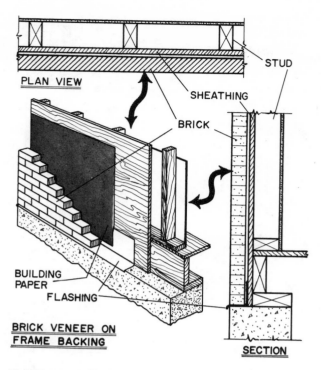

PLAN VIEW

STUD

SHEATHING

BRICK

BUILDING
PAPER

FLASHING

BRICK VENEER ON
FRAME BACKING

SECTION

Fig. 34-27. Brick-veneer plan and elevation.

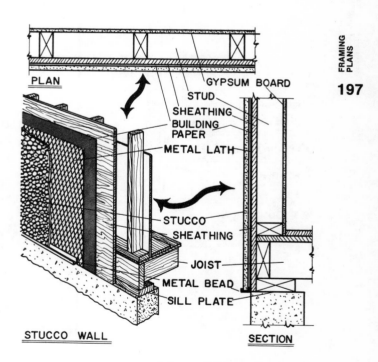

PLAN

GYPSUM BOARD

STUD

SHEATHING

BUILDING
PAPER

METAL LATH

STUCCO

SHEATHING

JOIST

METAL BEAD

SILL PLATE

STUCCO WALL

SECTION

Fig. 34-28. Stucco plan and elevation.

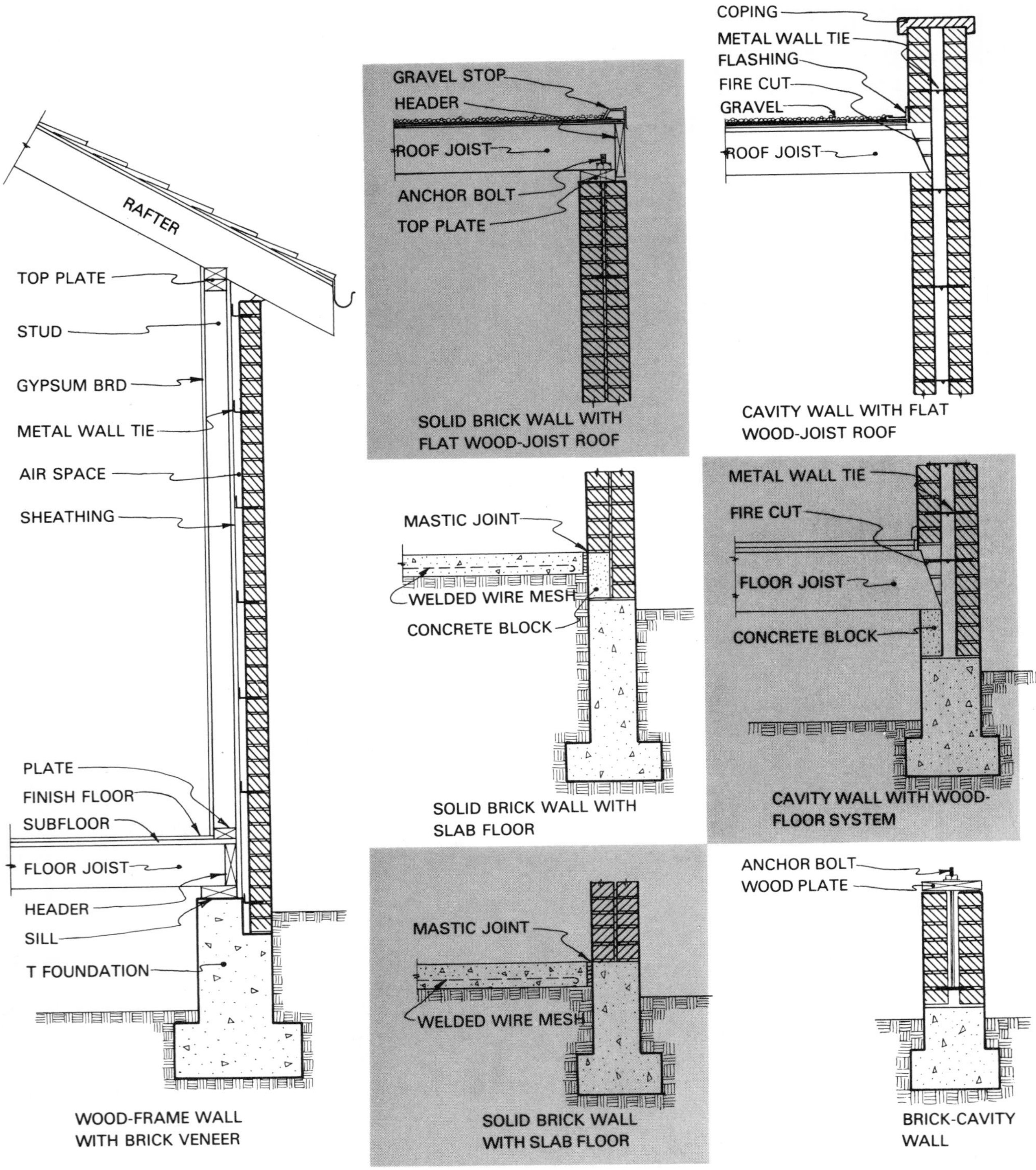

RAFTER

TOP PLATE

STUD

GYPSUM BRD

METAL WALL TIE

AIR SPACE

SHEATHING

PLATE
FINISH FLOOR
SUBFLOOR
FLOOR JOIST
HEADER
SILL
T FOUNDATION

WOOD-FRAME WALL
WITH BRICK VENEER

GRAVEL STOP
HEADER
ROOF JOIST
ANCHOR BOLT
TOP PLATE

SOLID BRICK WALL WITH
FLAT WOOD-JOIST ROOF

MASTIC JOINT
WELDED WIRE MESH
CONCRETE BLOCK

SOLID BRICK WALL WITH
SLAB FLOOR

MASTIC JOINT
WELDED WIRE MESH

SOLID BRICK WALL
WITH SLAB FLOOR

COPING
METAL WALL TIE
FLASHING
FIRE CUT
GRAVEL
ROOF JOIST

CAVITY WALL WITH FLAT
WOOD-JOIST ROOF

METAL WALL TIE
FIRE CUT
FLOOR JOIST
CONCRETE BLOCK

CAVITY WALL WITH WOOD-
FLOOR SYSTEM

ANCHOR BOLT
WOOD PLATE

BRICK-CAVITY
WALL

Fig. 34-29. Masonry wall details.

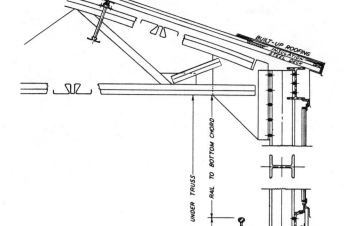

Fig. 34-30. Structural-steel wall details.

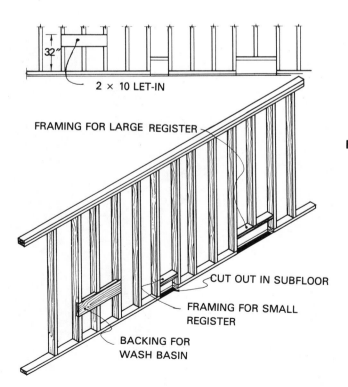

32"

2 × 10 LET-IN

FRAMING FOR LARGE REGISTER

CUT OUT IN SUBFLOOR

FRAMING FOR SMALL
REGISTER

BACKING FOR
WASH BASIN

**Fig. 34-31. Wall framing plan showing special framing
requirements.**

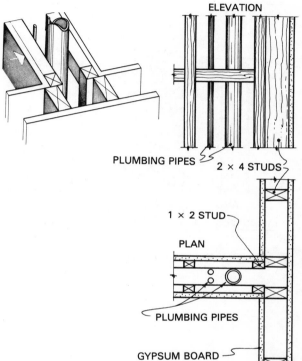

ELEVATION

PLUMBING PIPES   2 × 4 STUDS

1 × 2 STUD

PLAN

PLUMBING PIPES

GYPSUM BOARD

Fig. 34-32. Wall framing designed for pipe access.

FRAMING
PLANS

199

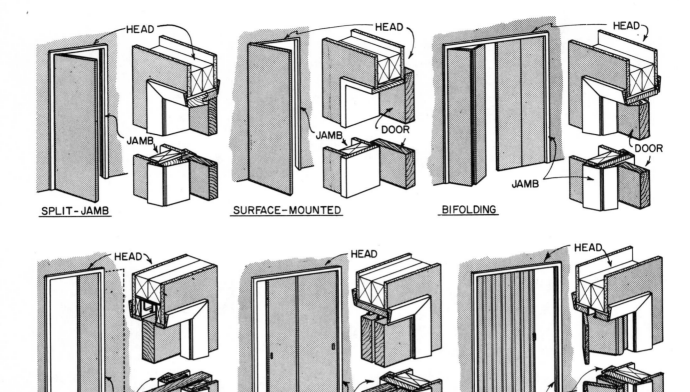

SPLIT-JAMB   SURFACE-MOUNTED   BIFOLDING

SLIDING-POCKET   SLIDING-BYPASS   FOLDING

Fig. 34-33. Interior door details.

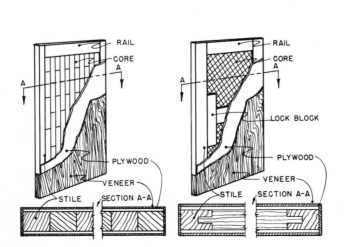

Fig. 34-34. Door construction details.

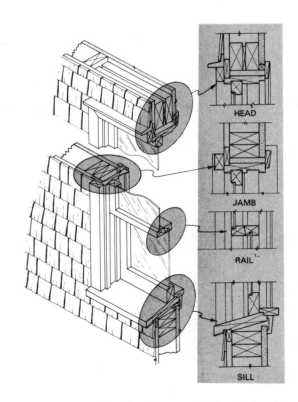

Fig. 34-35. Window details shown with head, jamb, and sill section.

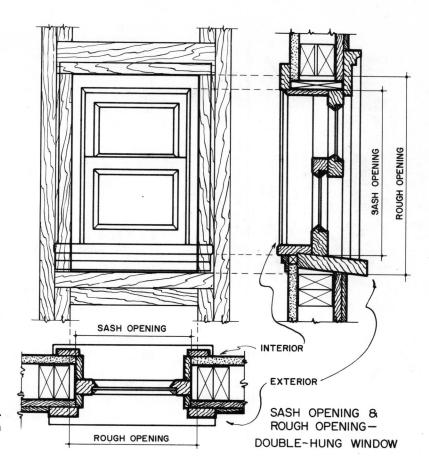

Fig. 34-36. Use of phantom lines to show rough opening for windows.

SASH OPENING &
ROUGH OPENING—
DOUBLE-HUNG WINDOW

INTERIOR

EXTERIOR

Fig. 34-37. Rough opening dimensioned for common sizes of windows.

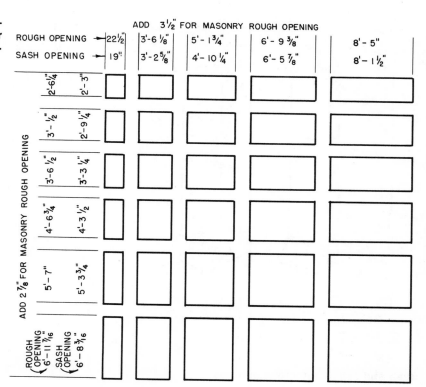

ADD 3½" FOR MASONRY ROUGH OPENING

| ROUGH OPENING → | 22½" | 3'-6⅛" | 5'-1¾" | 6'-9⅜" | 8'-5" |
| SASH OPENING → | 19" | 3'-2⅝" | 4'-10¼" | 6'-5⅞" | 8'-1½" |

ADD 2⅞" FOR MASONRY ROUGH OPENING

2'-6¼" / 2'-3"

3'-½" / 2'-9¼"

3'-6½" / 3'-3¼"

4'-6¾" / 4'-3½"

5'-7" / 5'-3¾"

ROUGH OPENING 6'-11⁷⁄₁₆" / SASH OPENING 6'-8³⁄₁₆"

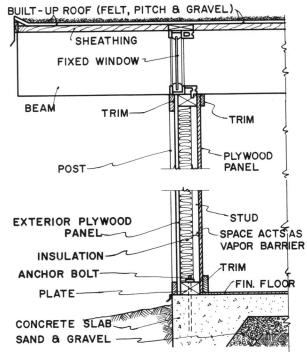

Fig. 34-38. Typical method of showing windows in exterior wall construction.

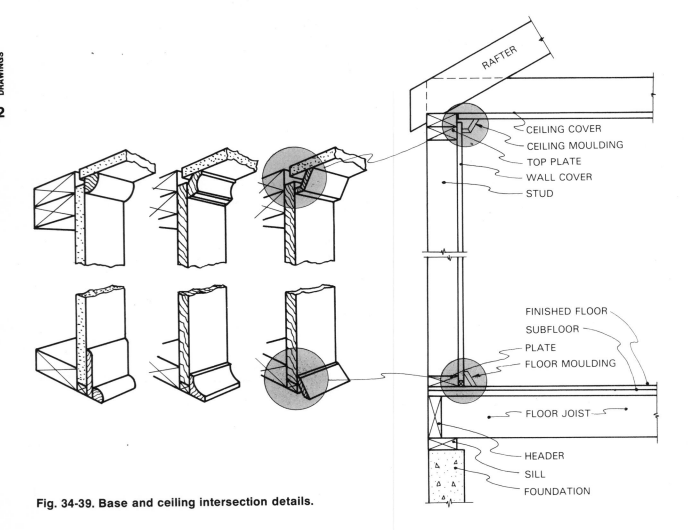

Fig. 34-39. Base and ceiling intersection details.

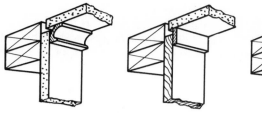
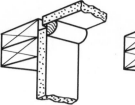
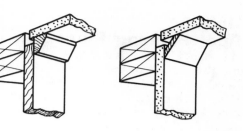

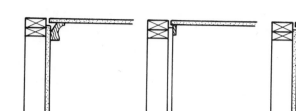

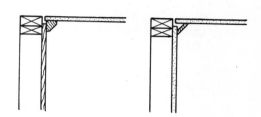

Fig. 34-40. Ceiling intersection details.

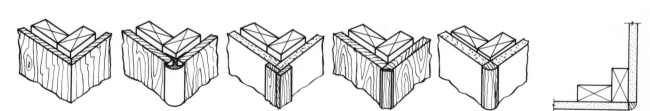

Fig. 34-41. Inside corner details.

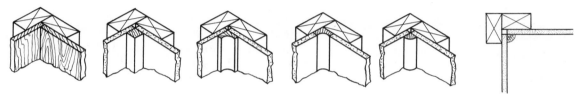

Fig. 34-42. Outside corner details.

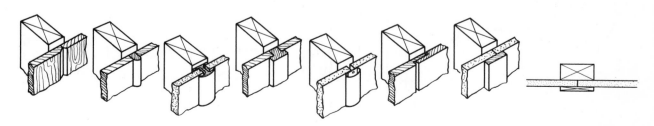

Fig. 34-43. Panel joist details.

Unit 35
Roof Framing Plans
A roof plan, as shown in Fig. 35-1, is a plan view of the roof showing the outline of the roof and the major object lines indicating ridges, valleys, hips, and openings. But a roof plan is not a roof framing plan. A roof framing plan shows the roof structure stripped of its covering to expose the position of each structural member, as shown in Fig. 35-2. The roof framing plan shows the exact position and spacing of each member. Figure 35-3 shows the relationship of the roof framing plan to a roof framing elevation drawing. Some reference to an elevation drawing is necessary to show the height relationships between various parts of the roof plan.

Types of Roof Framing Plans

SINGLE-LINE PLANS In the roof framing plan shown in Fig. 35-3, each structural member is represented by a single line. This method is used frequently in preparing roof framing plans and is quite usable when only the general relationship and center spacing of each member is needed.

COMPLETE PLAN When more details concerning the exact construction of intersections and joints are needed, a plan as shown in Fig. 35-4 is prepared. In this plan, the widths of the ridge boards, rafters, headers, and plates are shown to exact scale. When a complete roof framing plan of this type does not fully describe all framing details, additional pictorial drawings, such as the one shown at the top of Fig. 35-4, are sometimes added.

On roof framing plans, only the outline of the top of the rafters is shown. All areas underneath are shown by dotted lines.

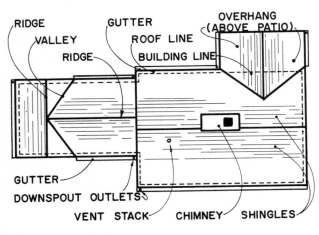

Fig. 35-1. Roof plan.

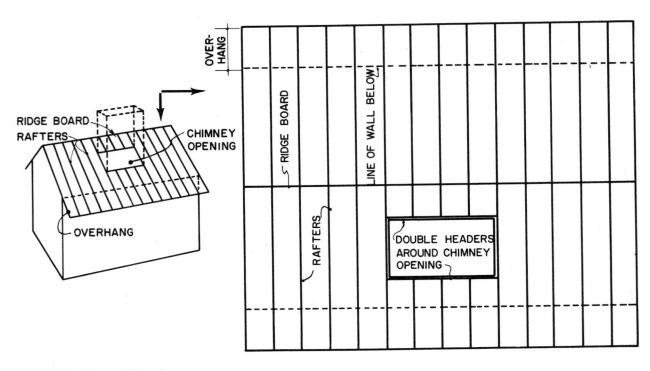

Fig. 35-2. Roof framing plan.

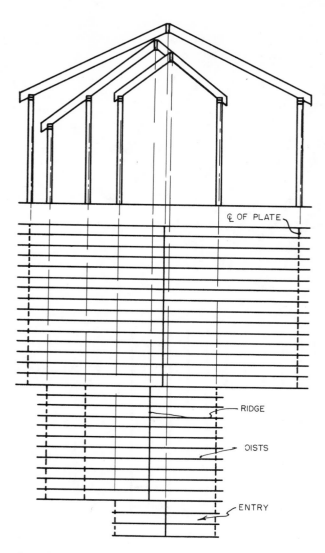

Fig. 35-3. Abbreviated roof framing plan projected from roof framing elevation.

Labels in figure: ℄ OF PLATE, RIDGE, JOISTS, ENTRY

Reading Roof Pitch

The pitch of a roof is the angle of the slope between the top ridge board and the top plate. Rise is the vertical distance from the top plate to the ridge. Run is the horizontal distance from the ridge to the top plate. The pitch is referred to as the rise over the run (rise/run). The run is, therefore, one-half the span, as shown in Fig. 35-5, and the run is always expressed in units of 12.

The pitch of a roof is referred to by one of three methods: the slope diagram, fractional pitch, or angular description. In a slope diagram description, a right triangle is made with the hypotenuse parallel to the slope of the sides of the roof. The other two sides represent the rise and the run, as shown in Fig. 35-5. The run in a slope diagram is always expressed in units of 12. Therefore, the rise is the resultant other leg of the triangle. The example in Fig. 35-5 shows roof pitches with slope diagrams of $^{18}/_{12}$, $^8/_{12}$, $^3/_{12}$, and $^1/_{12}$.

In using the fractional pitch method of pitch description, all rise values are converted into units of one. Slope diagrams are then converted to their lowest common denominator. The run of 12 units is doubled to the span of 24 units, as shown in Fig. 35-6. Thus, a slope diagram pitch of $^8/_{12}$ is converted to a $^1/_3$ pitch. A $^3/_{12}$ pitch is converted to a $^1/_8$ pitch, and a $^1/_{12}$ pitch is converted to a $^1/_{24}$ pitch. Figure 35-6 shows some typical roof pitches, comparing slope diagram designations with fractional pitches. The formula for determining pitches is therefore: pitch = rise/span.

The third method of indicating pitch is through using the angle of the slope. This is the angle between the horizontal surface and the hypotenuse of the slope diagram triangle.

Given the rise and run, the length of roof rafters is determined by using the following formula: $H^2 = b^2 + a^2$. H is the hypotenuse (rafter length); b is the base (run); a is the altitude (rise). Simply stated, this is: "The square of the hypotenuse is equal to the sum of the squares of the two sides." The carpenter's square is used, as shown in Fig. 35-7, to lay out roof framing angles according to the pitches shown.

Flat-Roof Framing Plans

A flat roof has no slope. Therefore, the roof rafters span directly from wall to wall or from wall to bearing partition. When rafters also serve as ceiling joists, the size of the rafter is designed to support the dead loads of both the roof and the ceiling. Figure 35-8 shows a typical flat roof construction. Large overhangs are possible on flat roofs since the roof joists can be extended past the top plates far enough to provide sun protection and not block the view. This extension is possible because there is no slope. When overhangs exist on all sides, lookout rafters as shown in Fig. 35-9 are used to extend the overhang on the side of the building perpendicular to the rafter direction. In post and beam construction, roof beams directly support roof covering materials and also function as exposed ceiling beams. Flat roof construction is usually shown through the use of a vertical section through the cornice (eave) area, as shown in Fig. 35-10.

Since there is no slope on a flat roof, drainage must be provided for by downspouts extending through the overhang. Flat roofs must be designed for maximum snow load, since snow will not slide from the roof but must completely melt and drain away.

Steel construction methods are especially applicable to flat roofs. The simplicity of erecting steel roofs results from the great strength of steel joists. The cross-bridging between the width of steel purlins (horizontal members), shown in Fig. 35-11, provides the purlin with a strength comparable to that of a truss.

Fig. 35-4. Complete roof framing plan.

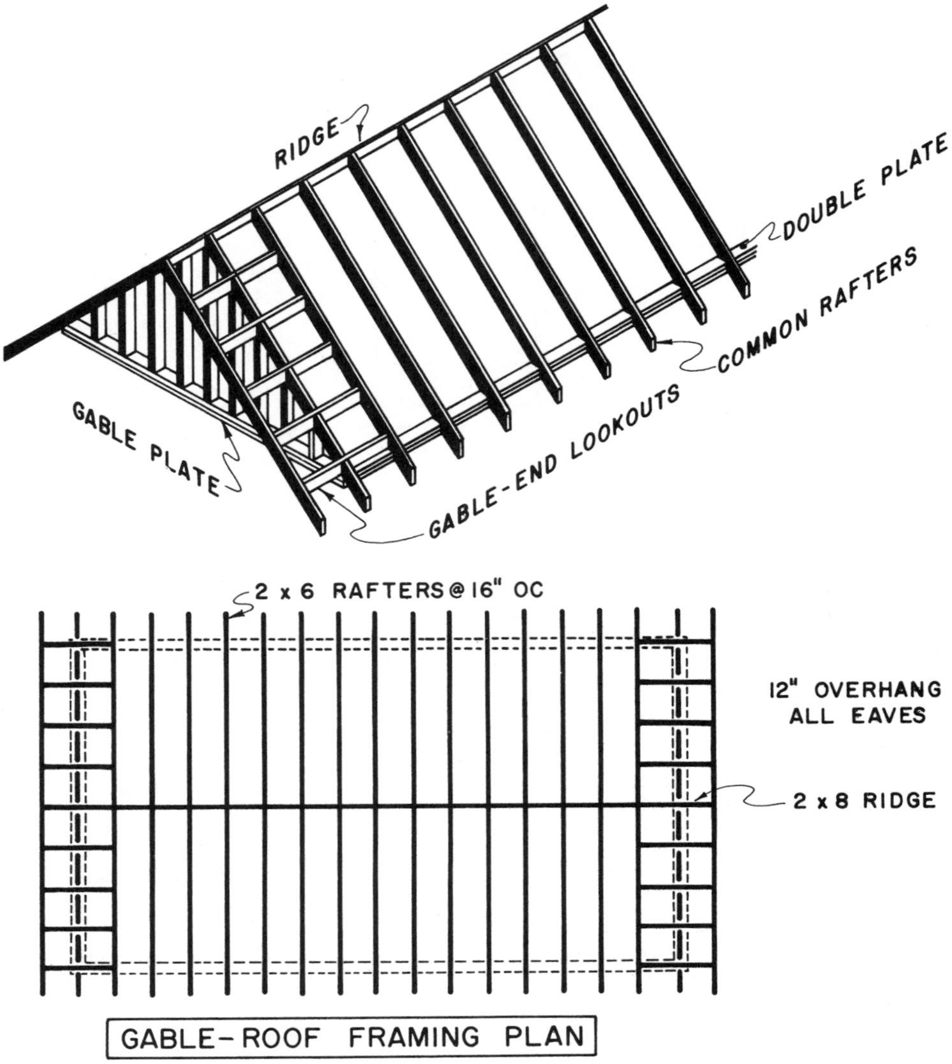

RIDGE

DOUBLE PLATE

COMMON RAFTERS

GABLE PLATE

GABLE-END LOOKOUTS

2 x 6 RAFTERS @ 16" OC

12" OVERHANG
ALL EAVES

2 x 8 RIDGE

GABLE-ROOF FRAMING PLAN

Fig. 35-5. Roof pitch is expressed as rise/run.

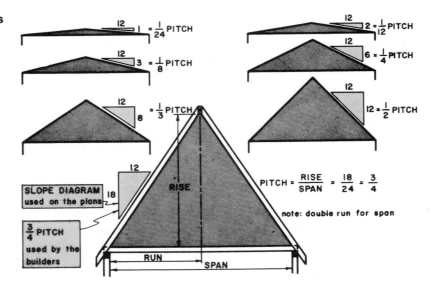

$\frac{12}{1} = \frac{1}{24}$ PITCH

$\frac{12}{2} = \frac{1}{12}$ PITCH

$\frac{12}{3} = \frac{1}{8}$ PITCH

$\frac{12}{6} = \frac{1}{4}$ PITCH

$\frac{12}{8} = \frac{1}{3}$ PITCH

$\frac{12}{12} = \frac{1}{2}$ PITCH

SLOPE DIAGRAM used on the plans

$\frac{3}{4}$ PITCH used by the builders

RISE

RUN   SPAN

$PITCH = \frac{RISE}{SPAN} = \frac{18}{24} = \frac{3}{4}$

note: double run for span

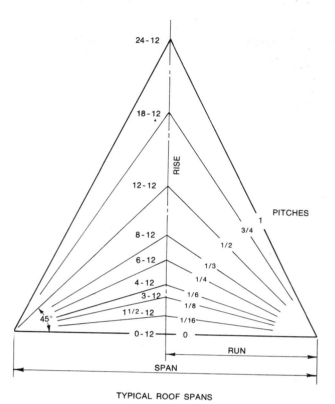

Fig. 35-6. Typical roof pitches.

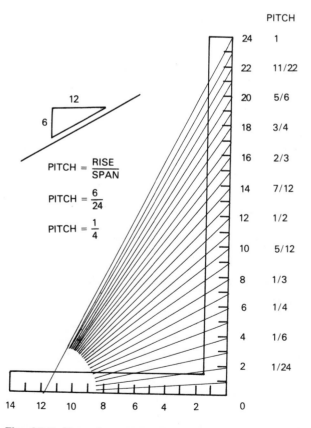

$$PITCH = \frac{RISE}{SPAN}$$

$$PITCH = \frac{6}{24}$$

$$PITCH = \frac{1}{4}$$

Fig. 35-7. Use of carpenter's square to lay out roof pitch.

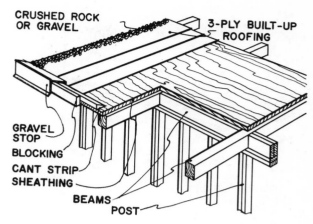

Fig. 35-8. Flat, built-up roof construction.

Gable Roof Framing Plans

Gable roofs are framed by one of three methods: conventional framing, truss framing, or post and beam framing. Figure 35-12 illustrates these three methods. In the conventional method, rafters and ceiling joists, usually spaced 16 in. on center, are covered with sheathing and shingles. An adaptation of this conventional method of constructing roofs is the use of roof trusses to replace conventional rafters and ceiling joists. Trusses provide a much more rigid roof but eliminate the use of space between the joists and rafters, which is often used for storage.

RIDGE BEAM The ridge board or ridge beam, as shown in Fig. 35-13, is the top member in the roof assembly. Rafters are fixed in their exact position by attachment to the ridge board. The ridge beam in a post and beam building may be part of the top plate assembly, as shown in Fig. 35-14.

GABLE END The gable end is the part of the exterior wall of a structure that extends from the top plate to the ridge (Fig. 35-15). In some cases, especially on low-pitch roofs, the entire gable-end wall from the floor to the ridge is panelized with varying lengths of studs. However, it is more common to prepare a conventional wall panel from the sill to the top plate and erect separate studs that project from the top plate of the panel to the roof rafters. Sheathing and siding are then added to the entire gable end of the building. When post and beam construction is used, windows in the gable end can be larger since studs are unnecessary in the gable end. This increasing use of windows in gable ends has necessitated the use of larger overhangs on gable-end walls. Gable-end lookouts are framed from the first or second rafters on the gable plate, as shown in Fig. 35-16. Figure 35-17 shows a plan and elevation drawing of a gable-end lookout construction. In this drawing, the plan view is shown exactly as

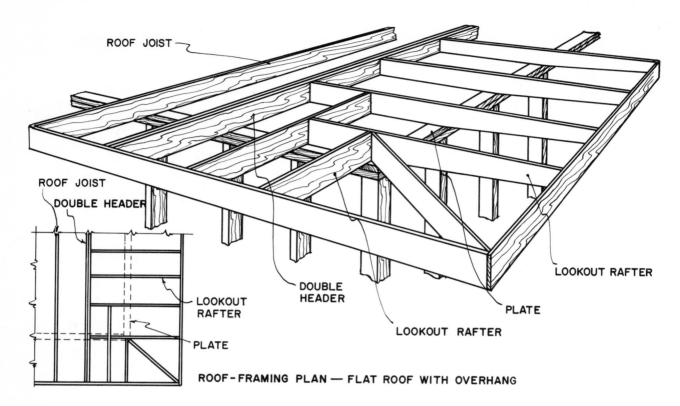

ROOF JOIST

ROOF JOIST
DOUBLE HEADER

LOOKOUT
RAFTER

PLATE

DOUBLE
HEADER

LOOKOUT RAFTER

LOOKOUT RAFTER

PLATE

ROOF-FRAMING PLAN — FLAT ROOF WITH OVERHANG

Fig. 35-9. Overhang for flat roof construction.

projected from the top, which reveals the bottom of each of the lookouts, which are placed on angles. Plan views are seldom prepared in this manner—they are usually prepared without showing the full depth of the lookout. In this case, only two lines representing the top of the rafters is drawn.

OVERHANGS Post and beam construction allows for larger overhangs since larger members are used and rafters are quite often exposed. Some variations of cuts are used to finish the rafter end. A comparison of the plumb, level, combination, and square cuts is shown in Fig. 35-18. Notice also that the rafters are notched with a seat (birds mouth) cut to provide a level surface for the intersection of the rafters on the top plate. The area of the rafter bearing on the plate should not be less than 3 in. (75 mm), as shown in Fig. 35-19.

COLLAR BEAMS Collar beams provide a tie between rafters. They may be placed on every rafter or on only every other rafter. Collar beams are used to reduce the rafter stress and sway that occurs between the top plate and the ridge. They also act as ceiling joists for finished attics, as shown in Fig. 35-20.

KNEE WALLS Knee walls are composed of vertical studs which extend from the floor level to the roof rafters, as shown in Fig. 35-21. Knee

walls add rigidity to the rafters and also provide wall framing for the interior partitions.

ROOF TRUSSES Figure 35-22 shows wood trusses that have become increasingly popular for homes and small buildings. Roof trusses allow more complete flexibility for interior partition spacing. They save approximately one-third on materials compared to the requirements of conventional roof framing. Trusses can be fabricated and erected in one-third the time required for rafter and ceiling joist construction. And truss con-

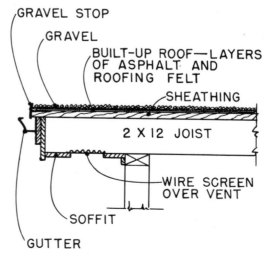

GRAVEL STOP

GRAVEL

BUILT-UP ROOF—LAYERS OF ASPHALT AND ROOFING FELT

SHEATHING

2 X 12 JOIST

WIRE SCREEN OVER VENT

SOFFIT

GUTTER

Fig. 35-10. Flat roof cornice (eave) section.

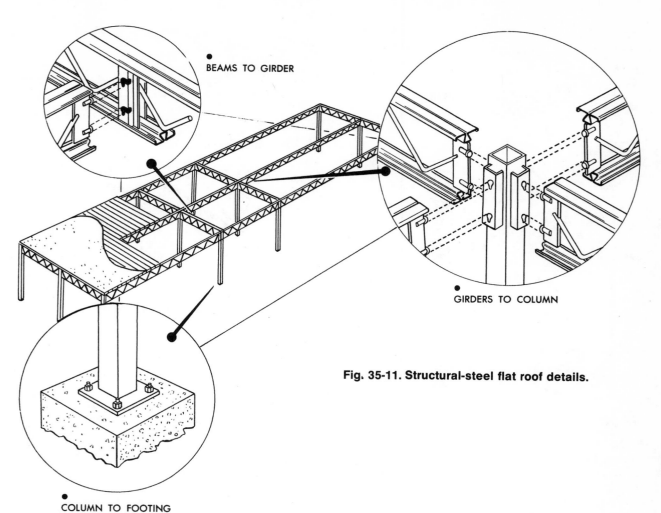

BEAMS TO GIRDER

GIRDERS TO COLUMN

COLUMN TO FOOTING

Fig. 35-11. Structural-steel flat roof details.

struction methods help bring the building under cover almost immediately. Trusses span their entire length without intermediate support; therefore, the use of trusses saves materials and erection time and eliminates the need for interior load bearing partitions.

Hip Roof Framing Plans

Hip roof framing is similar to gable roof framing except the roof slopes on four sides instead of intersecting a gable-end wall. The hip roof is shaped like a pyramid on square buildings. Where two adjacent slopes meet, a hip is formed on the external angle. A hip rafter extends from the ridge board over the corners of the top plate to the edge of the overhang.

The internal angle formed by the intersection of two slopes of the roof is known as the valley. A common rafter is a rafter that extends from the top plate of the exterior wall to the ridge board. A valley rafter is used on the internal angle in the same way a hip rafter is used on the external angle. Hip rafters and valley rafters are normally 2 in. (50 mm) deeper and 1 in. (25 mm) wider than common rafters for spans up to 12 ft. The larger

the span, the larger the rafter and the closer the spacing between rafters. Engineering tables should be consulted for exact size and spacing; however, for spans over 12 ft the rafter is usually doubled in width. Jack rafters are rafters that extend from the wall plate to the hip or valley rafters. They are parallel to the hip or valley rafters. They are parallel but always shorter than common rafters. Figure 35-23 illustrates the use of these various framing members in hip roof construction. Figure 35-24 shows a complete hip roof framing plan with rafter width and depth, the position of common rafters, hip rafters, and also the position of the wall plate, ridge board, and collar beams. Quite often single-line drawings are used to show hip roof construction, such as the one shown in Fig. 35-25. Compare and learn to visualize the pictorial drawing in Fig. 35-26 by studying the plan view of the hip framing plan.

In addition to the hip roof framing plans, details, usually of the elevation section type shown in Fig. 35-26, are prepared and indexed to the roof framing plan. Compare the construction detail shown in the pictorial drawing with the elevation and plan views shown in the hip roof framing plan in Fig. 35-26.

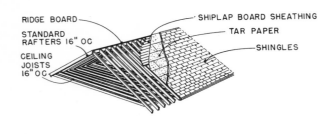

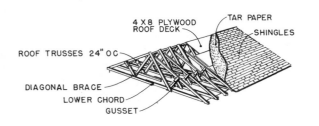

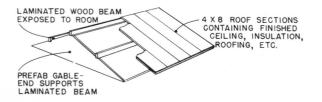

Fig. 35-12. Common types of gable roof construction.

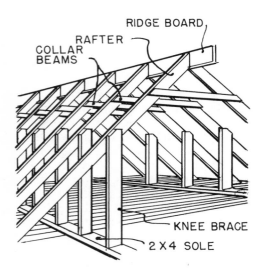

Fig. 35-13. Interior gable roof supports.

Roof Framing Details and Intersection Drawings

Where normal roof framing patterns are interrupted to accommodate the intersection of items such as chimneys, vent pipes, dormers, or sky-

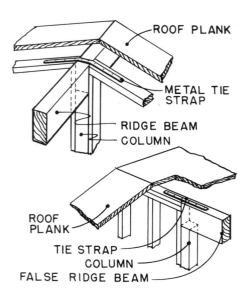

Fig. 35-14. Ridge beam assembly details.

lights, special framing details may be available, as shown in Figs. 35-27 through 35-30.

CHIMNEY DETAILS Chimney framing construction details are normally shown on detail drawings, such as the one shown in Fig. 35-27. Figure 35-27a shows a plan describing the intersection of a chimney with ceiling joist. Figure 35-27b shows the framing necessary for the intersection of the chimney and the roof rafters, and Fig. 35-27c shows a plan view and pictorial view of a saddle construction which is used to divert water away from a chimney.

DORMERS Parts of the roof framing plan which require special details include framing for dormer construction. The details of intersecting dormer roof framing with dormer walls are often shown on the roof framing plan, as in Fig. 35-28; or by the use of dormer elevations, which are shown in Fig. 35-29a and b. Figure 35-29b shows a side elevation of a shed dormer, and Fig. 35-29c shows a side elevation view of the framing for a gable dormer. The information found in Fig. 35-29a is sometimes incorporated into a roof framing plan. For this reason, the only additional detail added to the roof framing plan when dormers exist is a dormer elevation drawing, as shown in Fig. 35-29b and c.

CORNICE DETAILS Since neither a roof framing nor a roof plan can show cornice construction details, an elevation section of the cornice area is often prepared to describe this construction. Cornice details such as the one shown in Fig. 35-30 are used to show roof covering materials, as in a; gutter details, as in b; and soffit design and details, as in c.

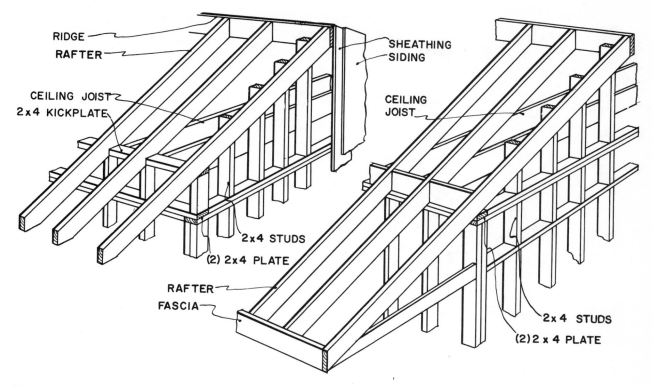

RIDGE
RAFTER

SHEATHING
SIDING

CEILING JOIST
2 x 4 KICKPLATE

CEILING JOIST

2 x 4 STUDS
(2) 2 x 4 PLATE

RAFTER
FASCIA

2 x 4 STUDS
(2) 2 x 4 PLATE

Fig. 35-15. Gable cornice details.

Steel Roof Construction

Roof framing plans for steel construction are similar to those for wood construction. However, in plans for structural steel, single-line drawings are almost universally used. A complete classification of each steel member is shown on the

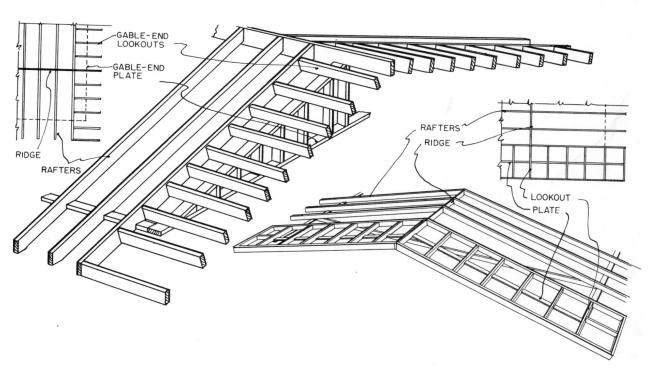

GABLE-END LOOKOUTS
GABLE-END PLATE

RAFTERS
RIDGE

RIDGE
RAFTERS

LOOKOUT PLATE

Fig. 35-16. Gable roof framing plan showing only top outline of each member.

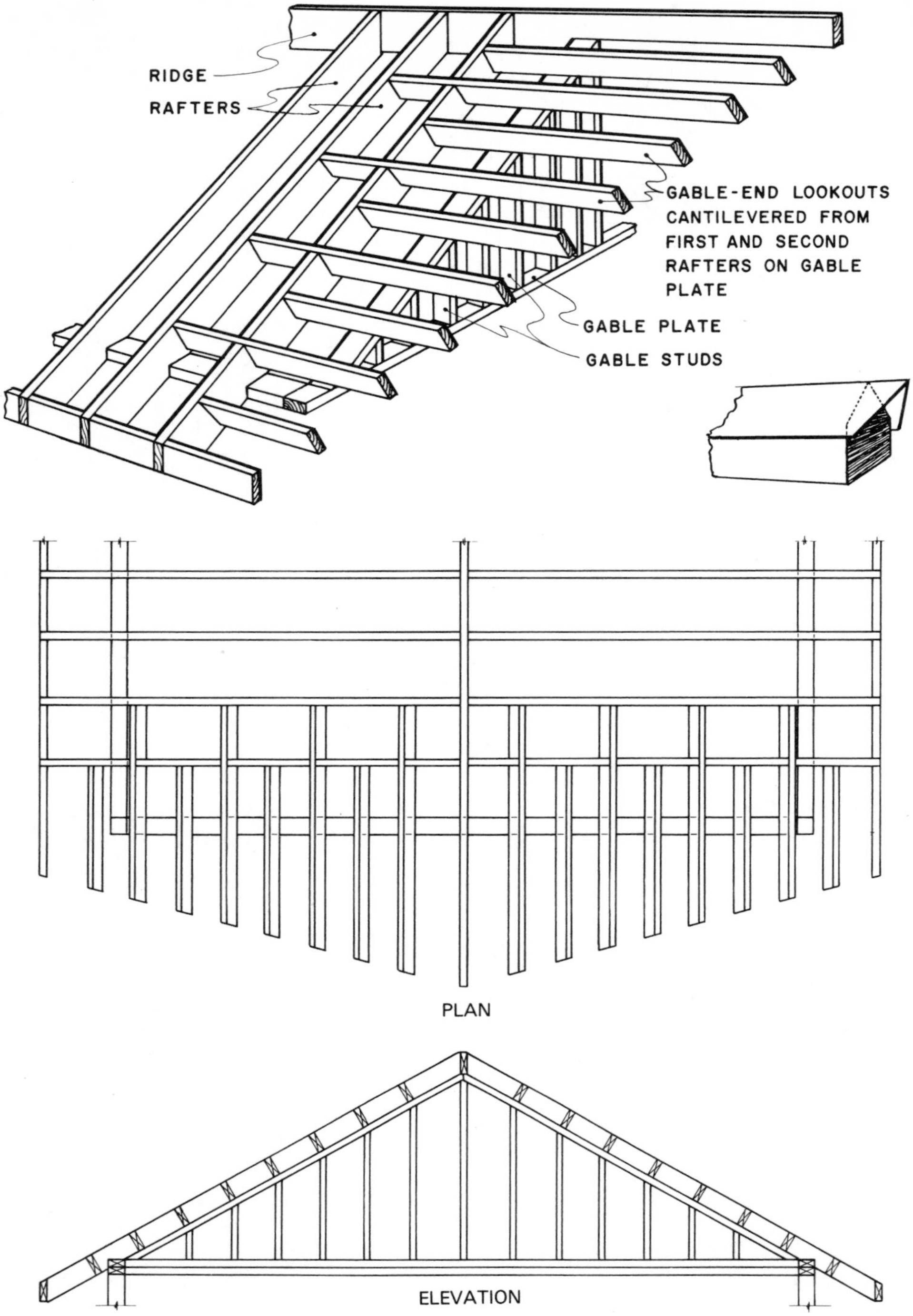

RIDGE

RAFTERS

GABLE-END LOOKOUTS CANTILEVERED FROM FIRST AND SECOND RAFTERS ON GABLE PLATE

GABLE PLATE

GABLE STUDS

PLAN

ELEVATION

Fig. 35-17. Winged gable details showing true projection of the lookout rafter depth.

drawing. This includes the size, type, and weight of the beams and columns. Each different type of member is shown by a different line weight, which relates to the size of the member. Struc-

tural steel roof framing plans frequently show bay areas (areas between columns), indicating the number of spaces and the spacing of each purlin between columns. Bays are frequently in-

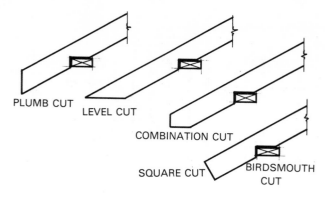

Fig. 35-18. Tail rafter cut details.

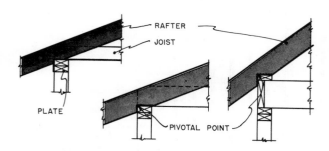

Fig. 35-19. Cornice (eave) intersection details showing seat cut.

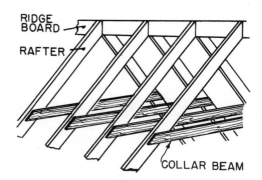

Fig. 35-20. Collar beams reduce rafter deflection.

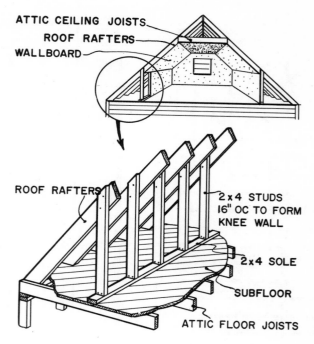

Fig. 35-21. Knee walls add rigidity to rafters.

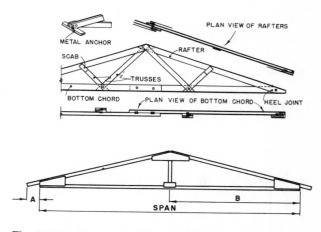

Fig. 35-22. Types of gable roof trusses.

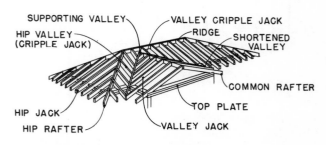

Fig. 35-23. Hip roof construction.

dexed in circles numerically on one side and alphabetically noted on the other side, as shown in Fig. 35-31a. Steel roof construction details are usually shown through the use of sectional drawings through the wall, eave, and roof line (Fig. 35-31b).

Roof Covering Materials

Roof coverings protect buildings from rain, wind, snow, heat, and cold. Materials used to cover

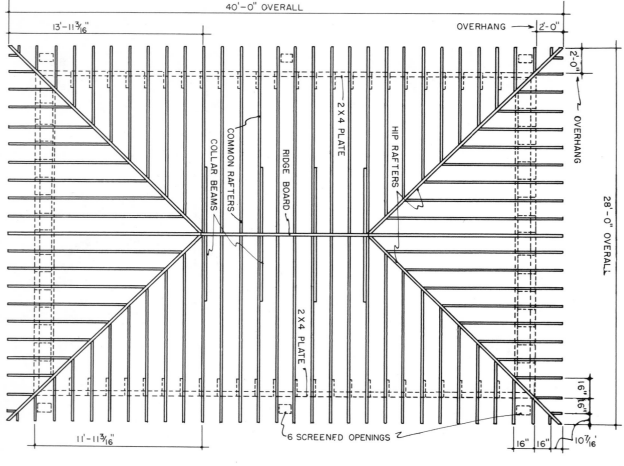

Fig. 35-24. Complete hip roof framing plan.

HIP-ROOF FRAMING PLAN

Fig. 35-25. Hip roof framing plan.

pitched roofs include wood shingles, asphalt shingles, and asphalt and asbestos shingles. On heavier roofs, tile or slate may also be used. Sheet materials, such as galvanized iron, aluminum, copper, and tin, are also used for flat or low-pitched roofs. A built-up roof consisting of layers of roofing felt covered with gravel topping is also used for low-pitched or flat roofs. Low-pitched

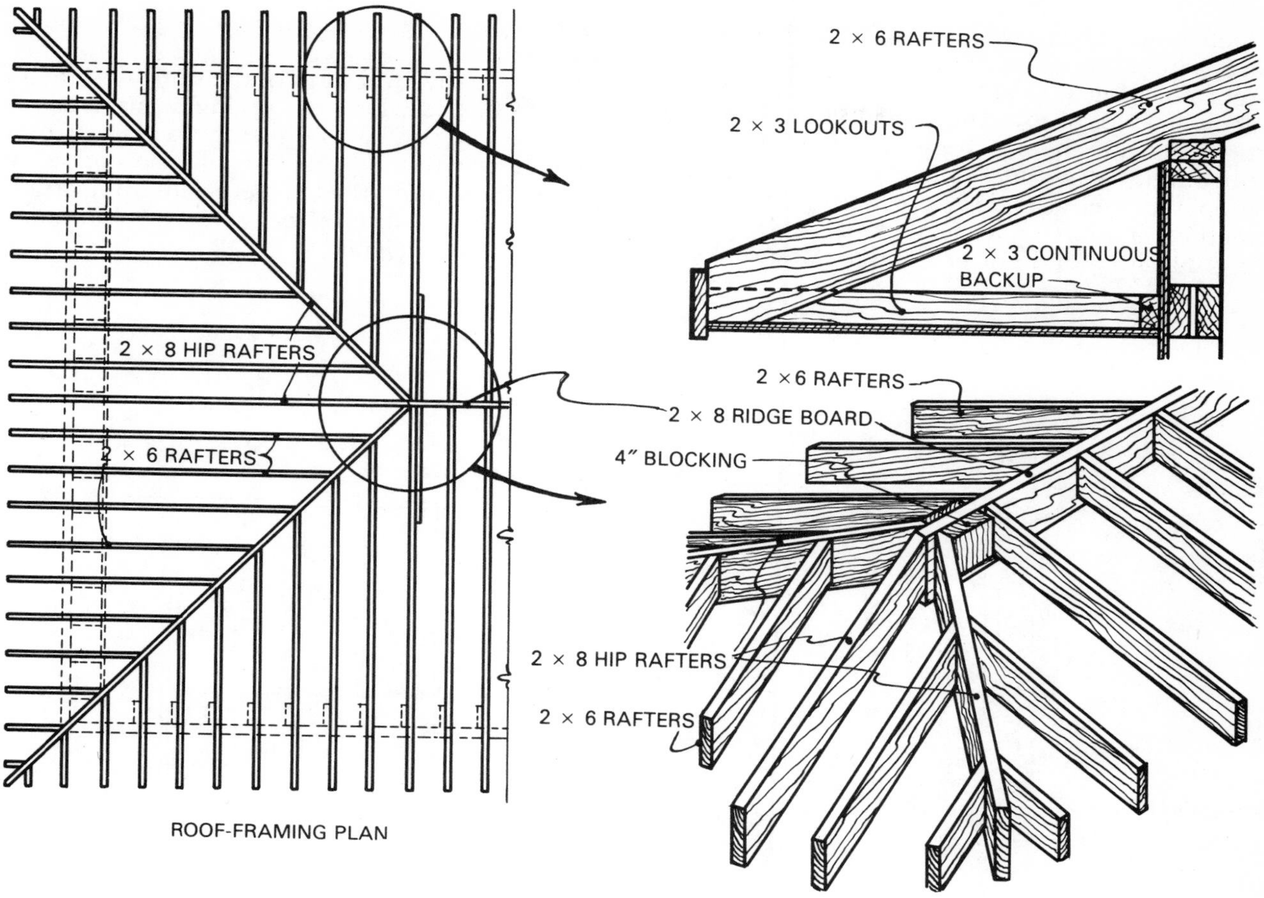

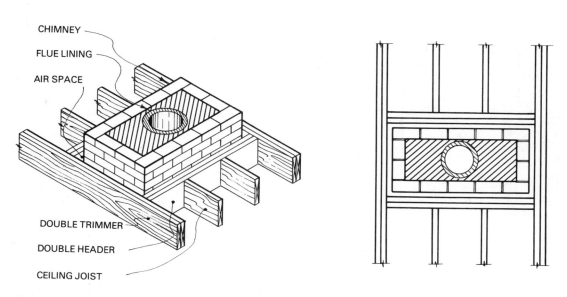

Fig. 35-26. Hip roof framing details.

2 × 6 RAFTERS

2 × 3 LOOKOUTS

2 × 3 CONTINUOUS BACKUP

2 × 6 RAFTERS

2 × 8 RIDGE BOARD

4" BLOCKING

2 × 8 HIP RAFTERS

2 × 6 RAFTERS

2 × 8 HIP RAFTERS

2 × 6 RAFTERS

ROOF-FRAMING PLAN

CHIMNEY

FLUE LINING

AIR SPACE

DOUBLE TRIMMER

DOUBLE HEADER

CEILING JOIST

Fig. 35-27a. Framing around chimney at ceiling joists.

roofs require heavier shingles to resist wind-lifting action. Separate plans are not prepared for roof covering materials, but elevation sections show the type of covering, and building specifications indicate the size, weight, and method of application. Sometimes pictorial drawings are prepared, such as Fig. 35-2a, b, and c, showing the method of applying a specific type of roof covering. Flashing and gutter details also are sometimes shown (Fig. 35-33) to indicate how water is diverted.

Where many structures of the same design are to be built and where the saving of materials is extremely critical, a roof sheathing plan such as the one shown in Fig. 35-34 is often prepared to guide the roofer in the most efficient application of sheathing to minimize waste and maximize construction efficiency.

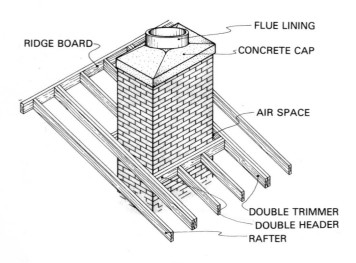
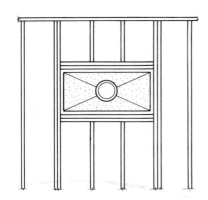

Fig. 35-27b. Framing around chimney at roof level.

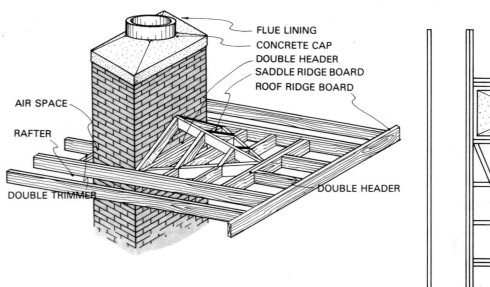
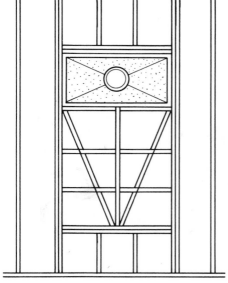

Fig. 35-27c. Saddle framing diverts water from chimney.

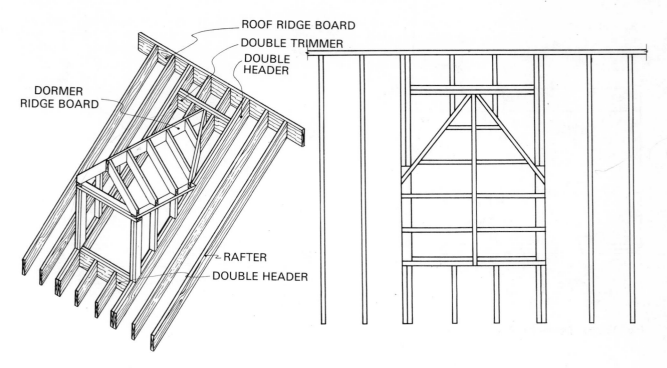

Fig. 35-28. Dormer framing plans.

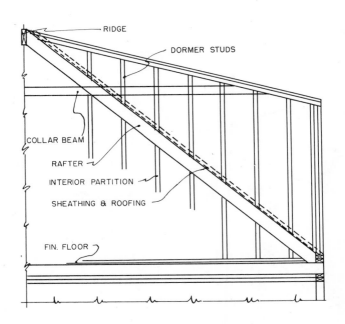

Fig. 35-29a. Shed dormer framing elevation.

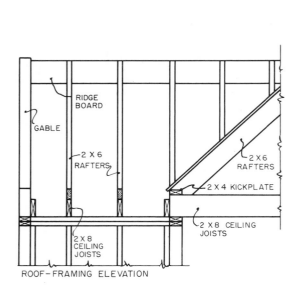

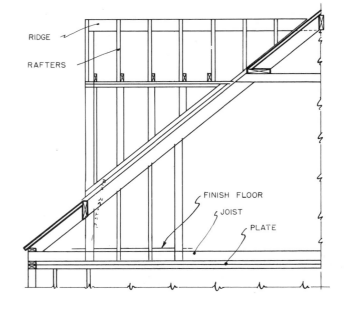

Fig. 35-29b. Dormer framing elevation.

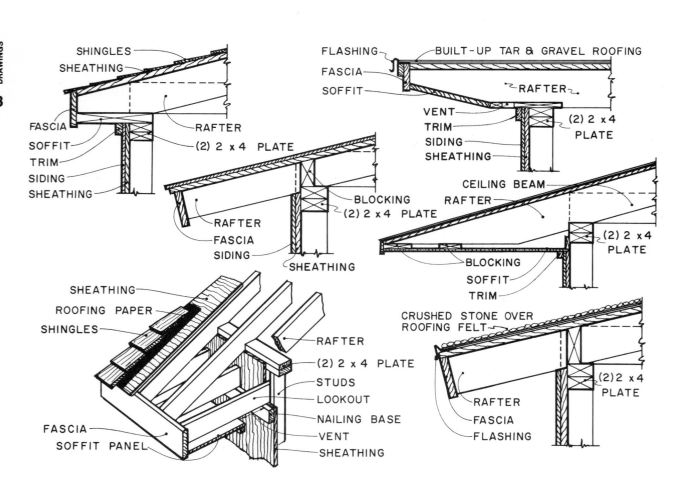

Fig. 35-30a. Cornice (eave) framing detail sections.

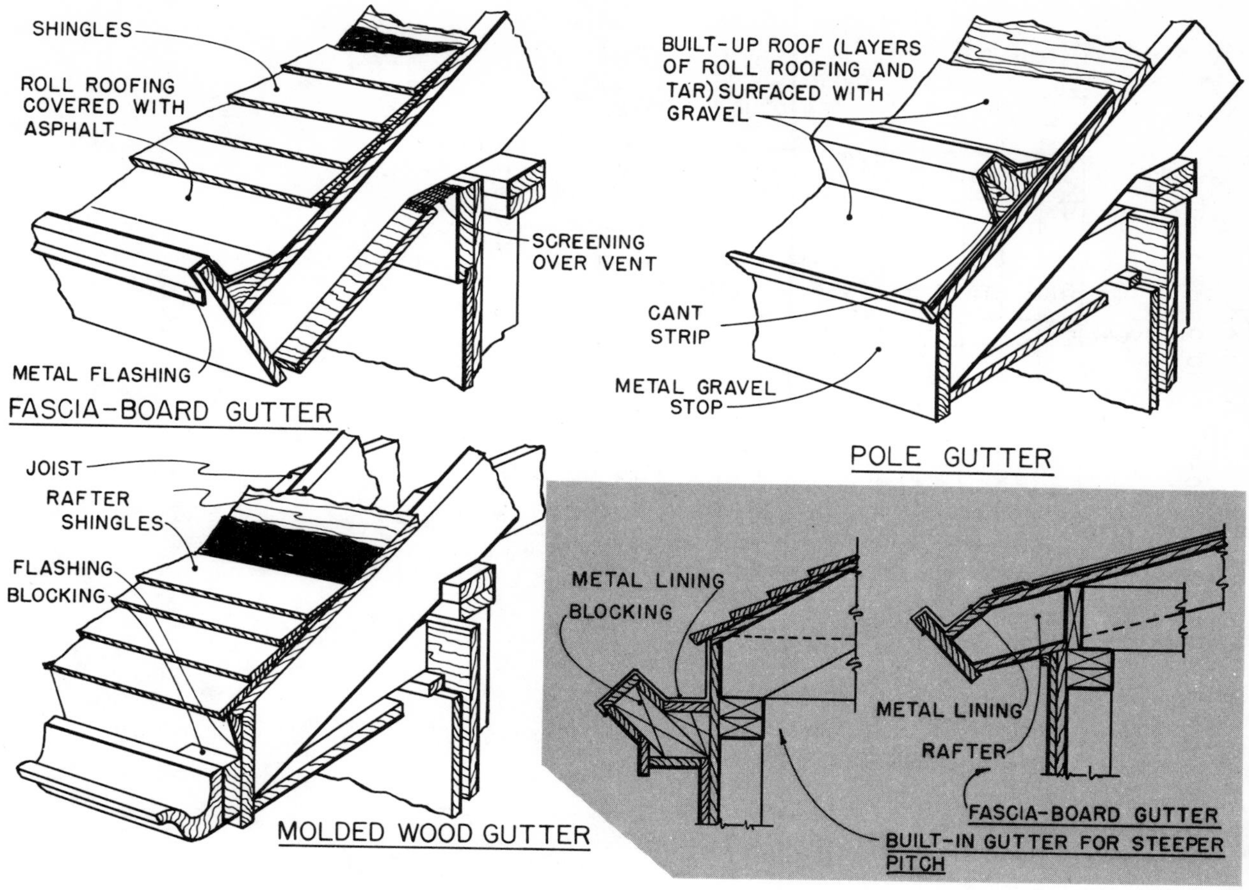

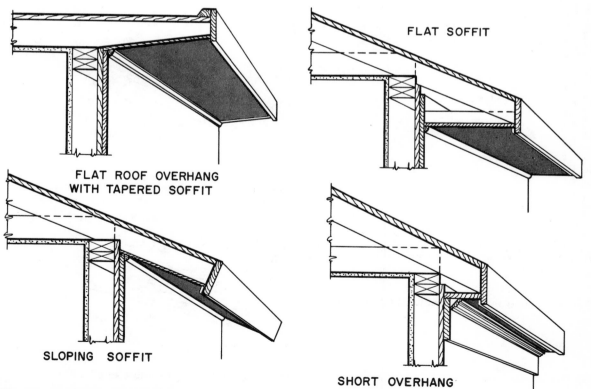

SHINGLES

ROLL ROOFING COVERED WITH ASPHALT

SCREENING OVER VENT

METAL FLASHING

FASCIA-BOARD GUTTER

BUILT-UP ROOF (LAYERS OF ROLL ROOFING AND TAR) SURFACED WITH GRAVEL

CANT STRIP

METAL GRAVEL STOP

POLE GUTTER

JOIST

RAFTER

SHINGLES

FLASHING

BLOCKING

MOLDED WOOD GUTTER

METAL LINING

BLOCKING

METAL LINING

RAFTER

FASCIA-BOARD GUTTER

BUILT-IN GUTTER FOR STEEPER PITCH

Fig. 35-30b. Built-in gutter details.

FLAT SOFFIT

FLAT ROOF OVERHANG WITH TAPERED SOFFIT

SLOPING SOFFIT

SHORT OVERHANG

Fig. 35-30c. Soffitt design details.

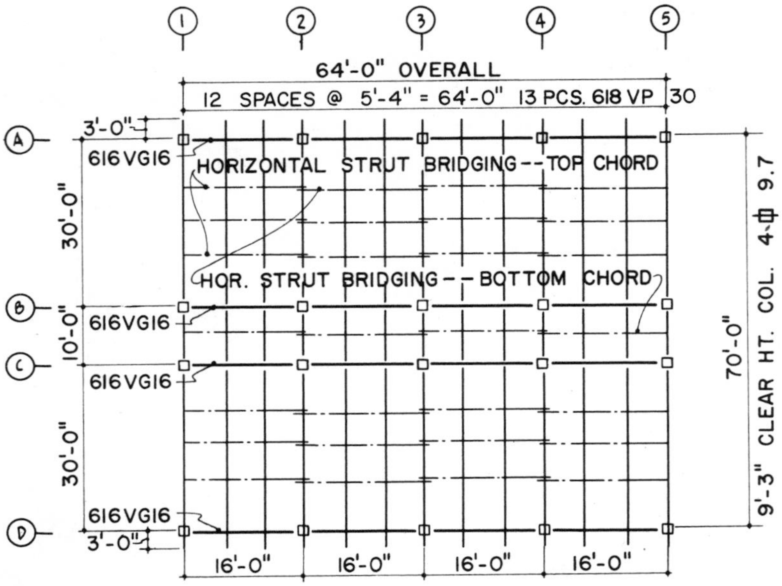

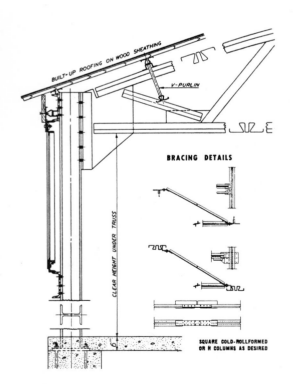

Fig. 35-31a. Structural-steel framing plan.

64'-0" OVERALL
12 SPACES @ 5'-4" = 64'-0" 13 PCS. 618 VP 30

616VG16

HORIZONTAL STRUT BRIDGING—TOP CHORD

HOR. STRUT BRIDGING—BOTTOM CHORD

70'-0" 9'-3" CLEAR HT. COL. 4-+ 9.7

BUILT-UP ROOFING ON WOOD SHEATHING

V-PURLIN

BRACING DETAILS

CLEAR HEIGHT UNDER TRUSS

SQUARE COLD-ROLLFORMED
OR H COLUMNS AS DESIRED

Fig. 35-31b. Section through structural-steel framed roof.

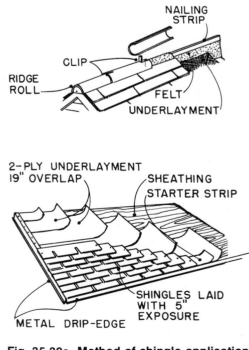

RIDGE SHINGLES

NAILING STRIP

CLIP

RIDGE ROLL

FELT UNDERLAYMENT

2-PLY UNDERLAYMENT 19" OVERLAP

SHEATHING STARTER STRIP

SHINGLES LAID WITH 5" EXPOSURE

METAL DRIP-EDGE

Fig. 35-32a. Method of shingle application.

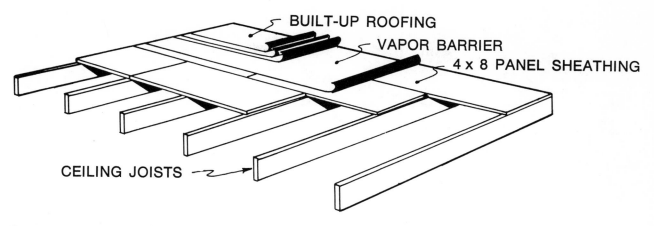

Fig. 35-32*b.* Built-up roof construction.

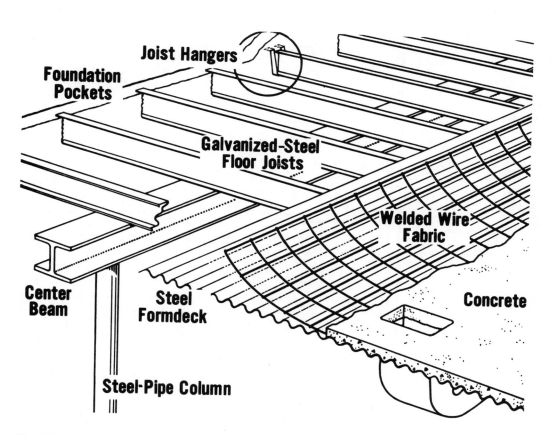

Fig. 35-32*c.* Steel roof framing.

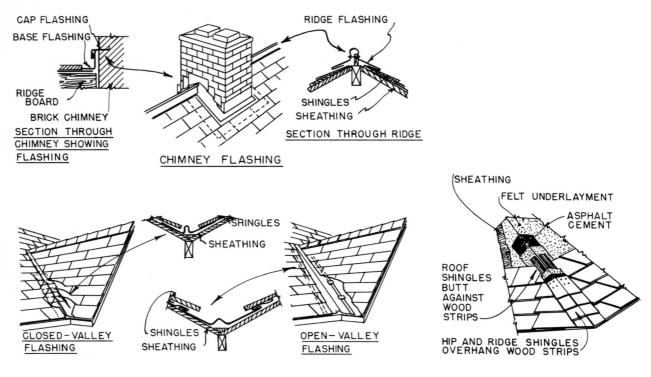

Fig. 35-33. Flashing details.

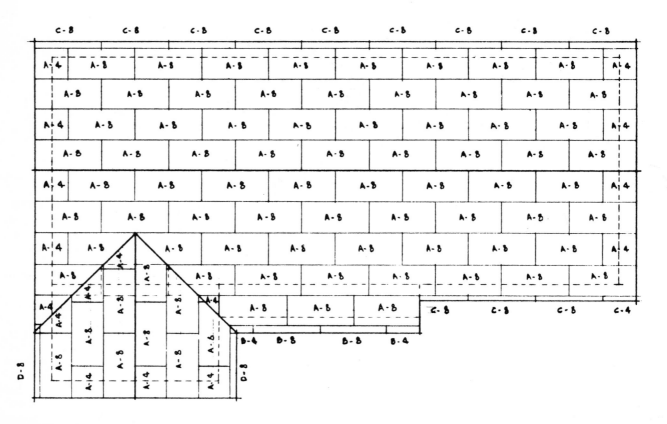

Fig. 35-34. Roof sheathing plan.

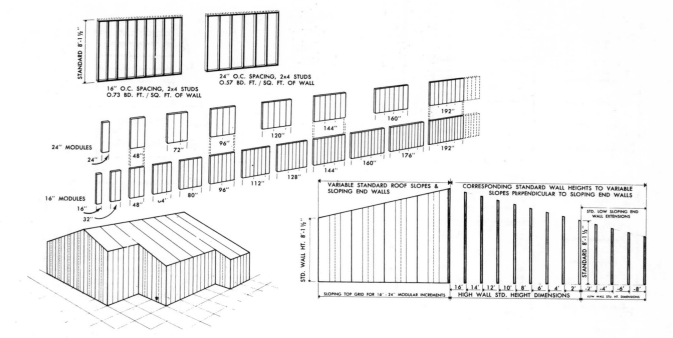

Fig. 36-1. Standard modular rhythms.

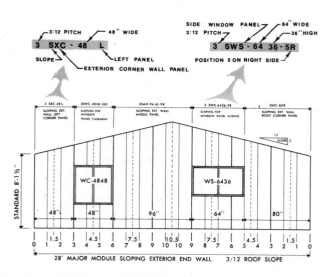

Fig. 36-2. Modular wall increments.

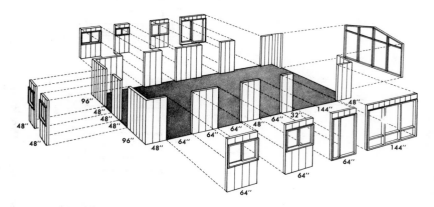

Fig. 36-3. Modular coding system.

Unit 36
Reading Modular Framing Plans

Modular plans are fitted to a modular grid. Thus, the modular relationship of foundations, floors, walls, windows, doors, partitions, and roofs is established. All dimensions and components precisely fall on modular increments, as shown in Figs. 36-1 and 36-2.

Some plans are completely modular and use a coding system similar to the one shown in Fig. 36-3 to describe the position of modular parts. Figure 36-4 shows a modular framing panel elevation drawing, which uses a coding system rather than numerical dimensions for size description. When constructing buildings of this type, modular components are installed according to the numbering system, thus eliminating the need for matching framing components di-

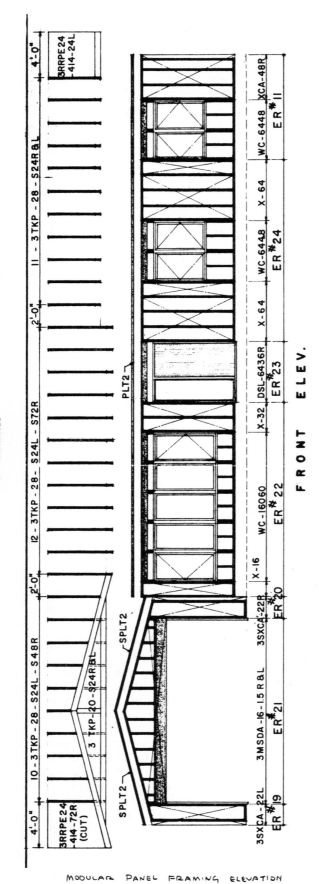

MODULAR PANEL FRAMING ELEVATION

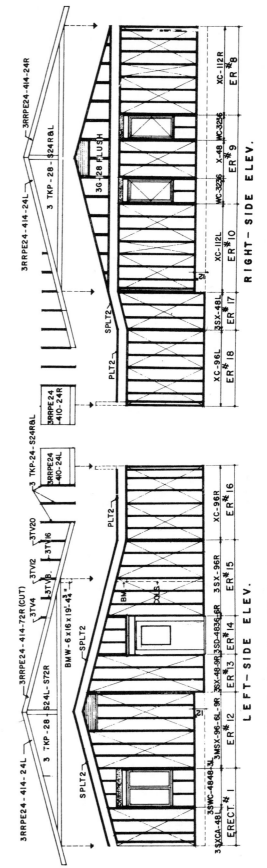

Fig. 36-4. Modular panel framing elevations. (National Lumber Mfg. Assc.)

mensionally. However, in this type of system, units must be installed precisely where located, or framing components will not match or align.

Where conventional framing and modular components are combined in the same plan, framing for the modular component is extremely critical since any variation in the framing opening, as shown in Fig. 36-5, will result in a misfit when the modular units are positioned into the framework. Figure 36-6 shows modular components for doors and windows that are used with modular component structures.

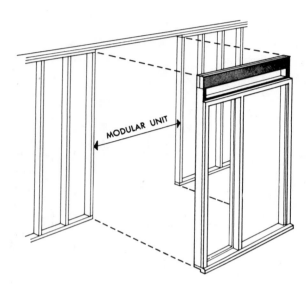

Fig. 36-5. Framing for modular components.

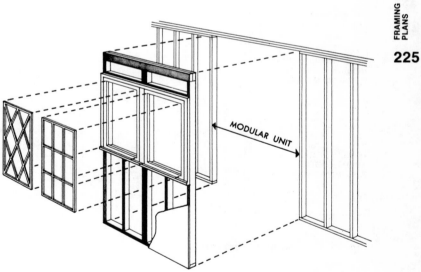

Fig. 36-6. Modular increments for doors and windows.

Unit 37

Reading Framing Dimensions
In reading framing dimensions, look for sizing and spacing of framing members and major distances between framing components. Figure 37-1 shows a typical framing plan dimensioned with overall dimensions and subdimensions and with sizes of all framing materials labeled. Regular interval spacing of structural members, such as roof rafters, floor joists, and wall studs, are not dimensioned if they fall on modular increments. Look for notes on framing drawings to determine overall spacing of members. On detail drawings, such as the one shown in Fig. 37-2, overall dimensions are not given; only key distances between structural levels or horizontal distances are shown. If material sizes are not given on the framing drawing, refer to the specifications list for materials description, which includes sizes.

On modular framing drawings, dimensions are sometimes omitted in lieu of using a modular grid. Framing members which fall on the center lines of the grid are modular and can be determined by counting the number of blocks from one member to another. However, framing members that do not fall on the modular grid are dimensioned separately.

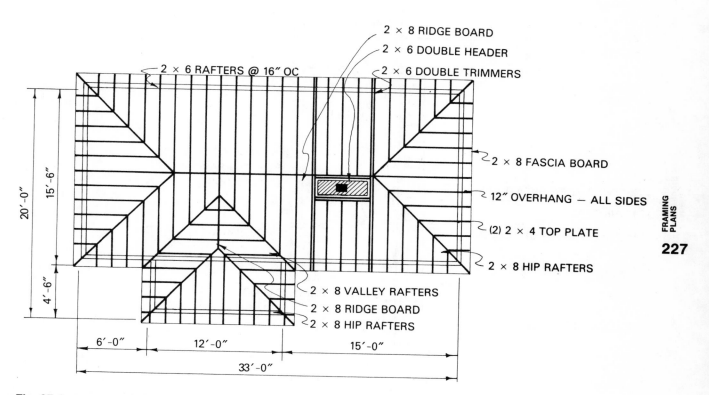

Fig. 37-1. A typical dimensioned framing drawing.

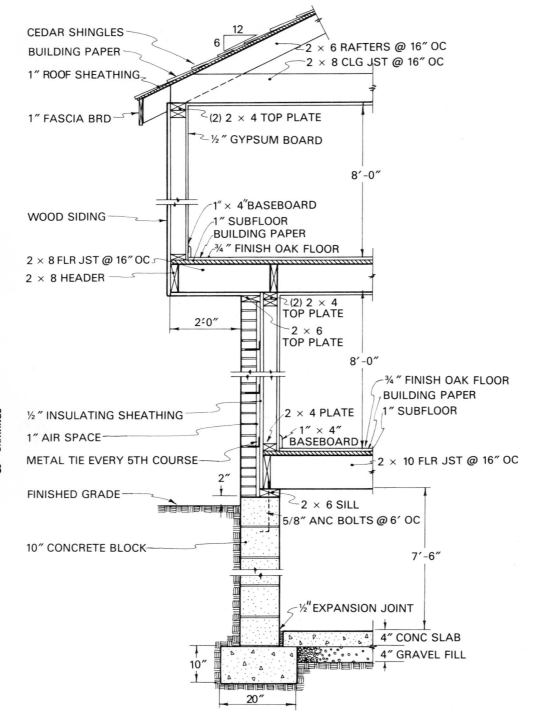

CEDAR SHINGLES
BUILDING PAPER
1″ ROOF SHEATHING

12
6

2 × 6 RAFTERS @ 16″ OC
2 × 8 CLG JST @ 16″ OC

1″ FASCIA BRD

(2) 2 × 4 TOP PLATE
½″ GYPSUM BOARD

8′-0″

WOOD SIDING

1″ × 4″ BASEBOARD
1″ SUBFLOOR
BUILDING PAPER
¾″ FINISH OAK FLOOR

2 × 8 FLR JST @ 16″ OC
2 × 8 HEADER

2′-0″

(2) 2 × 4 TOP PLATE
2 × 6 TOP PLATE

8′-0″

¾″ FINISH OAK FLOOR
BUILDING PAPER
1″ SUBFLOOR

½″ INSULATING SHEATHING
1″ AIR SPACE
METAL TIE EVERY 5TH COURSE

2 × 4 PLATE
1″ × 4″ BASEBOARD

2 × 10 FLR JST @ 16″ OC

2″

FINISHED GRADE

10″ CONCRETE BLOCK

2 × 6 SILL
5/8″ ANC BOLTS @ 6′ OC

7′-6″

½″ EXPANSION JOINT

4″ CONC SLAB
4″ GRAVEL FILL

10″

20″

Fig. 37-2. Dimensioned framing details.

CHAPTER 10

Reading Location Plans

Location plans provide the builder with essential information about the property. There are four basic types of location plans: the survey plan, the plat plan, the plot plan, and the landscape plan. The survey plan shows the geographical figures of the property. The plat plan shows the relationship of the property to the community. The plot plan shows the location of all structures on the property, and the landscape plan shows how the various features of the terrain are used in the overall design.

Unit 38
Survey Plans
A survey is a map (drawing) which shows the exact size, shape, and level of a lot. When prepared by a licensed surveyor, the survey is used as a legal document and is filed with the deed to the property. A lot survey includes the length of each boundary line, tree locations, corner elevations, contour of the land, and positions of streams, rivers, roads (or streets) and utility lines. It also lists the name of the owner of the lot and owner or title of adjacent lots.

A survey drawing is a complete, accurate description of the features of the lot. Symbols are used extensively to describe the features of the terrain. Figure 38-1 shows survey symbols. Some symbols depict the appearance of the feature, but most survey plan symbols are schematic representations of a feature.

The numbered arrows in Fig. 38-2 correspond to the following guides for reading a survey drawing:

1. The elevation above the datum of the lot is recorded in one corner of the drawing.
2. The size and location of streams and rivers are shown by wavy lines. These are blue lines on geographical survey maps.
3. A cross is used to show the position of existing trees. The elevation at the base of the trunk is shown.
4. The compass direction is given of each property line by degrees, minutes, and seconds. Figure 38-3 shows how the compass direction of each property line relates to compass orientation of the lot.
5. A north arrow is used to show general compass direction and orientation of the drawing.

6. Contour lines show the elevation of each line above the datum plane. The heights of the contour lines above the datum is shown at a break in each contour line.
7. Lot corners are identified by small circles surrounding the intersection of the corner lines.
8. Property lines are shown by a heavy line with two dashes repeated throughout the length of the line.
9. The elevations above datum of contour lines is the same at every point on the line. Brown lines are used for contour lines on geographical survey maps. Contour lines result from imaginary cuts made through the terrain at regular intervals. Figure 38-4 shows how this cut forms contour lines. The contour interval or the vertical distance between contour lines can be any convenient distance. It is usually an increment of five. Contour intervals of 5 ft, 10 ft, 15 ft, and 20 ft are common on large survey plans. The use of smaller contour intervals (1 to 4 ft) provides a more accurate description of the slope and shape of the terrain than does the use of larger intervals. Land on any part of a contour line has the same altitude (height) above the datum. Contour lines are therefore always continuous. The areas covered by contour lines may be so vast that the lines extend off the drawing sheet. However, if a larger geographical area is shown, the lines would ultimately meet and connect. Figure 38-5 shows how contours are projected from a profile of a hill. Contour lines that are very close together indicate a very steep slope. Contour lines that are far apart indicate a more gradual slope.
10. Proposed changes in the original grade line are shown by dotted contour lines. Figure 38-6 shows how the contour of a lot is planned for recontour and how the recontour lines are established to show the new heights above the datum. Figure 38-7 shows another method of indicating height points above datum and the direction of slope. To help read drawings where this type of indication is used, the points of the arrows having similar distances above the datum can be connected, creating a contour line.
11. Lot dimensions are shown directly on the property line. The distance on each line indicates the distance between corners or intersections. Dimensions relate only to straight-line

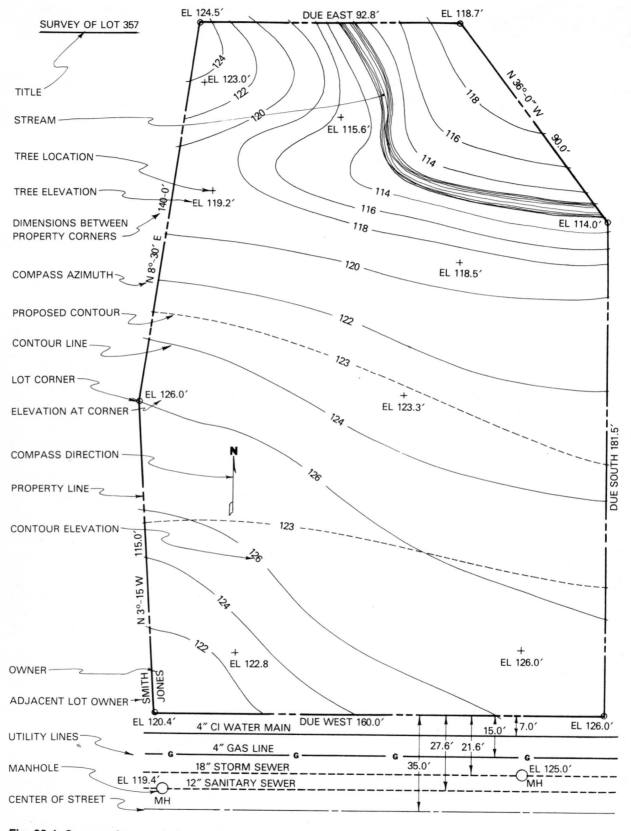

SURVEY OF LOT 357

TITLE

STREAM

TREE LOCATION

TREE ELEVATION

DIMENSIONS BETWEEN
PROPERTY CORNERS

COMPASS AZIMUTH

PROPOSED CONTOUR

CONTOUR LINE

LOT CORNER

ELEVATION AT CORNER

COMPASS DIRECTION

PROPERTY LINE

CONTOUR ELEVATION

OWNER

ADJACENT LOT OWNER

UTILITY LINES

MANHOLE

CENTER OF STREET

EL 124.5′   DUE EAST 92.8′   EL 118.7′

124
+EL 123.0′
122
120

N 36°-0″ W
90.0′

118
116
114

114
116
118

EL 115.6′ +

EL 119.2′
140-0′

N 8°-30′ E

120   EL 118.5′ +

122

123

EL 126.0′
124   EL 123.3′ +

126

123

126

124

N 3°-15′ W   115.0′

122   EL 122.8 +

EL 126.0′ +

SMITH   JONES

EL 120.4′   DUE WEST 160.0′

4″ CI WATER MAIN

4″ GAS LINE — G — G — G — G —

18″ STORM SEWER

12″ SANITARY SEWER

EL 119.4′
MH

EL 125.0′
MH

15.0′   7.0′

27.6′   21.6′

35.0′

EL 114.0′

DUE SOUTH 181.5′

EL 126.0′

Fig. 38-1. Survey plan symbols.

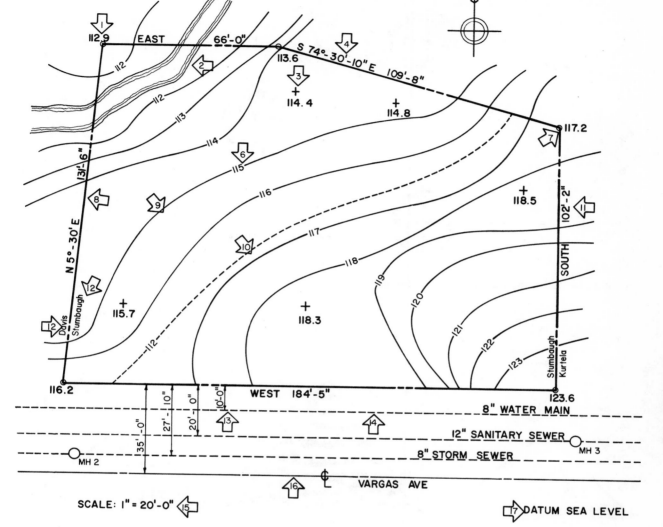

Fig. 38-2. Guides for reading survey plans.

distances. Each corner, regardless of how slight, is the end of the dimensioned line. Figure 38-8 shows how dimensions are placed on the perimeter of a lot. Notice that each time the compass direction changes, a new dimension is supplied. Thus, you will always find a dimension on top of the property line accompanying the compass direction below the line. Corners are circled to make slight angle changes more apparent. Curved property lines, as shown in Fig. 38-9, are dimensioned from the point of tangency to the next point of tangency. Points of tangency are indicated by dots or circles. A point of tangency is the terminal point of an arc and indicates the point where the curve becomes a straight line.

12. The names of the owners of adjacent lots are placed outside the property line. The name of the lot owner is shown on the inside of the property line.

13. The distance from the property line to all utility lines is dimensioned.

14. The position of the utility lines is shown by dotted lines. Utility lines are labeled according to their function.

15. Surveys are prepared using an engineer's scale. Common scales for surveys are $1/10'' = 1'-0''$ and $1/20'' = 1'-0''$.

16. Existing streets and roads are shown either by center lines or by curb and sidewalk outlines.

17. Datum base is shown as sea level or an established local plane.

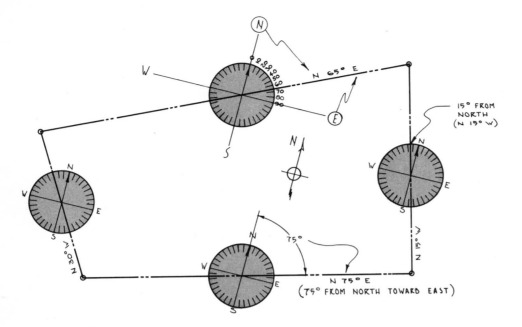

Fig. 38-3. Azimuth readings on property lines.

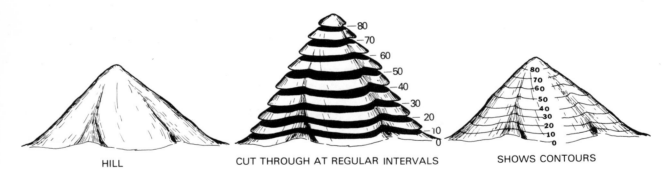

HILL

CUT THROUGH AT REGULAR INTERVALS

SHOWS CONTOURS

Fig. 38-4. Contour lines result from imaginary lines
cut at regular intervals horizontally through the terrain.

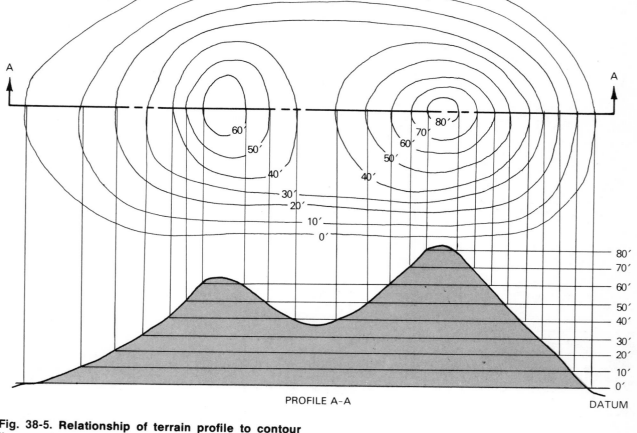

CONTOUR LINE

PROFILE A-A

DATUM

Fig. 38-5. Relationship of terrain profile to contour lines.

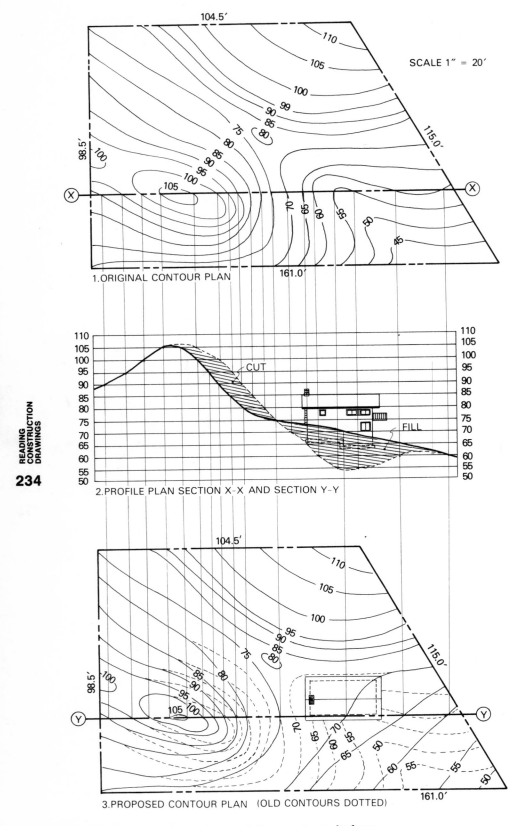

SCALE 1″ = 20′

1. ORIGINAL CONTOUR PLAN

2. PROFILE PLAN SECTION X-X AND SECTION Y-Y

3. PROPOSED CONTOUR PLAN (OLD CONTOURS DOTTED)

Fig. 38-6. Dotted lines show existing contours before grading.

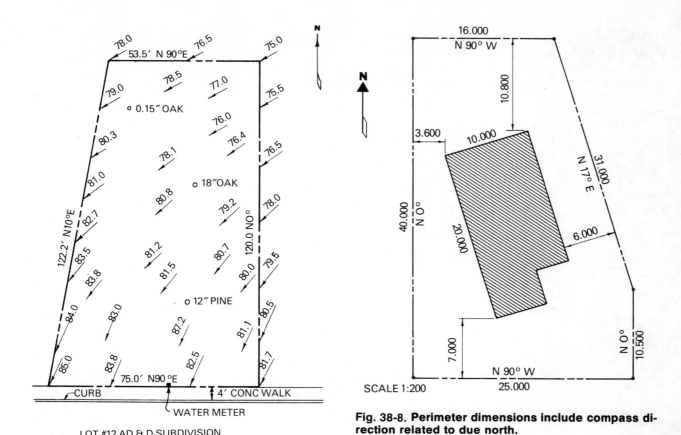

Fig. 38-7. Elevation of terrain shown by height dimensions and slope direction.

Fig. 38-8. Perimeter dimensions include compass direction related to due north.

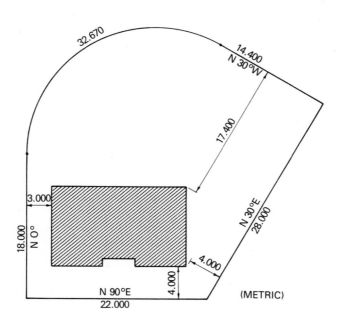

Fig. 38-9. Length of property line curves is shown from point of tangency.

Unit 39

Plat Plans Plat plans are outlines of land subdivisions. They may be of home developments, industrial parks, or urban neighborhoods. They show the size and shape of each parcel of property in the area.

The entire world is divided into geographical survey regions, even though not all of these regions have been surveyed. Geographical survey maps show the general contour of the area, nature of the terrain, and constructed features. Figure 39-1 shows a portion of a geographical survey plan. Geographical survey regions are further subdivided into township grids. A township grid is 24 sq mi. Each township grid is divided into 16 townships, each 6 sq mi; and each township is divided into 36 sections, each 1 sq mi. Figure 39-2

shows the township grid subdivided into townships and sections.

Plat plans are further subdivisions of township sections and are identified by section. Plat plans are identified in the following order:

1. Name of plat
2. Section
3. Township
4. County
5. State

Plat plans are located from a specific corner of a particular section in order to identify the plat in relation to the entire township grid. The dimensions of properties within the plat are described in the identical manner as survey plans describe individual lots (Fig. 39-3).

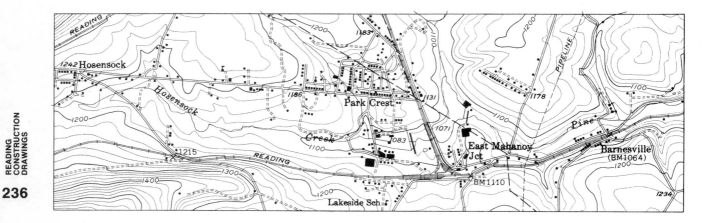

Fig. 39-1. Contour lines shown on geographical survey plans.

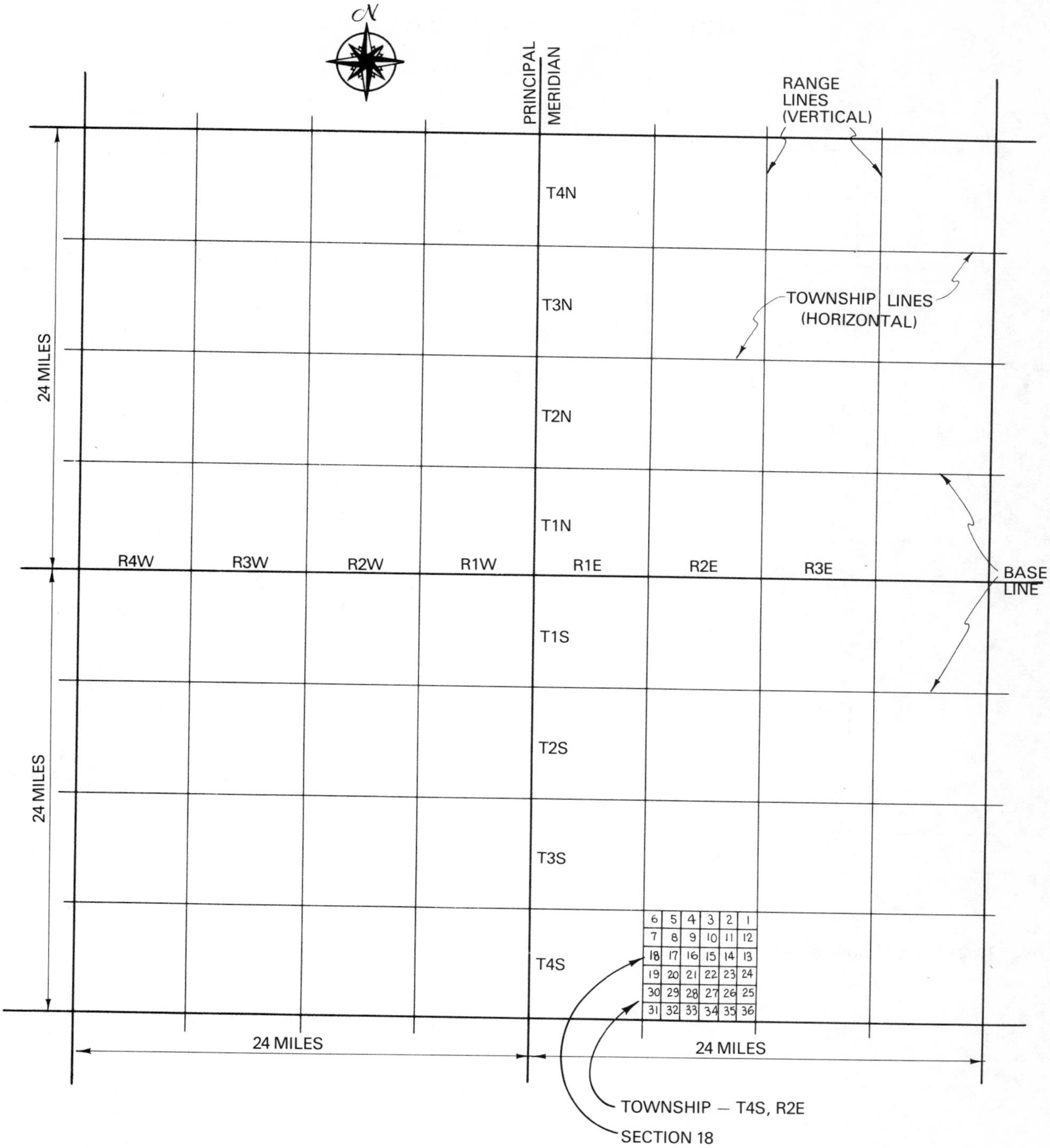

N

PRINCIPAL MERIDIAN

RANGE LINES (VERTICAL)

T4N

T3N

TOWNSHIP LINES (HORIZONTAL)

T2N

T1N

24 MILES

R4W R3W R2W R1W R1E R2E R3E

BASE LINE

T1S

T2S

24 MILES

T3S

T4S

6	5	4	3	2	1
7	8	9	10	11	12
18	17	16	15	14	13
19	20	21	22	23	24
30	29	28	27	26	25
31	32	33	34	35	36

24 MILES

24 MILES

TOWNSHIP — T4S, R2E

SECTION 18

Fig. 39-2. The township grid is 24 sq mi.

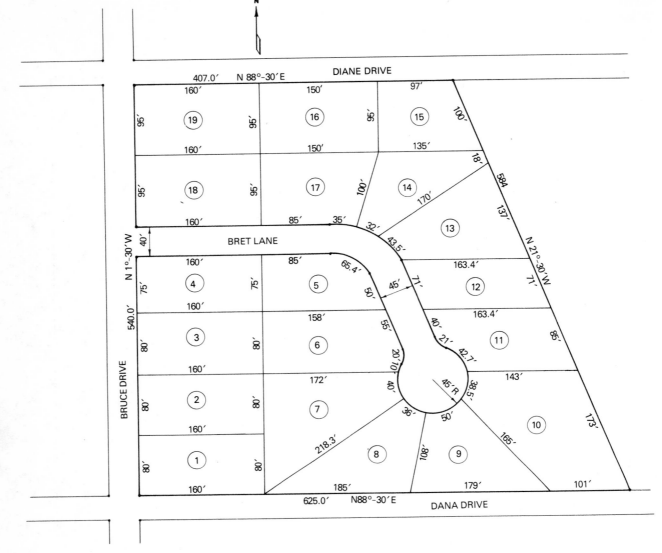

238

Fig. 39-3. Typical subdivision plat plan.

Unit 40

Plot Plans

Plot plans are used to show the location and size of all buildings on a lot. Overall building dimensions and lot dimensions are shown on plot plans. The position and size of walks, drives, patios, and courts are also shown. Compass orientation of the lot is given, and contour lines are sometimes shown. Figure 40-1 shows the key features and symbols commonly used on plot plans.

The numbered arrows shown in Fig. 40-2 illustrate the following guides for reading plot plans:

1. Only the outline of a structure is shown. Cross-hatching is frequently used in place of interior walls, although on some plot plans an abbreviated floor plan may be shown.
2. Outlines of other buildings on the lot are shown and dimensioned from the main building or from property lines.
3. Plot plan dimensions include the overall size of buildings, the distance from each building to at least two property lines, and the dimension and compass azimuth (direction) of each property line (Fig. 40-3).
4. Each building in the lot is dimensioned from the property line to the outside wall of the building. Property lines show the legal limits of the lot on all sides. Most building codes require that buildings be located (set back) specific minimum distances from property lines.

Figure 40-4 shows typical property lines and building line setback requirements. If setback is not specified, check local codes before beginning construction of any facility.

5. The position and size of all driveways is shown.
6. Location and size of walkways is shown.
7. Grade elevation of key surfaces, such as patios, driveways, and courts, are shown as are heights above datum.
8. Surface materials used on patios, decks, and driveways are either labeled or shown with a plan symbol for that material.
9. Streets adjacent to the property are labeled to help identify and orient the property to its surroundings.
10. Overall lot dimensions are placed either on extension lines outside the property line or directly on the property line.
11. Size and location of all courts are shown and dimensioned.
12. Size and locations of pools, ponds, and other bodies of water are outlined and labeled.
13. The compass orientation of the lot is identified by the use of a north arrow.
14. Decimal scales such as $1/10'' = 1'-0''$, $1/20'' = 1'0''$ or metric scales of $1:100$ or $1:200$ are used on plot plans.
15. Utility lines are labeled by function. Their relationship to the property and to structures is shown in Fig. 40-5.

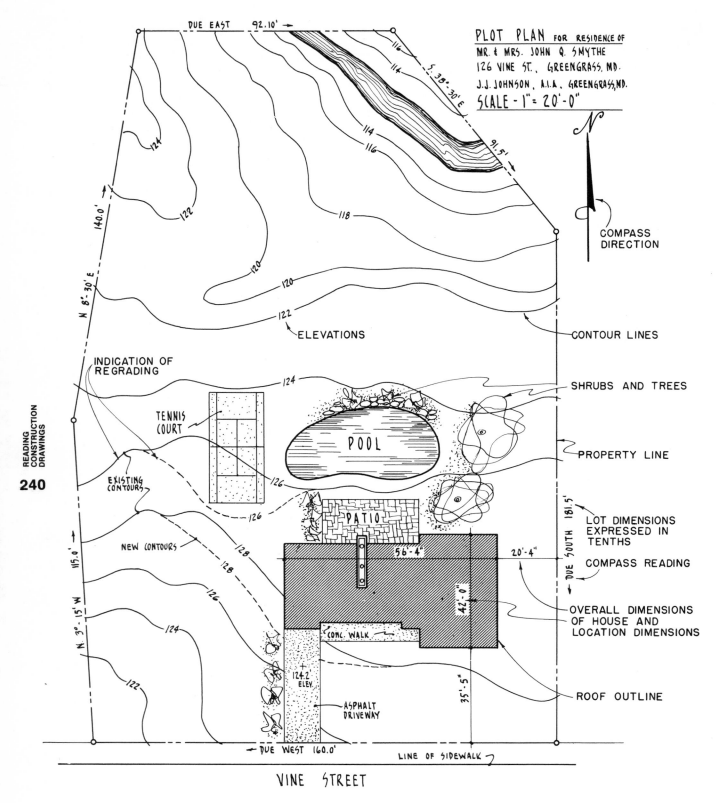

Fig. 40-1. Plot plan symbols.

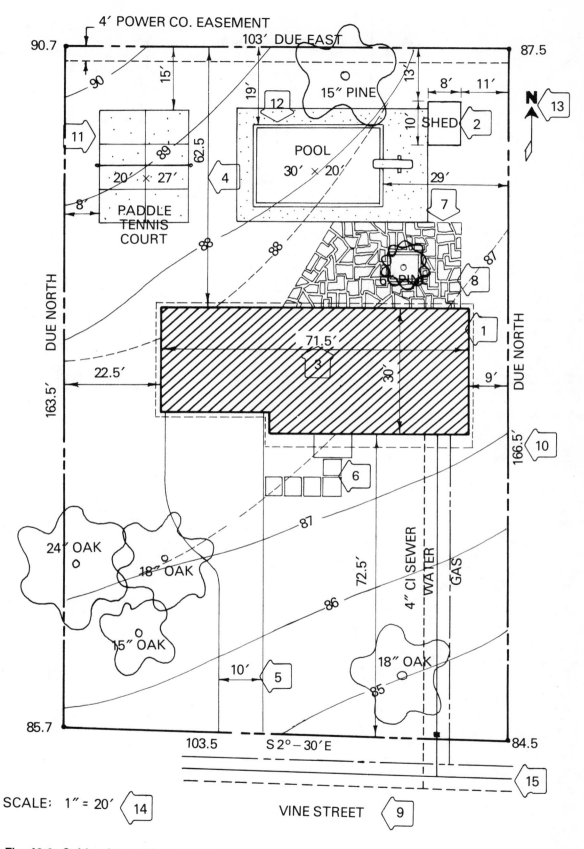

Fig. 40-2. Guides for reading plot plans.

4' POWER CO. EASEMENT

103' DUE EAST

90.7

87.5

15'

90

15" PINE

19'

12

13'

8'   11'

11

SHED   2

N   13

89

20' × 27'

62.5

POOL
30' × 20'

10'

4

8'

PADDLE
TENNIS
COURT

29'

7

88

88

87

8

PINE

DUE NORTH

1

71.5'

DUE NORTH

22.5'

3

30'

9'

163.5'

166.5'

10

6

87

24" OAK

18" OAK

4" CI SEWER

WATER

GAS

15" OAK

72.5'

86

18" OAK

85

10'

5

85.7

84.5

103.5

S 2° – 30' E

15

SCALE: 1" = 20'   14

VINE STREET   9

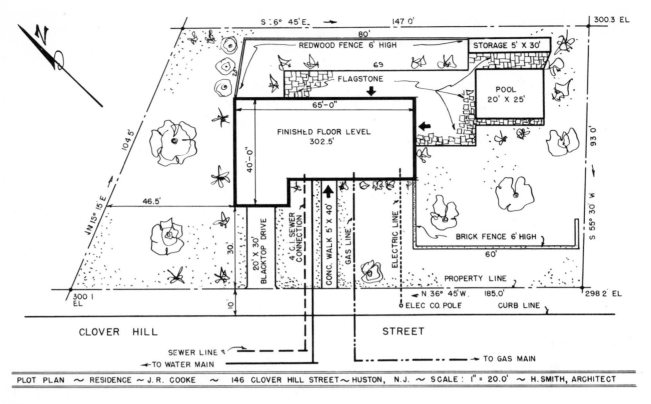

Fig. 40-3. Plot plan dimensions.

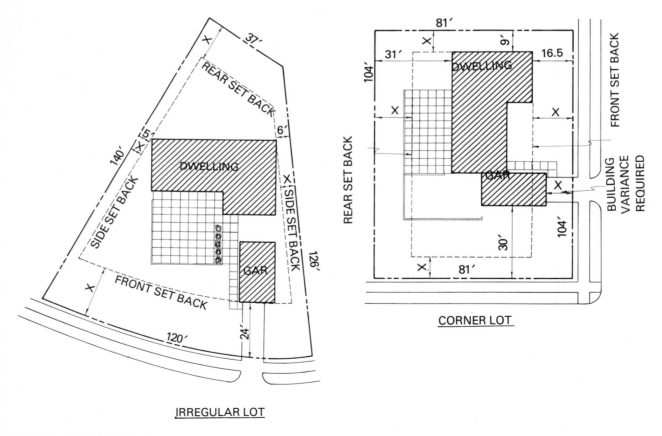

JRREGULAR LOT

CORNER LOT

Fig. 40-4. Property and building lines shown on plot plans.

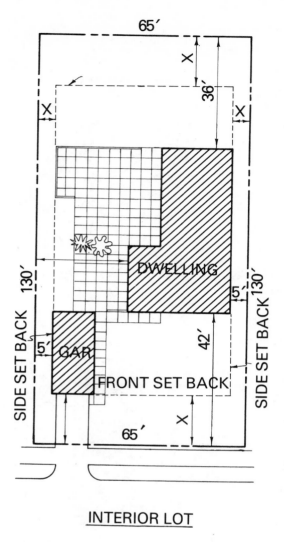

INTERIOR LOT

Fig. 40-4 (con't)

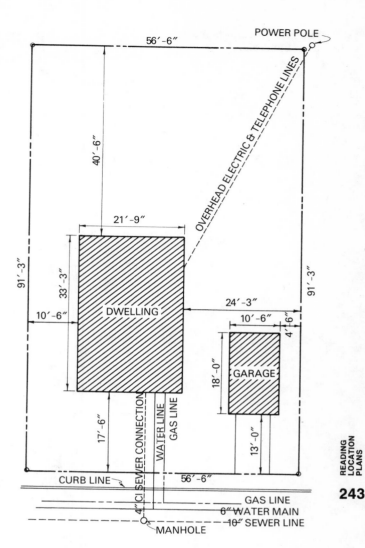

Fig. 40-5. Utility lines shown on plot plans.

Unit 41
Landscape Plans

Landscape plans show the type and location of vegetation existing or planned for a lot. They may also show the contour of the land and the position of buildings. Such features are often necessary to make the placement of vegetation meaningful. Symbols are used on landscape plans to show the position of trees, shrubbery, flowers, vegetable gardens, hedge rows, and lawns. Figure 41-1 shows some common symbols used on landscape plans.

The following guides for reading landscape plans are illustrated by the numbered arrows shown in Fig. 41-2.

1. The elevation of all trees is noted to show the datum level.
2. Vegetable gardens are shown by an outline of the planting furrows.
3. Orchards are shown by an outline of each tree in the pattern.
4. The property line is shown to define the limits of the lot.
5. Trees are located with crosses.
6. Shrubbery is outlined and used to define boundaries, to outline walks, to conceal foundation walls, and to balance irregular contours.
7. The outline and subdivisions of courts are shown.

Fig. 41-1. Landscape plan symbols.

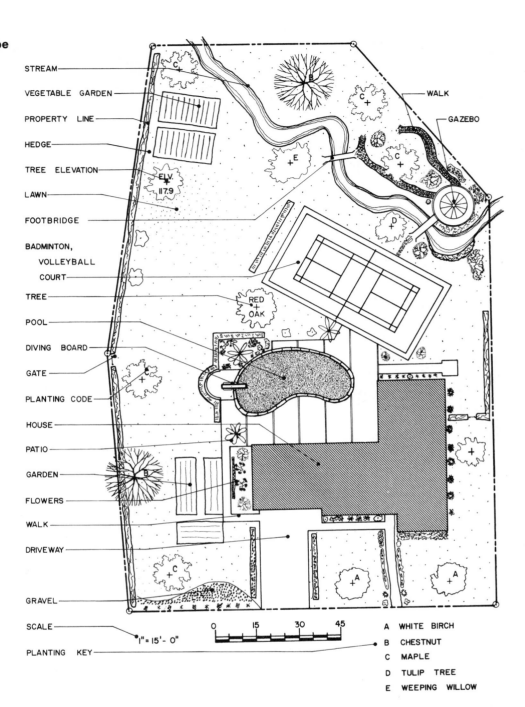

STREAM

VEGETABLE GARDEN

PROPERTY LINE

HEDGE

TREE ELEVATION

LAWN

FOOTBRIDGE

BADMINTON,

VOLLEYBALL

COURT

TREE

POOL

DIVING BOARD

GATE

PLANTING CODE

HOUSE

PATIO

GARDEN

FLOWERS

WALK

DRIVEWAY

GRAVEL

SCALE

PLANTING KEY

WALK

GAZEBO

ELV.
117.9

RED
OAK

1" = 15'- 0"

0   15   30   45

A   WHITE BIRCH
B   CHESTNUT
C   MAPLE
D   TULIP TREE
E   WEEPING WILLOW

8. Flower gardens are shown by the outline of their shapes.
9. Lawns are shown by small, sparsely placed dots or vertical lines.
10. The outlines of all walks and planned paths are shown.
11. Conventional map symbols are used for small features such as bridges.
12. The outline and surface covering of all patios and terraces are indicated.

13. The name of each tree and shrub is labeled on the symbol or keyed to a planting schedule.
14. All landscaping changes are noted with dotted lines.
15. Flowers are labeled when clustered into beds.
16. The house is outlined, crosshatched, or shaded. In some cases, the outline of the floor plan is shown in abbreviated form. This helps to show the relationship of the outside to the inside living areas.

Fig. 41-2. Guides to read landscape plans.

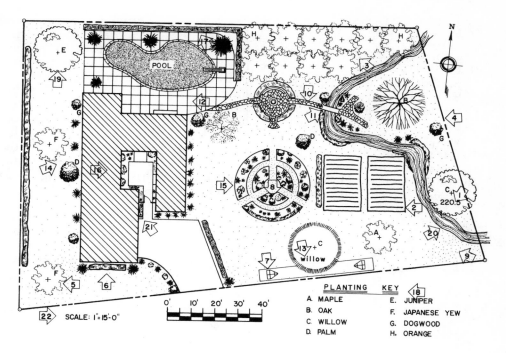

PLANTING KEY

A. MAPLE E. JUNIPER
B. OAK   F. JAPANESE YEW
C. WILLOW G. DOGWOOD
D. PALM   H. ORANGE

SCALE: 1"=15'-0"

ESTATE PLAN

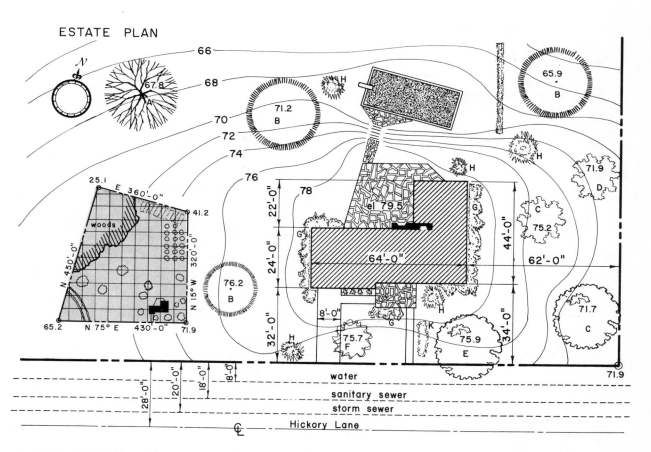

Fig. 41-3. Estate plan with enlarged partial detail.

17. Hedge is outlined and is used as a screening device to provide privacy, to divide an area, to control traffic, or to serve as a windbreak.

18. Each tree or shrub may be indexed to a planting schedule, if there are too many to be labeled on the drawing, as suggested in rule 13.

19. A tree is shown by drawing an outline of the area covered by its branches. This symbol varies from a perfect circle to irregular lines representing the appearance of branches. A cross (+) indicates the location of the trunk.

20. Water is indicated by irregular parallel lines.

21. Shrubbery which is to be planted in phases is labeled as phase 1, phase 2, phase 3, and so forth by the use of different shading or cross-hatching.

22. Because surveyors use this measure, an engineer's scale is used to prepare landscape plans.

Often the lot or estate is too large to be shown accurately on a standard landscape plan. A scale such as 1″ = 20′-0″ may not show the entire estate. Or a scale must be used that is so small that the features cannot readily be identified, labeled, and dimensioned. In this case a plan, such as the one shown in Fig. 41-3, is prepared showing the total estate with only one portion enlarged. This portion is indexed to the drawing of the immediate area surrounding the main structure. Figure 41-3 shows the total estate plan as an insert in the drawing, showing the lower right corner of the estate developed in more detail.

It is sometimes desirable or necessary to combine all of the features of the survey plot plan and landscape plan into one drawing. In such a combination, all the symbols and dimensions are incorporated into one location plan.

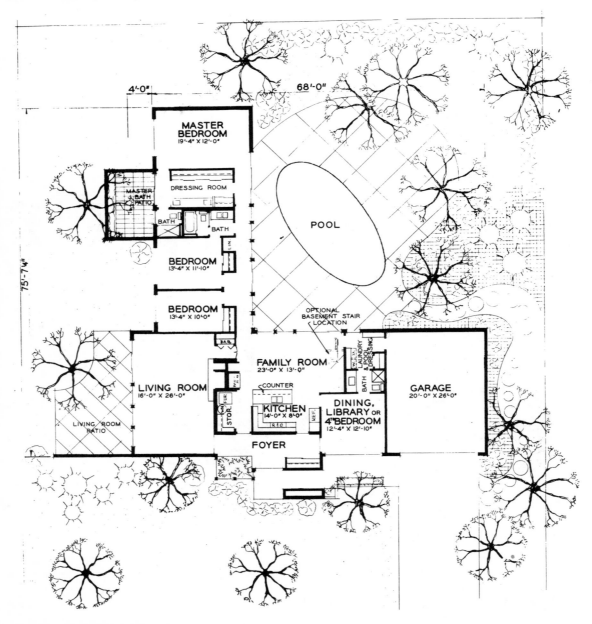

Figure 41-4. Symbolic landscape plan.

CHAPTER 11

Reading Electrical Plans

Electrical plans are floor plans and sometimes elevation drawings to which symbols indicating the position of electrical fixtures, outlets, and switches have been added.

Unit 42
Electric Circuits

To adequately read and understand electrical plans, some basic knowledge of electric circuits is necessary. Electrical energy is brought to a building by service entrance wires, as shown in Fig. 42-1. The size of the service entrance wire determines the amount of electricity that can safely enter the building wiring system. Service wires are connected to a watt meter, then to a distribution panel (Fig. 42-2), which distributes the electricity throughout the building through branch circuits.

Between the watt meter and the distribution panel (branch circuit box) is a main fuse or circuit breaker. If too much current is drawn from outside sources and heats the wires, the circuit breaker will disconnect (open) the circuit. In addition to this protection to the main source of power, each branch circuit is also protected with circuit breakers. Branch circuit breakers trip (open) when the wires in that circuit become overloaded and too hot.

Fig. 42-1. Service entrance equipment.

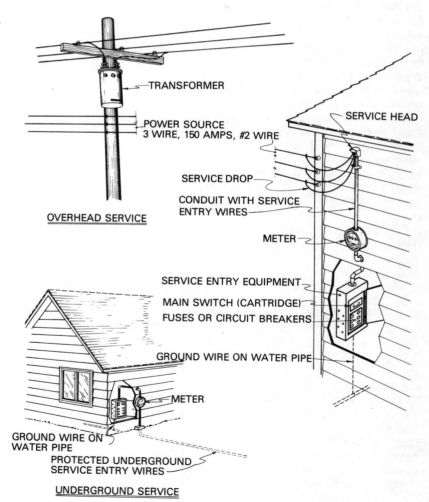

TRANSFORMER

POWER SOURCE
3 WIRE, 150 AMPS, #2 WIRE

OVERHEAD SERVICE

SERVICE HEAD

SERVICE DROP

CONDUIT WITH SERVICE
ENTRY WIRES

METER

SERVICE ENTRY EQUIPMENT

MAIN SWITCH (CARTRIDGE)
FUSES OR CIRCUIT BREAKERS

GROUND WIRE ON WATER PIPE

METER

GROUND WIRE ON
WATER PIPE

PROTECTED UNDERGROUND
SERVICE ENTRY WIRES

UNDERGROUND SERVICE

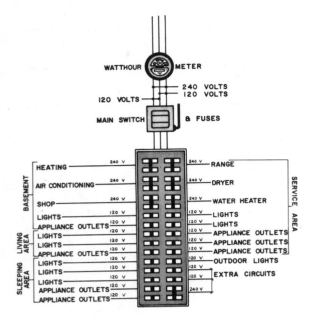

Fig. 42-2. Branch circuit distribution panel.

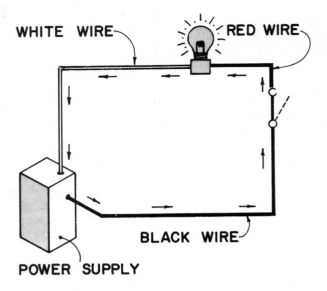

Fig. 42-3. Simple electric circuit.

Branch Circuits

Each branch circuit delivers electricity to one or more outlets. It is necessary to divide the amount of electricity that enters the building into branch circuits so that one line will not carry all the energy (voltage). If the entire building is placed on one circuit, overloading will leave the entire building without power.

A circuit is a path which electricity follows from the power supply source to a light, appliance, or electrical device and back to the power supply source again, as shown in Fig. 42-3. Each branch circuit functions like this simple electric circuit. Branch circuits are divided into three groups: lighting circuits, small-appliance circuits, and individual circuits.

Lighting circuits, as shown in Fig. 42-4, provide lighting outlets for the building. Different lights in each room are usually on different circuits so that if one circuit breaker trips, the room will not be in total darkness.

Small-appliance circuits provide power only to convenience outlets, as shown in Fig. 42-5. These circuits are designed to provide electricity where small low-wattage appliances will be connected.

Individual circuits such as the circuits shown in Fig. 42-6 are designed to serve only one piece of electrical equipment. Appliances and devices that are served by individual circuits include: electric ranges, automatic heating units, water heaters, clothes dryers, air conditioners, built-in electric heaters, workshop outlets, and large, motor-driven appliances, such as washers, disposals, and dishwashers. When a motor starts, it needs an extra serge of power to bring it to full

speed. This is called a starting load. Individual circuits are designed to provide sufficient power for starting loads.

Where low voltage is required, usually for decorative outdoor lighting, a low-voltage control circuit such as the one in Fig. 42-7 is used.

Circuit Switch Control

Small-appliance circuits and individual circuits are usually hot; that is, there is electricity available in the outlet at all times. Lighting circuits, however, may be hot or may be controlled with switches. Figure 42-8 shows several types of switches used to control the flow of electricity to an outlet. Figure 42-9 shows a three-way switching circuit and its application to lighting.

There are many different kinds of switches used to control circuits: toggle switches, such as the ones shown in Fig. 42-8, are available in single-pole, three-way, or four-way types. Mercury switches, which are silent, shockproof, easy to wire and install, last longer than toggle switches and are also available in single-pole, three-way, and four-way types. Automatic cycle controls, as used on washers, are installed on appliances to

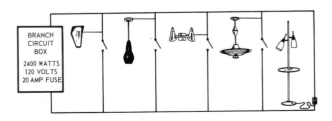

Fig. 42-4. Typical lighting circuit.

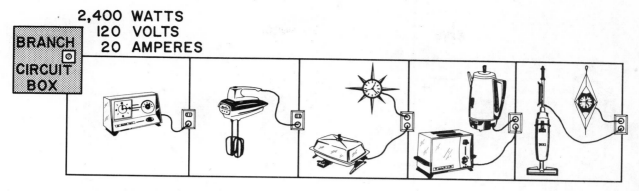

Fig. 42-5. Small appliance circuit.

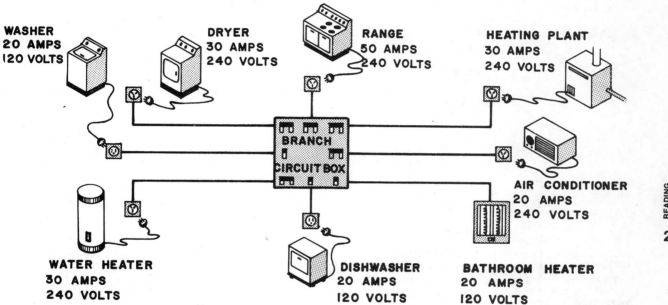

Fig. 42-6. Individual circuits.

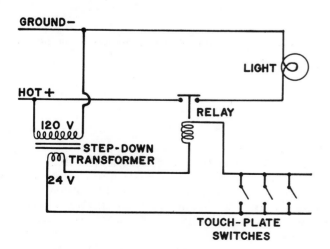

Fig. 42-7. Low-voltage control systems.

make them perform their functions on a time cycle. Photoelectric cells control switching by blocking a beam of light. Automatic controls adjust heating and cooling systems to desired temperatures. Clock thermostats adjust heating and cooling by adjusting both temperature and time. Aquastats keep water heated or cooled to selected temperatures. Dimmer switches control intensity of light. Time switches cause lights or devices to switch on and off at prespecified time intervals. Safety alarm systems and switches activate bells or lights when a circuit, usually on a door or window, is broken. Master switches control circuits throughout an entire building from one location. Low-voltage switching systems provide economical long runs for low-voltage lighting.

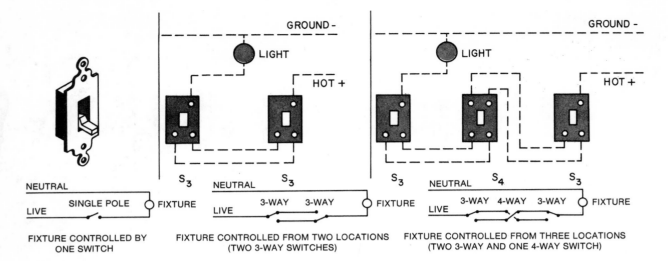

Fig. 42-8. Switching controls.

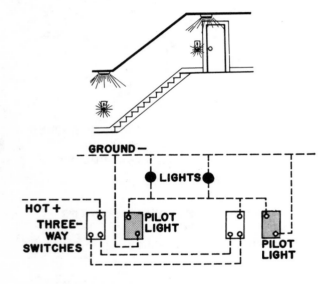

Fig. 42-9. Three-way switches with pilot light.

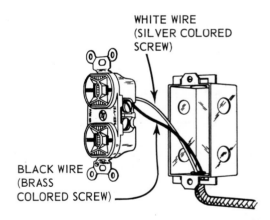

Fig. 42-10b. Wiring for duplex convenience outlets.

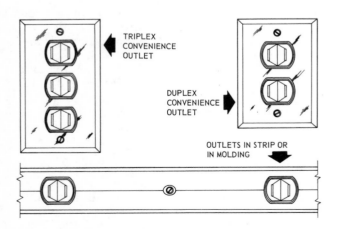

Fig. 42-10a. Electrical convenience outlets.

Electrical Outlets

There are several kinds of electrical outlets: the convenience outlet (Fig. 42-10a and b), lighting outlets (Fig. 42-10c), and special purpose outlets (Fig. 42-10d). Special purpose outlets and convenience outlets are connected to hot circuits, while lighting outlets are controlled with a switching device. Figure 42-11 shows electrical wiring symbols used on construction drawings.

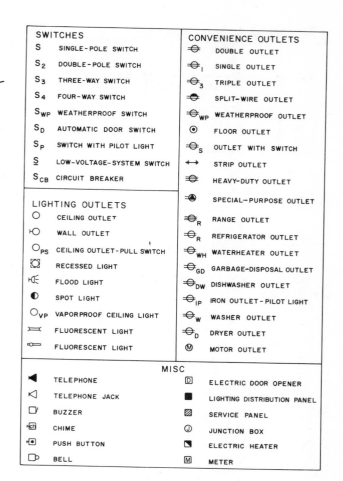

Fig. 42-10c. Electrical lighting outlets.

SWITCHES		CONVENIENCE OUTLETS	
S	SINGLE-POLE SWITCH	DOUBLE OUTLET	
S₂	DOUBLE-POLE SWITCH	SINGLE OUTLET	
S₃	THREE-WAY SWITCH	TRIPLE OUTLET	
S₄	FOUR-WAY SWITCH	SPLIT-WIRE OUTLET	
S_WP	WEATHERPROOF SWITCH	WEATHERPROOF OUTLET	
S_D	AUTOMATIC DOOR SWITCH	FLOOR OUTLET	
S_P	SWITCH WITH PILOT LIGHT	OUTLET WITH SWITCH	
S	LOW-VOLTAGE-SYSTEM SWITCH	STRIP OUTLET	
S_CB	CIRCUIT BREAKER	HEAVY-DUTY OUTLET	
		SPECIAL-PURPOSE OUTLET	
LIGHTING OUTLETS		RANGE OUTLET	
O	CEILING OUTLET	REFRIGERATOR OUTLET	
O	WALL OUTLET	WATERHEATER OUTLET	
O_PS	CEILING OUTLET-PULL SWITCH	GARBAGE-DISPOSAL OUTLET	
	RECESSED LIGHT	DISHWASHER OUTLET	
	FLOOD LIGHT	IRON OUTLET-PILOT LIGHT	
	SPOT LIGHT	WASHER OUTLET	
O_VP	VAPORPROOF CEILING LIGHT	DRYER OUTLET	
	FLUORESCENT LIGHT	MOTOR OUTLET	
	FLUORESCENT LIGHT		

Switch subscripts: S_2, S_3, S_4, S_{WP}, S_D, S_P, S_{CB}
Outlet subscripts: Range $_R$, Refrigerator $_R$, Waterheater $_{WH}$, Garbage-disposal $_{GD}$, Dishwasher $_{DW}$, Iron $_{IP}$, Washer $_W$, Dryer $_D$

MISC			
	TELEPHONE		ELECTRIC DOOR OPENER
	TELEPHONE JACK		LIGHTING DISTRIBUTION PANEL
	BUZZER		SERVICE PANEL
	CHIME		JUNCTION BOX
	PUSH BUTTON		ELECTRIC HEATER
	BELL		METER

Fig. 42-11. Electrical wiring symbols.

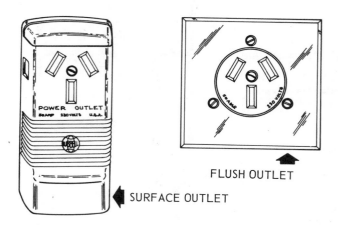

POWER OUTLET
50 AMP 230 VOLTS U.S.A.

FLUSH OUTLET

SURFACE OUTLET

Fig. 42-10d. Special purpose outlets.

Unit 43
Wiring Plans The position of all outlets and controls are shown on the electrical plan by using electrical wiring symbols (Fig. 42-11); however, the entire circuit is not drawn on the electrical plan. A true wiring diagram shows the manner in which a light or fixture is wired to the switch and how the wire in the switch is actually broken. In the architectural abbreviated method, shown in Fig. 43-1, only the position of the fixture and the switch is found on the drawing with a dotted line connecting the outlet with the switch which controls it. The dotted line does not represent the path of the actual wire. Figure 43-2, for example, is a complete wiring plan which shows the position of all switches, outlets, and fixtures. This plan also shows the position of each lighting fixture with a dotted line connecting that fixture to the switch used to control each fixture. Three-way switches indicate that the fixture is controlled by two separate switches, such as the hall light shown in Fig. 42-9, which is connected to the switch at the entry to the living room and also connected to another switch at the entry to bed-

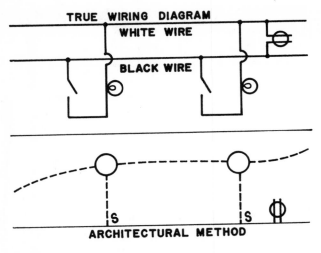

Fig. 43-1. True wiring diagram compared to architectural method.

room number 1. Figure 43-3 shows the relationship between actual symbols on an electrical plan, the appliance they represent, and the circuit needed to supply the necessary electrical power for that appliance.

Figures 43-4 through 43-13 show some typical wiring diagrams of various specific rooms of a residence. Refer to Unit 2 to relate the symbols found on these plans to the fixtures, devices, and types of switches they represent. Trace the control of each fixture to a switch. Notice and compare the complexity of the circuits designed for the kitchen and laundry with the circuits used on other rooms. This difference is because of the need for more special circuits to serve specific appliances and devices in addition to lighting circuits in kitchens and laundries. Notice also the use of three-way and four-way switches in halls and other traffic areas to provide flexibility in control.

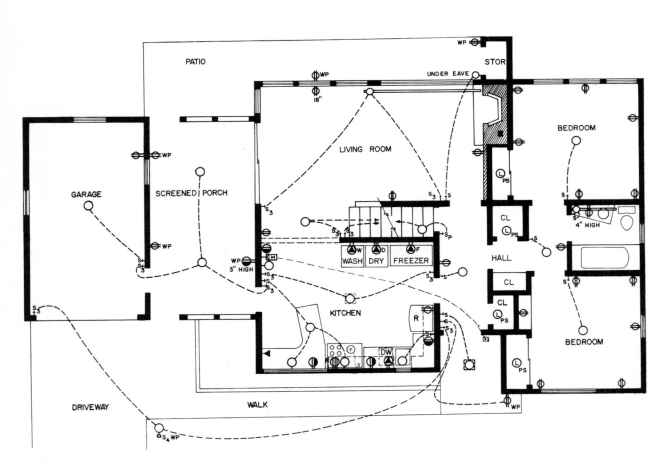

Fig. 43-2. Complete wiring plan.

Fig. 43-3. Relationship of wiring plan to fixture requirements.

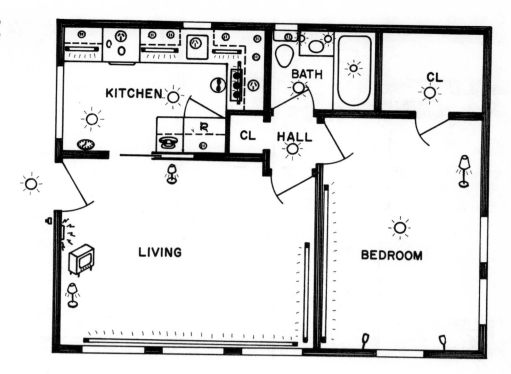

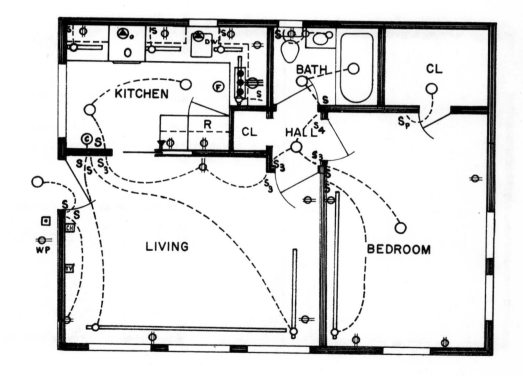

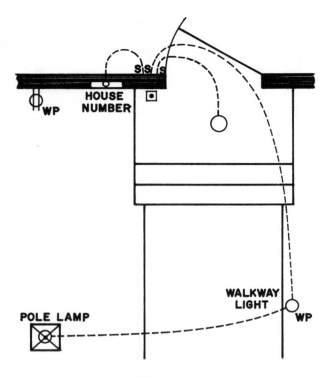

Fig. 43-4. Entrance wiring plan.

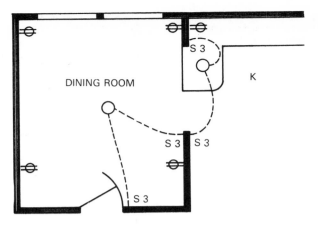

Fig. 43-6. Dining room wiring plan.

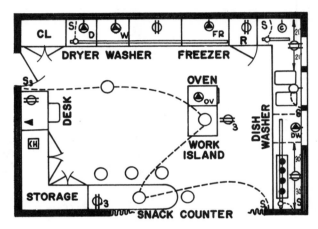

Fig. 43-7. Kitchen wiring plan.

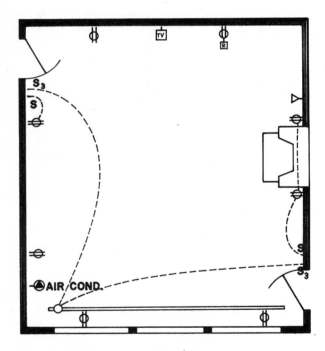

Fig. 43-5. Living room wiring plan.

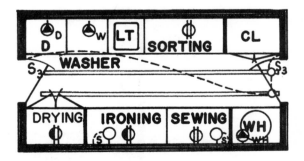

Fig. 43-8. Utility room wiring plan.

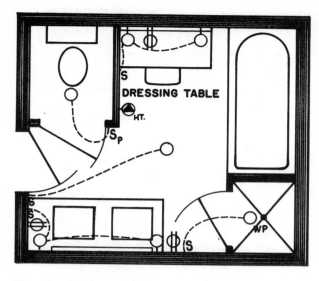

Fig. 43-9. Bathroom wiring plan.

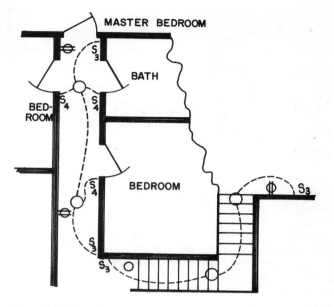

Fig. 43-12. Traffic area wiring plan.

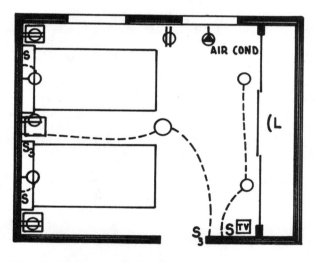

Fig. 43-10. Bedroom wiring plan.

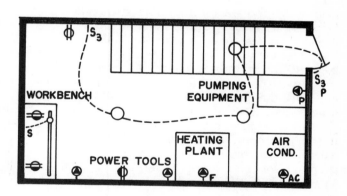

Fig. 43-13. Basement-shop wiring plan.

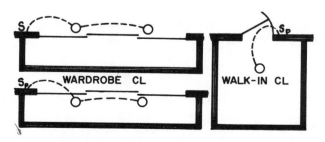

Fig. 43-11. Wiring plan for closets.

CHAPTER 12

Reading Plumbing Diagrams

Plumbing diagrams are drawings which describe piping systems that supply water to and drainage of waste material from buildings. A plumbing system (Fig. 44-1) consists of supply pipes that carry fresh water under pressure from a public water supply or individual well to building fixtures and pipes, which in turn carry wastes to a disposal system by gravity drainage (see Fig. 44-9). Plumbing diagrams are shown in both plan and elevation form. Figure 44-2 shows plumbing symbols used on both plumbing plans and elevations.

Unit 44
Schematic Plumbing Plans

Plumbing symbols are either added to an existing floor plan, or a separate plumbing plan is prepared as shown in Figure 44-3. Sometimes plumbing symbols, electrical symbols, heating and air-conditioning symbols are all combined in one plan. When this is done, dimensions and other construction notes are eliminated to facilitate reading these symbols without interference. This is because the plumbing contractor is concerned solely with the placement of plumbing fixtures and with the length of piping runs, and not with other construction details.

Since most plumbing fixtures are concentrated in the bathroom, the laundry, and kitchen areas, a detailed schematic plumbing plan is sometimes prepared for these rooms as shown in Figs. 44-4, 44-5, and 44-6. When reading these plans, visualize the plumbing symbols as they appear in Fig. 44-7.

Water Supply Systems

Fresh water is brought to all plumbing fixtures under pressure. This water is supplied from either a public water supply or from private wells. Because this water is under pressure, the pipes may run in any convenient direction after leaving the main control pipe, as shown in Fig. 44-8. Water lines require shutoff valves at the property line and at the entrance of the building. A water meter is located at the shutoff valve near the building.

Hot water is obtained by routing water through a hot-water heater. The hot water is then directed, under pressure, to appropriate fixtures. The hot-water valve is always on the left of each fixture. Placing insulation around hot-water lines conserves hot water and reduces the total cost of fuel for heating water. This is usually included in a note on the plumbing diagram, which specifies the type and thickness of the insulation material.

Waste Discharge Systems

Waste water is discharged through the disposal system by gravity drainage, as shown in Fig. 44-9. All pipes in this system must slant in a downward direction so that the weight of the waste will cause it to move down toward the main disposal system and away from the structure. Because of this gravity flow, waste lines that connect to sewerage systems are much larger than the water supply lines, in which the water is under pressure, as shown in Fig. 44-10. Waste lines are concealed in walls and under floors. The vertical lines are called stacks, and the horizontal lines are called branches. There are also vents which provide for the circulation of air and permit sewer gases to escape through the roof to the outside, and equalize the air pressure in the drainage system. Fixture traps stop gases from entering the building. Each fixture has a separate trap (seal) to prevent backflow of sewer gas. Fixture traps are exposed for easy maintenance; however, water closet traps are built into the fixture.

The flow of waste water starts at the fixture trap. It flows through the fixture branches to the soil stack. It then continues through the building drain to the sewer drain, and finally reaches the main sewer line.

Waste stacks carry only waste water. The lines which carry solid waste are called soil lines. Soil lines therefore are the largest in the system and are flushed with water after each use.

Fresh water supply pipes are full of water (wet pipes) and under pressure all the time. The waste and soil pipes are wet pipes which have water in them only when waste water is flushed through them. The vent pipe system is composed of dry pipes which never contain water.

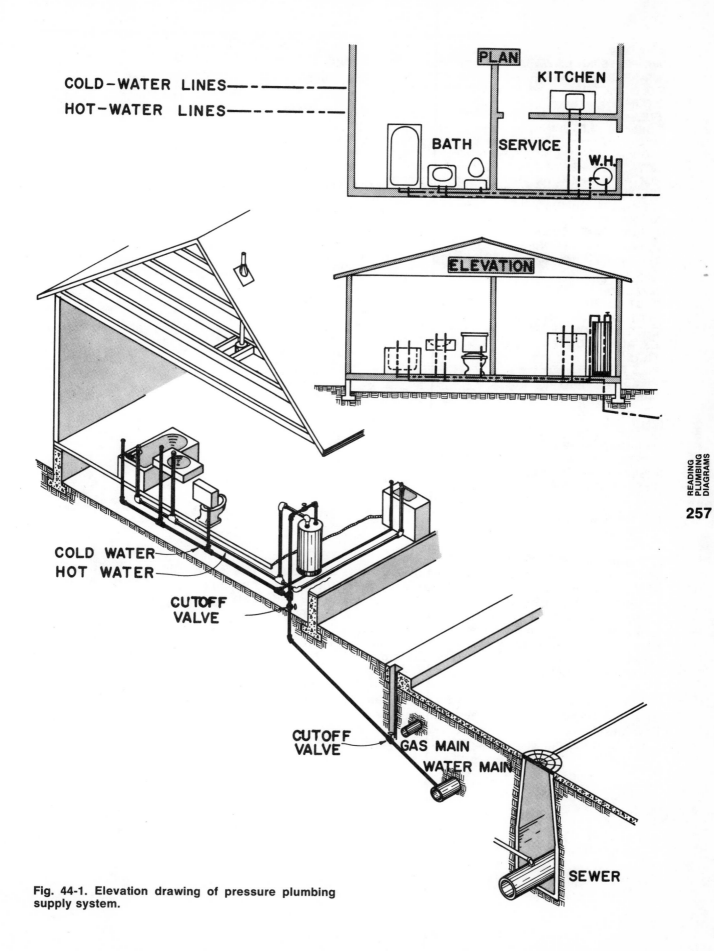

COLD-WATER LINES————————

HOT-WATER LINES————————

PLAN

KITCHEN

BATH SERVICE

W.H.

ELEVATION

COLD WATER

HOT WATER

CUTOFF
VALVE

CUTOFF
VALVE

GAS MAIN

WATER MAIN

SEWER

Fig. 44-1. Elevation drawing of pressure plumbing
supply system.

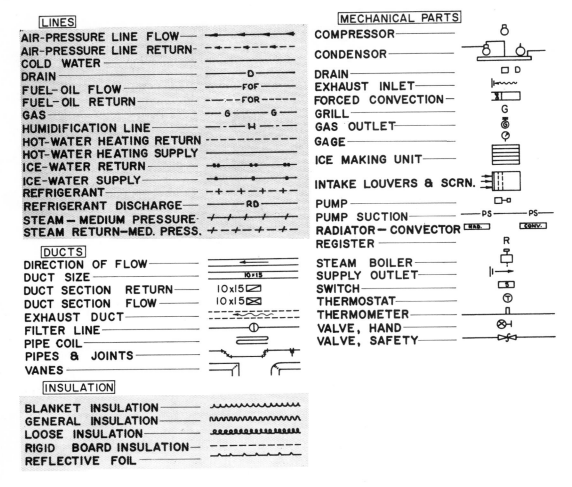

Fig. 44-2. Plumbing symbols.

READING CONSTRUCTION DRAWINGS

Outside Waste Systems

When municipal sewer disposal facilities are available, connection of waste pipes to the public system are shown on the plot plan. However, when public systems are not available, a septic tank system is used and either a separate drawing is provided or the location of the system is indicated on the plot plan. This location drawing is usually required by the building code and supervised by the local board of health.

A septic tank system converts waste solids into liquid by bacterial action. The building wastes flow into a septic tank buried some distance from the building. The lighter part of the liquid flows out of the septic tank into drainage fields through porous pipes spread over an area to allow wide distribution of liquids, as shown in Fig. 44-11.

The size and type of the septic system varies according to the number of occupants of a building, the contour of the terrain, and soil type. Figure 44-12 shows several types of drainage systems. The size of the lines and the distance of the septic tank and drainage fields from the building depends on the building codes of the community. Figure 44-13 shows the type of septic system drawing required by most boards of health.

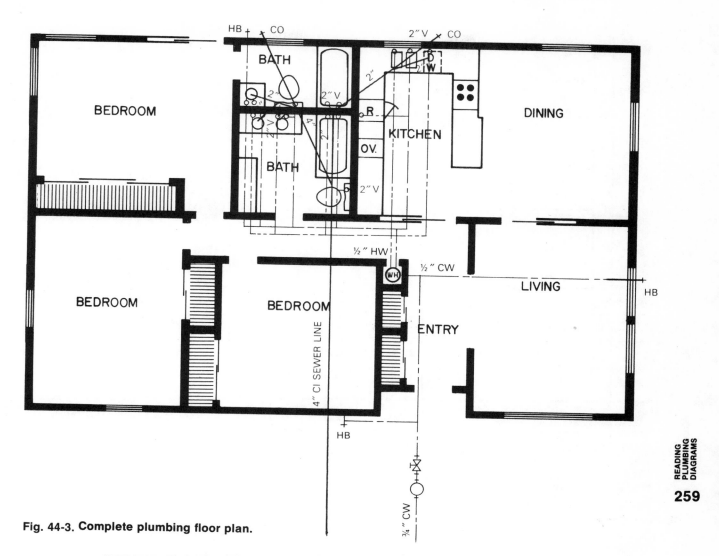

Fig. 44-3. Complete plumbing floor plan.

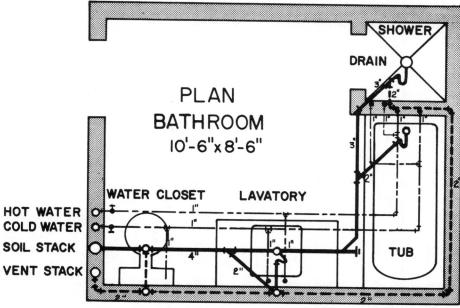

Fig. 44-4. Schematic bathroom plumbing plan.

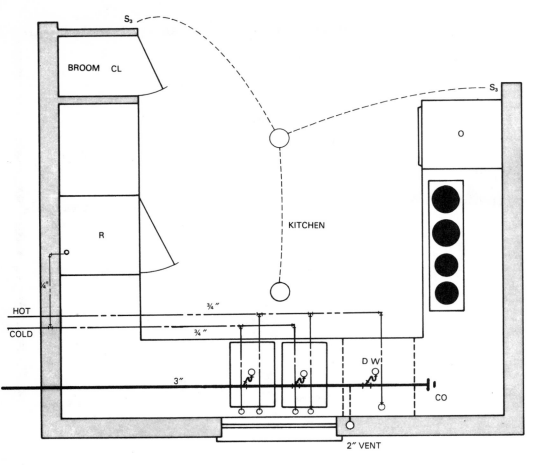

Fig. 44-5. Schematic kitchen plumbing plan.

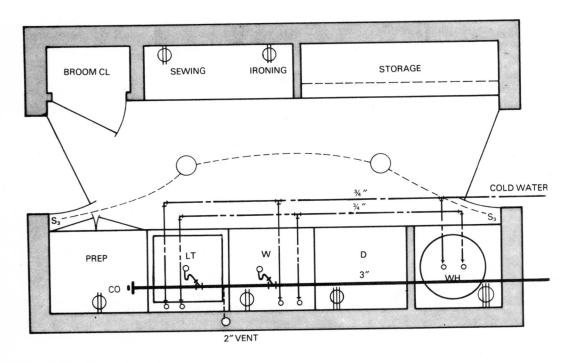

Fig. 44-6. Schematic utility room plumbing plan.

Fig. 44-7. Visualizing a plumbing plan.

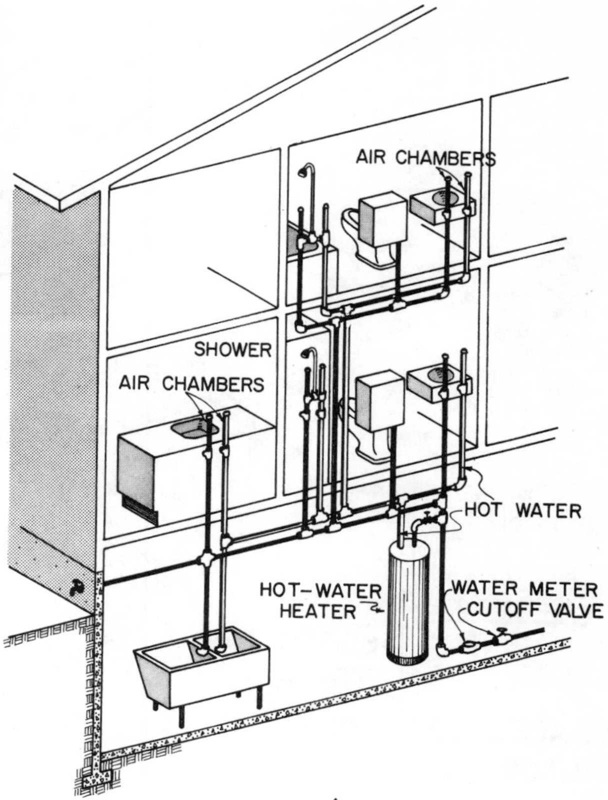

**Fig. 44-8.
Water distribution system.**

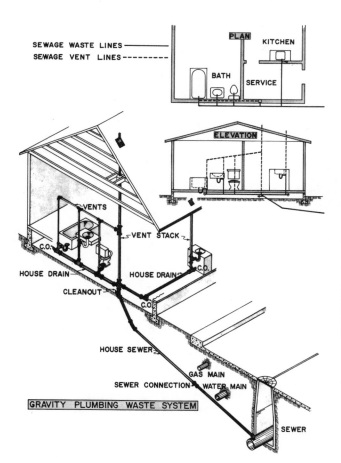

Fig. 44-9. Gravity waste discharge system.

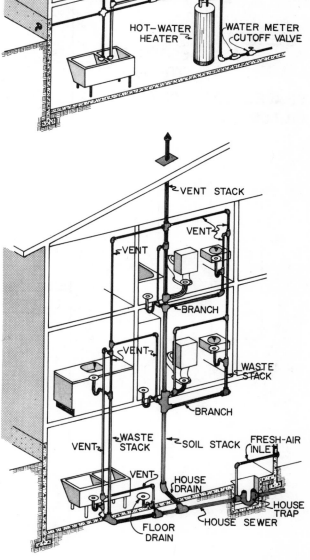

Fig. 44-10. Pictorial isolation of a gravity waste system.

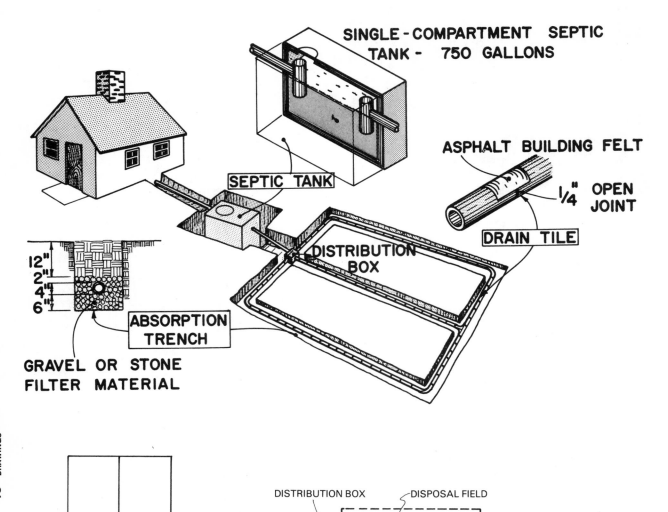

SINGLE - COMPARTMENT SEPTIC TANK - 750 GALLONS

ASPHALT BUILDING FELT

1/4" OPEN JOINT

SEPTIC TANK

DRAIN TILE

DISTRIBUTION BOX

12"
2"
4"
6"

ABSORPTION TRENCH

GRAVEL OR STONE FILTER MATERIAL

DISTRIBUTION BOX

DISPOSAL FIELD

EFFLUENT SEWER

HOUSE SEWER

CI SEWER

SEPTIC TANK

Fig. 44-11. Septic distribution system.

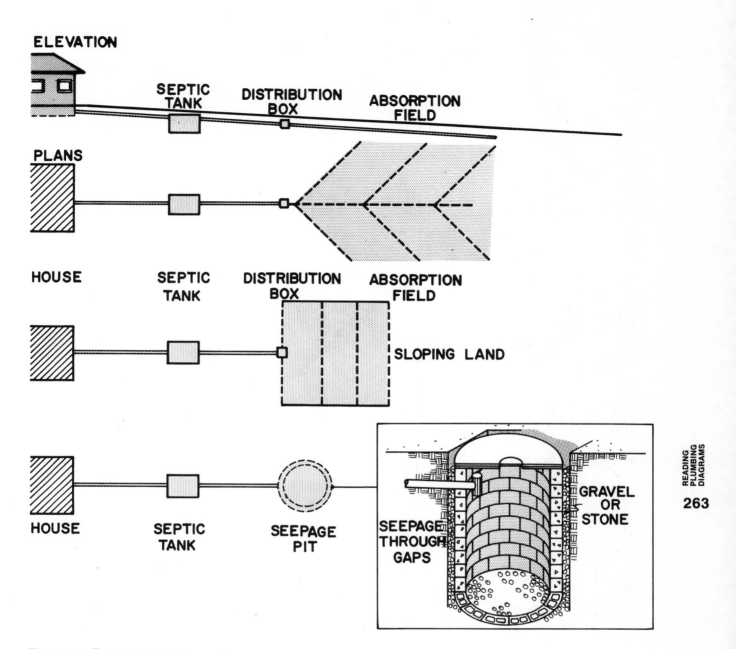

ELEVATION

SEPTIC TANK   DISTRIBUTION BOX   ABSORPTION FIELD

PLANS

HOUSE   SEPTIC TANK   DISTRIBUTION BOX   ABSORPTION FIELD

SLOPING LAND

HOUSE   SEPTIC TANK   SEEPAGE PIT

SEEPAGE THROUGH GAPS   GRAVEL OR STONE

Fig. 44-12. Types of drainage systems.

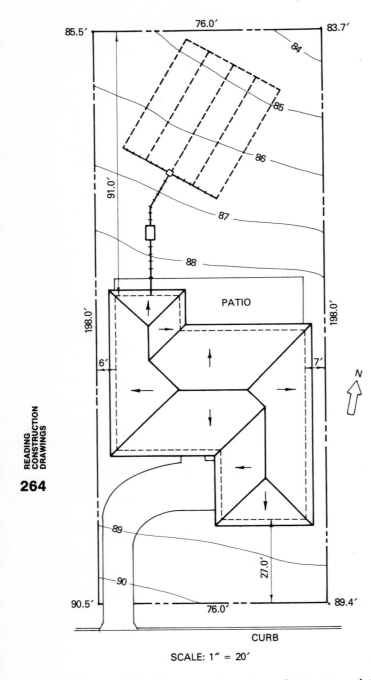

SCALE: 1″ = 20′

Fig. 44-13. Septic drainage system shown on a plot plan.

Unit 45
Schematic Plumbing Elevations

Since schematic plans show only the horizontal positioning of pipes, the amount of rise above floor level and the flow of fresh water and wastes between levels is difficult to read. Elevation drawings such as the water distribution system elevation, shown in Fig. 45-1, provides this vertical orientation. Elevation drawings are also used for some details such as the positioning of shutoff valves (Fig. 45-2).

Since most plumbing work is concentrated in the kitchen, the bathroom, and the utility room, separate schematic elevations are sometimes prepared for only these rooms, as shown in Fig. 45-3.

Unlike plumbing plans, which combine both the water supply and sewage disposal system on one drawing, plumbing elevations usually include only one system on each drawing. Figure 45-4 shows only the sewage disposal system in elevation form from the second floor to the public sewer drain.

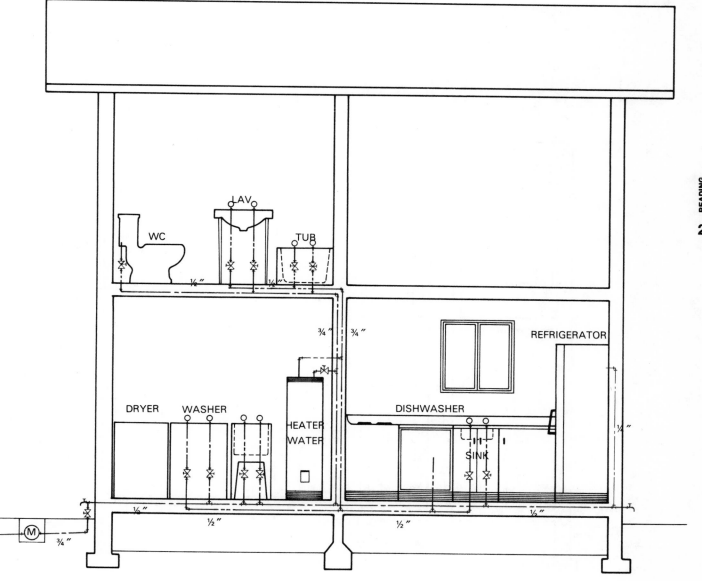

Fig. 45-1. Water distribution system elevation.

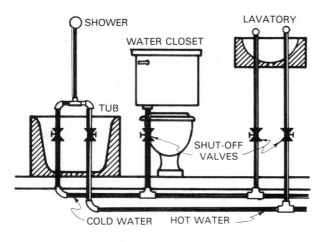

Fig. 45-2. Elevation drawing showing position of shut-off valves.

Fig. 45-3. Schematic elevation of bathroom plan.

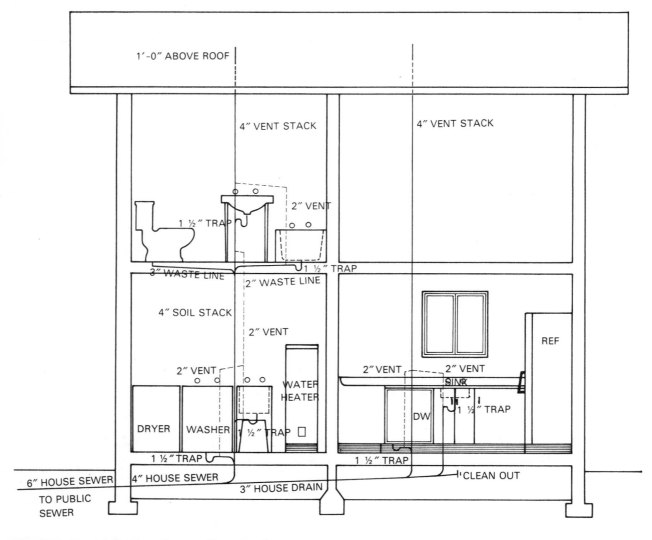

1'-0" ABOVE ROOF

4" VENT STACK

4" VENT STACK

1 ½" TRAP

2" VENT

1 ½" TRAP

3" WASTE LINE

2" WASTE LINE

4" SOIL STACK

2" VENT

REF

2" VENT

2" VENT

2" VENT

SINK

WATER
HEATER

1 ½" TRAP

DRYER   WASHER   1 ½" TRAP

DW

6" HOUSE SEWER

4" HOUSE SEWER

3" HOUSE DRAIN

1 ½" TRAP

CLEAN OUT

TO PUBLIC
SEWER

Fig. 45-4. Complete elevation section showing sewage disposal system.

CHAPTER 13

Reading Climate Control Plans

Climate control plans show systems of maintaining specific degrees of temperature, amounts of moisture (humidity), and the exchange of odorless air. This is achieved through heating systems, cooling systems, air filters, and humidifiers.

Unit 46
Heating System Plans
Many different types of systems are used to heat buildings. The effective use of insulation, ventilation, roof overhang, caulking, weather stripping, and solar orientation are also used to increase the efficiency of heating systems. Heat is transferred from a warm surface to a cold surface by three methods: radiation, convection, and conduction. Figure 46-1 illustrates these methods.

In radiation, heat flows to the cooler surface through space in the same way light travels through space. The warm air contacts objects, and the cooler object it strikes becomes warm. The object in turn warms the air that surrounds it. In convection, a warm surface heats the air around it. The warmed air rises and cool air moves in to take its place, causing a convection current. In conduction, heat moves through and heats a solid material; the denser the material, the better it will conduct heat.

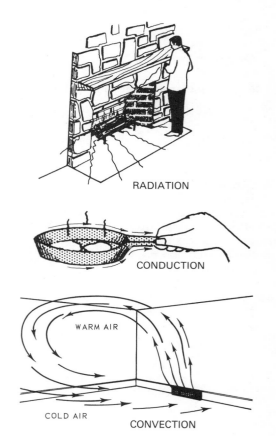

Fig. 46-1. Types of heat transfer.

Heating Delivery Systems
The two most popular and flexible types of heating systems for the delivery of heat from the source to the outlets are the perimeter heating and radiant heating systems, as shown in Fig. 46-2. In perimeter heating, the heat outlets are located on the outside walls of rooms usually under windows. The heat emitted from outlets rises and covers the coldest area in each room. Warm air rises, passes along the ceiling, and returns to an air-return duct. Baseboards, convectors, and radiators are used as devices to project this heat through the perimeter system. In this system the main loss of heat is through windows and outside walls.

Radiant heating functions by heating an area of a wall, ceiling, or floor. These warm surfaces in turn radiate heat to cooler objects. The heating surfaces may be lined with pipes containing hot water, hot air, or with electrical resistance wires.

Heating Devices
Devices that produce the heat used in various heating delivery systems include warm-air, hot-water (hydronic), steam, electrical, and solar devices.

Symbols for various heating devices are usually shown on the floor plan, or they may be shown on a separate floor plan devoted exclusively to heating and air conditioning or combined with plumbing and electrical plans. Figure

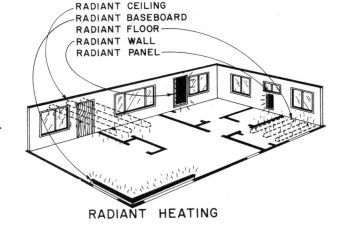

Fig. 46-2. Perimeter and radiant heating transmission.

46-3 shows climate control symbols used on floor plans.

WARM-AIR UNITS In a warm-air unit, the air is heated in a furnace (Fig. 46-4) and air ducts distribute this heated air to outlets throughout the building. The air supply is taken from the outside, from the furnace room, or from return air ducts in heated rooms. Warm-air units operate either by gravity or forced air. In forced air systems, the air is blown through the duct by use of a fan in the furnace. Gravity systems rely on allow-

ing the warm air to rise naturally to higher levels without the use of a fan. Therefore, the furnace in a gravity system must be located on a level lower than the area to be heated. Warm air, often called hot air, provides almost instant heat. Air filters and humidity control devices may be combined with the heating and cooling system. Warm-air system plans include the location of each heating outlet and the location of all duct work from the furnace to the outlet, as shown in Fig. 46-5.

Duct work for warm-air systems falls into several categories: individual duct systems, as shown in Fig. 46-5; extended plenum systems, as shown in Fig. 46-6; perimeter loop systems, such as the one shown in Fig. 46-7; and perimeter radial systems, as shown in Fig. 46-8.

When only an abbreviated heating and air-conditioning plan is available, only the location of warm-air outlets (Fig. 46-9) are shown on the

(T)	THERMOSTAT
▭	RADIATOR/CONVECTOR
⊠ ▭ ○	HEAT REGISTERS
〰	RADIANT SYSTEM
⊢———⊣	AIR DUCT
⊢——←——⊣	DUCT AIRFLOW (DIRECTION)
⊢ 10 × 18 ⊣	DUCT SIZE (INCHES)
⊏———⊐	DUCT SIZE CHANGE
— · — · —	COLD WATER
— — — —	HOT WATER
————————	STEAM/HOT-WATER SUPPLY
— — — —	STEAM/HOT-WATER RETURN
————————	HIGH-PRESSURE STEAM SUPPLY
— — — —	HIGH-PRESSURE STEAM RETURN

Fig. 46-3. Common climate control symbols.

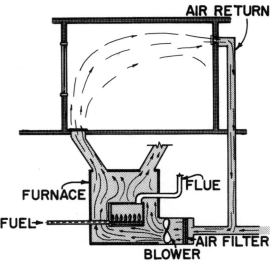

Fig. 46-4. Forced warm-air system.

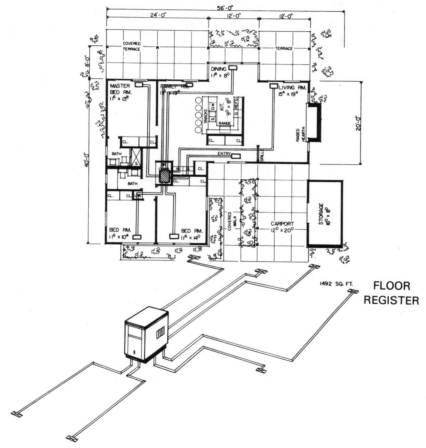

FLOOR
REGISTER

Fig. 46-5. Individual duct system plan. (Home Planners, Inc.)

plan. Figure 46-10 shows an abbreviated heating and air-conditioning plan showing only the position of outlets. When this type of plan is provided, the builder must determine the type, size, and location of all duct work connecting the furnace with the room outlets.

HOT-WATER UNITS A hot-water (hydronic) unit (Fig. 46-11) uses a boiler to heat water and a water pump to send the heated water to radia-

tors, thin tubes, convectors, or baseboard units located throughout the building. Forcing the water through the system with a pump is, of course, faster than allowing gravity to provide the flow. Hot-water heating units provide even heat and keep heat in the outlets longer than warm-air units. The hot-water boiler is smaller than warm-air furnaces, and hot-water pipes are smaller and easier to install than warm-air ducts. However, hot-water units are incompatible with

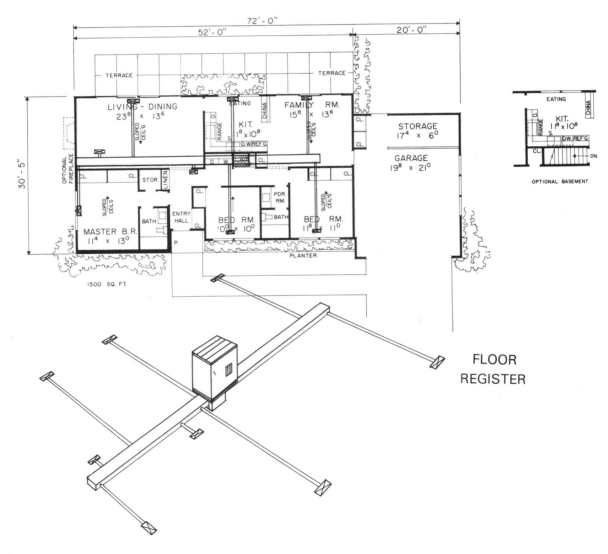

Fig. 46-6. Extended plenum trunk system plan. (Home Planners, Inc.)

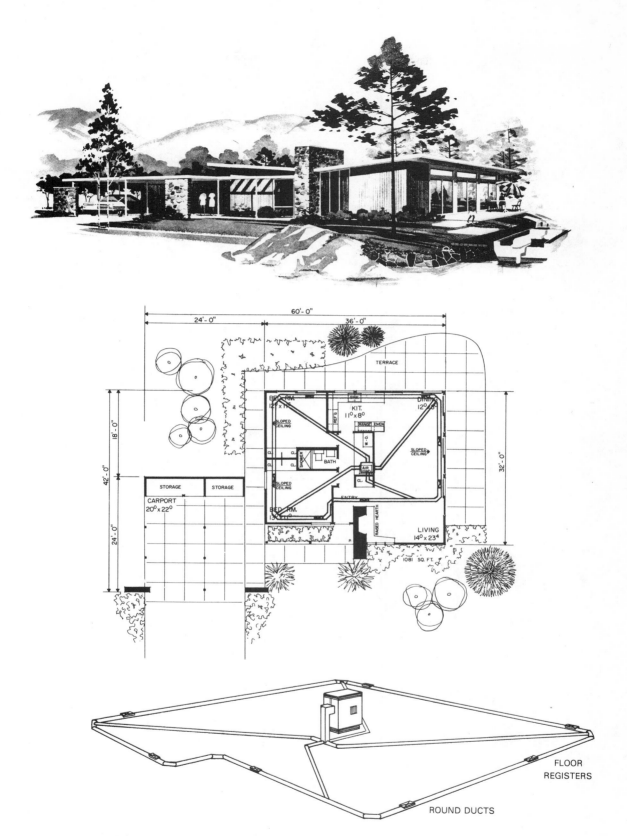

FLOOR
REGISTERS

ROUND DUCTS

Fig. 46-7. Perimeter loop system. (Home Planners, Inc.)

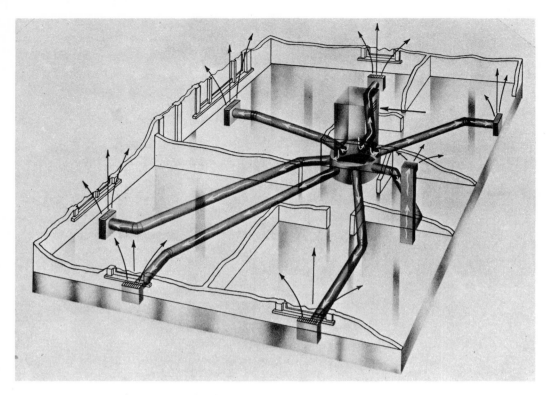

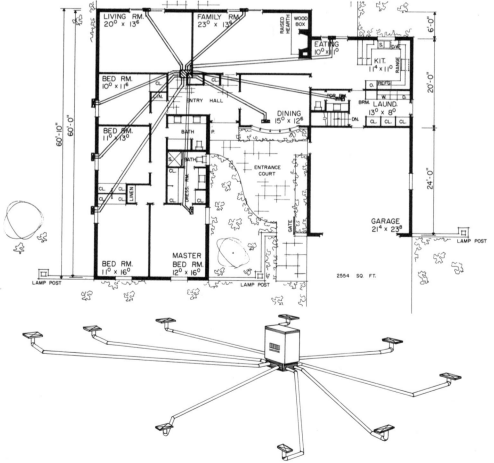

Fig. 46-8. Perimeter radial system. (Home Planners, Inc.)

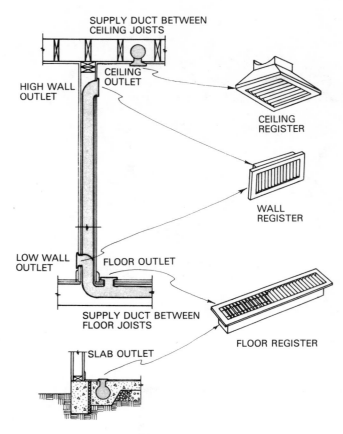

Fig. 46-9. Warm-air system outlets.

air-conditioning systems which require the installation of air ducts. There are several types of hot-water systems used to supply heat from the boiler to heating units: the series loop system, the one-pipe system, the two-pipe system, and the radial system.

The *series loop system*, as shown in Fig. 46-12, is a continual loop of pipes containing hot water which passes through baseboard units. Hot water flows continually from the boiler through the baseboard units and back again to the boiler for reheating. The heat in a series loop system cannot be controlled except at the source of the loop. Thus it is only effective for small areas where radiators or convectors produce heat at the same temperature throughout the system. The only way to vary temperature in this type of system is to increase the number of loops in a building. The temperature for each loop can then be varied at each individual loop boiler.

In *one-pipe systems*, heated water is circulated through pipes which are connected to radiators or convectors by means of bypass pipes. This allows each radiator to be individually controlled by valves. Water flows from one side of each radiator to the main line and returns to the boiler for reheating. Figure 46-13 includes a floor plan with a one-pipe hot-water system. The relation-

ship to the complete system is shown in the pictorial insert.

In a *two-pipe system*, as shown in Fig. 46-14, there are two parallel pipes; one for the supply of hot water from the boiler to each radiator, and the other for the return of cooled water from each radiator back to the boiler. The heated water is directed from the boiler to each radiator but returns from each radiator through the second pipe to the boiler for reheating. In this system, all radiators receive water at the same temperature.

Hot-water heating plans are sometimes shown without piping details. If only the position of outlets is shown, as in Fig. 46-15, then the plumbing contractor must determine the exact position of piping and also ensure that the radiators or convectors are located as determined on the piping plan.

The *radiant hot-water heating system* distributes hot water through a series of pipes in floors or ceilings. The warm surfaces of the floor or ceiling radiate heat to cooler objects. Ceilings are often used for radiant hot-water heating since furniture and rugs restrict the distribution of heat in floors and walls. Figure 46-16 shows a radiant heating system together with a floor plan arrangement for radiant steam heating systems.

STEAM HEATING SYSTEMS   Although *steam heating systems* function on water vapor rather than hot water, drawings for steam systems are identical to those prepared for hot-water systems. Steam systems are easy to install and maintain, but they are not suitable for use with most conventional convector-type radiators. They are most popular for large apartments, commercial buildings, and industrial complexes where separate steam generation facilities are provided. Steam heat is delivered through either perimeter or radial systems.

ELECTRIC HEAT Electric heat is produced when electricity passes through resistance wires. This heat is usually radiated or fan-blown. Resistance wires are placed either in panel heaters (radiant heat) built into walls or ceilings, placed in baseboards, or set in plaster to heat walls, ceilings, or floors. Units for distribution of electric heat, which is very clean, use very little space, require no air for combustion, and require no fuel storage facilities or duct work. Electric heat tends to be very dry, consequently complete ventilation and humidity control must accompany electric heat devices, since no air circulation is generalized. Figure 46-17 shows the installation of resistance wires in floors and ceilings. No plans are drawn specifically for electric heat but notations are made on the floor plans concerning the location of either resistance wires or electric panels. On electrical plans the location of facilities for power supply and thermostating is shown.

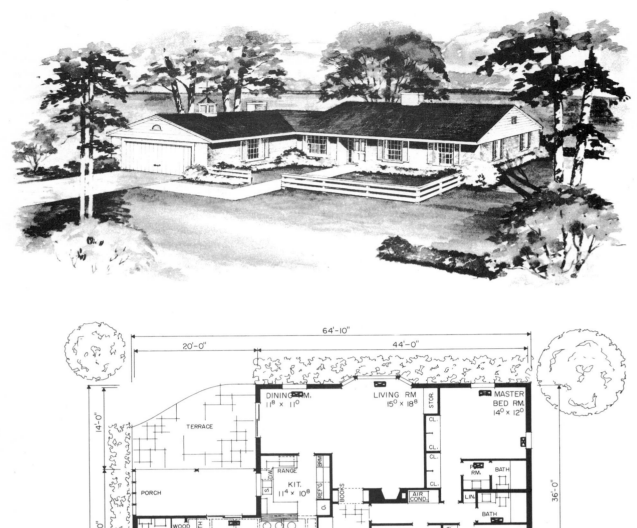

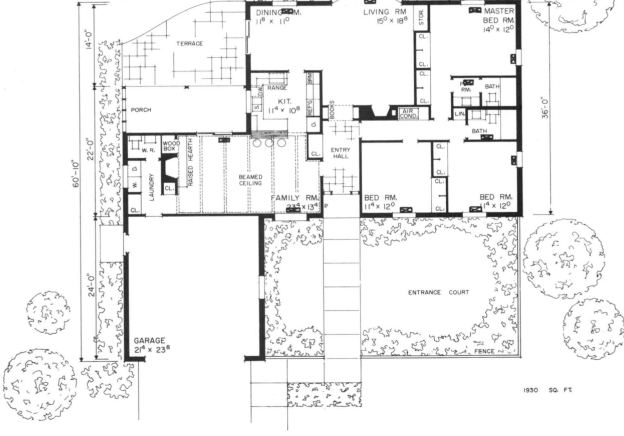

Fig. 46-10. Abbreviated heating and air-conditioning plan showing only the position of outlets. (Home Planners, Inc.)

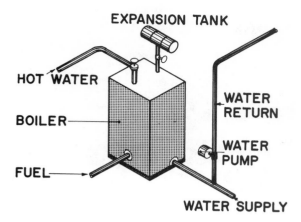

Fig. 46-11. Hot-water unit.

Solar heating is, of course, used without equipment by utilizing the relationship between the roof overhang and large glass areas on the south side of a building. Overhangs are designed so that the summer sun, which is more directly overhead, is blocked off and the winter sun, which is lower on the southern horizon, is allowed to contact and heat large window areas. Summer control on the east and west sides of a building is also provided through the use of protective window coverings, trees, and shrubbery. In summer the trees provide shade, in winter they lose their leaves and allow the sun to enter the structure. Figure 46-19 shows a solar heat installation plan currently used to heat a structure.

Another electrical device used in supplying heat (and cooling) for structures is the heat pump, shown in Fig. 46-18. The heat pump is a year-round air conditioner. In winter it takes heat from the outside and pumps it into the building. There is always some heat in the air regardless of the temperature. In the summer the pump is reversed and the heat in the structure is pumped outside as the pump works like a reversible refrigerator. Drawings for heat pump duct work layouts are identical to those used for forced warm-air systems.

SOLAR HEAT Solar heating uses the sun to help heat structures. There are several methods of harnessing solar energy to heat and cool buildings. One procedure is through the use of coated aluminum plates under two transparent covers that absorb thermal energy and transfer it to a fluid stored in an insulated tank. The heated fluid is then pumped throughout the structure for heating or cooling a hot-water delivery system.

Another solar procedure uses the sun's heat for heating, cooling, and for converting sunlight into electricity to run home appliances. This system uses large panels (collectors) which consist of a number of solar cells made of sandwiches of cadmium sulfide and copper sulfide placed between thin layers of glass. These solar cells produce electric current upon exposure to sunlight. Part of the electric current produced in this matter is fed immediately into the structures' electrical system to operate lights and appliances. The remainder is used to charge a series of batteries which provide energy when the sun is not shining on the panels.

A third method is the use of a solar furnace. The solar furnace is a collection of mirrors that focuses the sun's heat on a concentrated area. Temperatures as high as 3500° (1926°C) are obtained by this means.

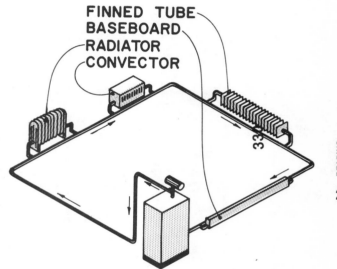

FINNED TUBE
BASEBOARD
RADIATOR
CONVECTOR

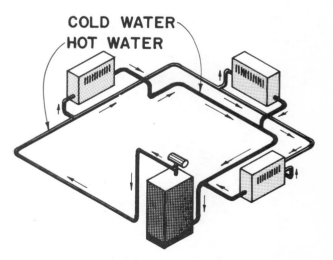

COLD WATER
HOT WATER

Fig. 46-12. Series loop of hot-water system.

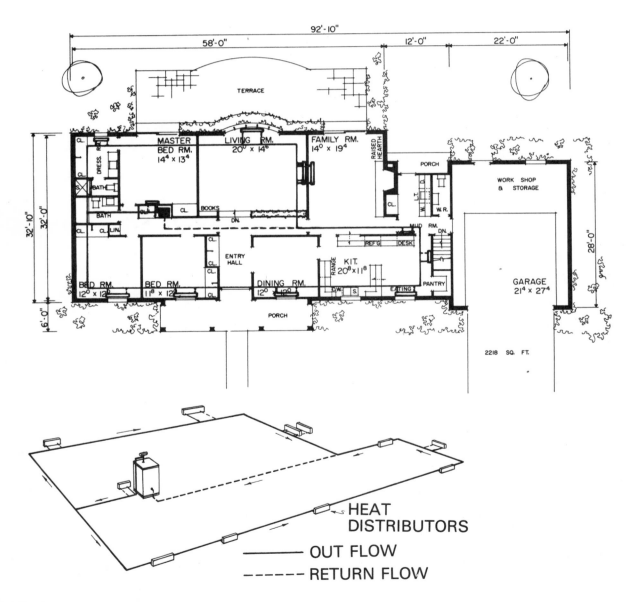

92'-10"

58'-0" **12'-0"** **22'-0"**

TERRACE

MASTER
BED RM.
14⁴ x 13⁴

LIVING RM.
20⁰ x 14⁶

FAMILY RM.
14⁰ x 19⁴

RAISED HEARTH

PORCH

WORK SHOP
& STORAGE

DRESS RM.

CL.

CL.

BATH

S

BATH

CL.

BOOKS

DN.

CL.

W. T. D.

W. R.

32'-10"

32'-0"

CL.

CL. LIN

MUD RM.

DN.

REFG

DESK

28'-0"

BED RM.
12⁰ x 12⁰

BED RM.
11⁸ x 12⁸

CL.

CL.

CL.

ENTRY
HALL

CL.

DINING RM.
12⁰ x 12⁰

RANGE

KIT.
20⁸ x 11⁸

EATING

PANTRY

GARAGE
21⁴ x 27⁴

6'-0"

DW.

S

PORCH

2218 SQ. FT.

HEAT
DISTRIBUTORS

———— OUT FLOW

-------- RETURN FLOW

**Fig. 46-13. One-pipe hot-water heating system.
(Home Planners, Inc.)**

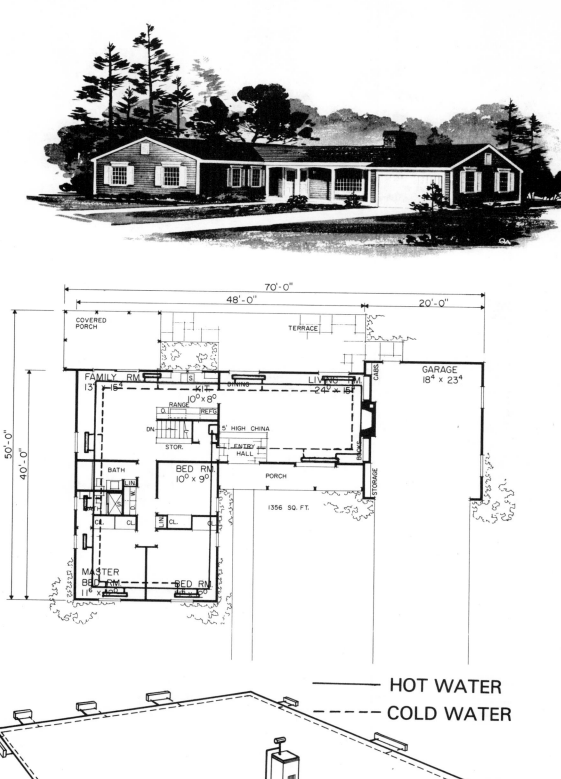

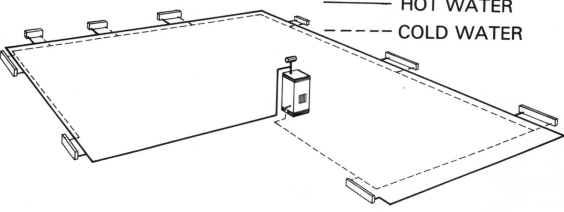

Fig. 46-14. Two-pipe hot-water system. (Home Planners, Inc.)

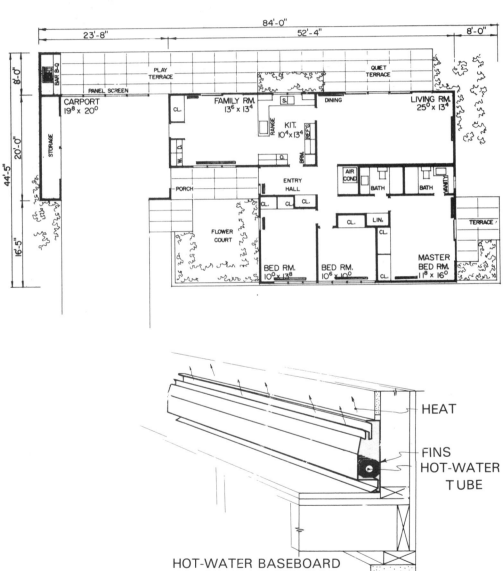

Fig. 46-15. Abbreviated plan with only hot-water outlets shown. (Home Planners, Inc.)

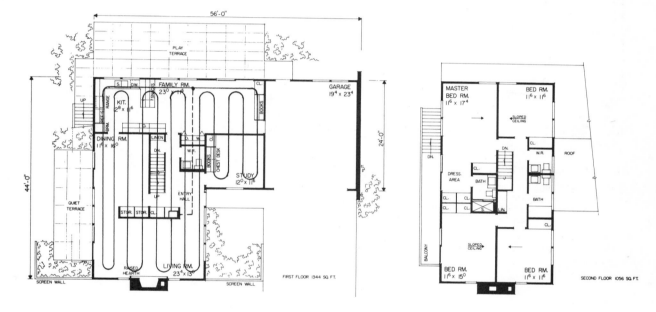

FIRST FLOOR 1344 SQ. FT.

SECOND FLOOR 1056 SQ. FT.

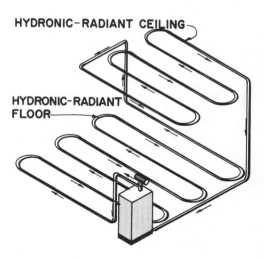

HYDRONIC-RADIANT CEILING

HYDRONIC-RADIANT FLOOR

Fig. 46-16. Radiant hot-water system plan between floors. (Home Planners, Inc.)

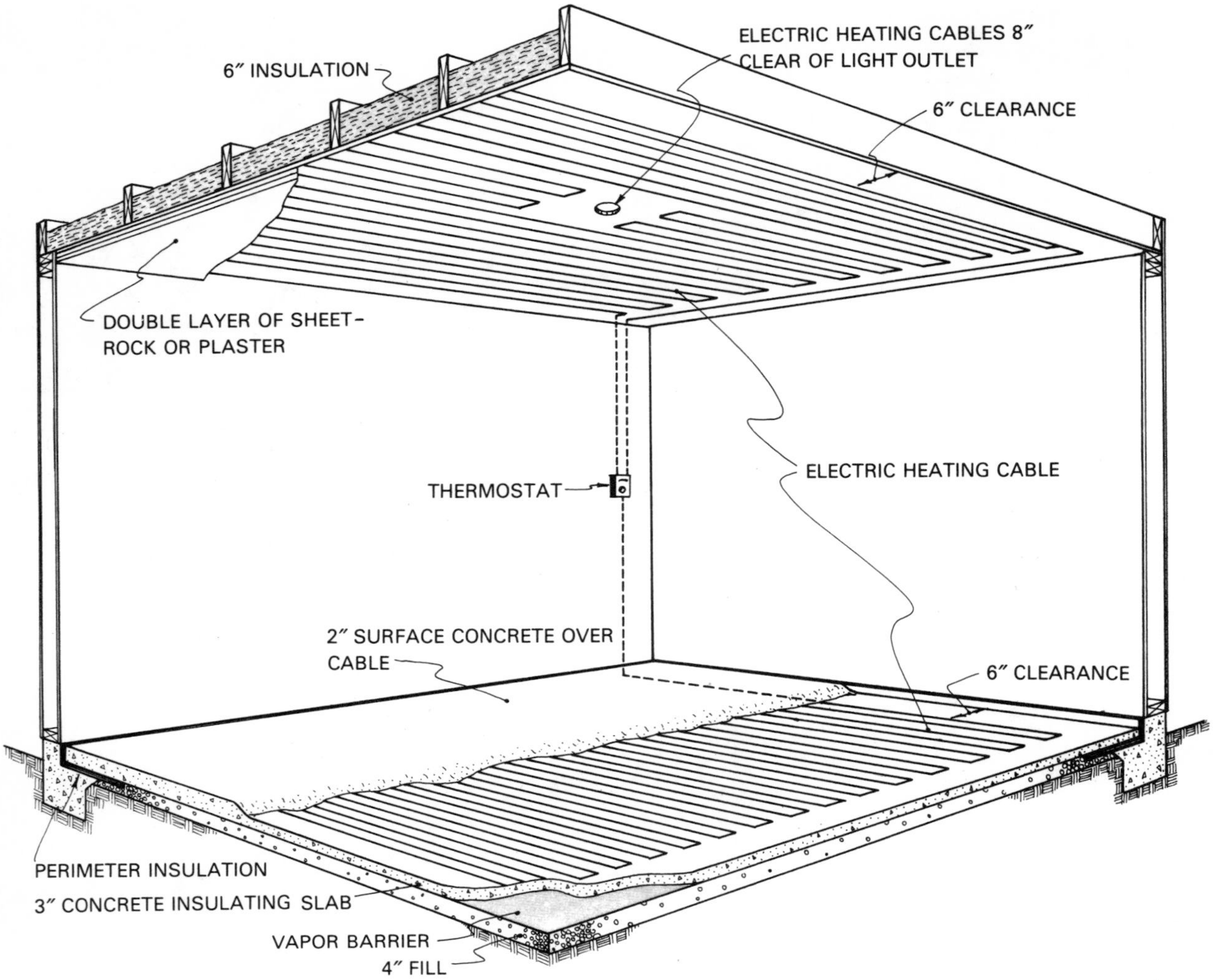

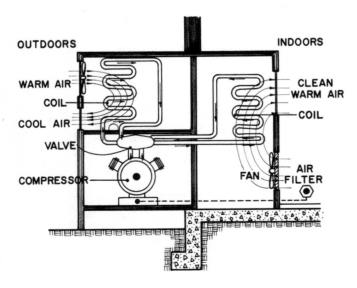

6" INSULATION

ELECTRIC HEATING CABLES 8"
CLEAR OF LIGHT OUTLET

6" CLEARANCE

DOUBLE LAYER OF SHEET-
ROCK OR PLASTER

ELECTRIC HEATING CABLE

THERMOSTAT

2" SURFACE CONCRETE OVER
CABLE

6" CLEARANCE

PERIMETER INSULATION

3" CONCRETE INSULATING SLAB

VAPOR BARRIER

4" FILL

Fig. 46-17. Electric radiant heating cables in floors and ceilings.

OUTDOORS

INDOORS

WARM AIR

COIL

COOL AIR

VALVE

COMPRESSOR

CLEAN
WARM AIR

COIL

FAN

AIR
FILTER

Fig. 46-18. Operation of heat pump.

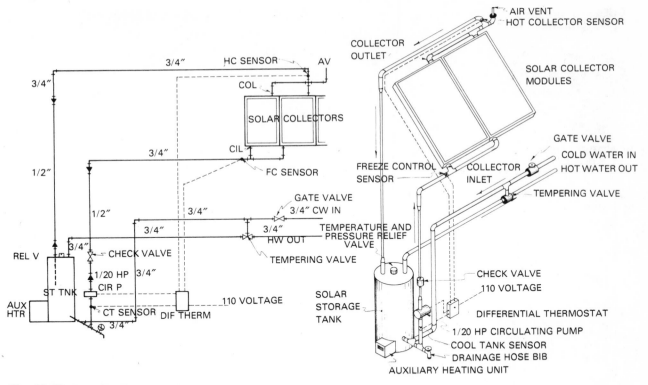

Fig. 46-19. Solar heating system plan.

Unit 47
Cooling System Plans

A building is cooled by removing the heat. Heat can be transferred in one direction only: from a warmer object to a cooler object. Therefore, to cool a building a central air-conditioning system absorbs the heat from the building and transfers it to a liquid refrigerant (usually freon). Warm air is carried away from rooms through ducts to the air-conditioning unit. A filter removes dust and other impurities from the air. A cooling coil containing the refrigerant then absorbs heat from air passing through it, and a blower which pulls the heat-laden air from rooms pushes heat-free air (cool air) back to the rooms. There are four main methods of cooling structures, as shown in Fig. 47-1: the waste-water method, the cooling tower method, the evaporation method, and the air-cool method.

Regardless of the type, the cooling unit is shown exactly the same on floor plans. Classification of units are found in the building specifications. The cooling unit is either part of the heating unit using the same blower and vents, or the heating and the cooling systems can be separate as shown in Figs. 47-2 and 47-3.

When cooling systems are combined with heating systems, a combined heating and cooling systems plan is usually prepared as shown in Figure 47-4. This plan is read exactly the same as a warm-air heating duct plan except for the addition of the cooling unit.

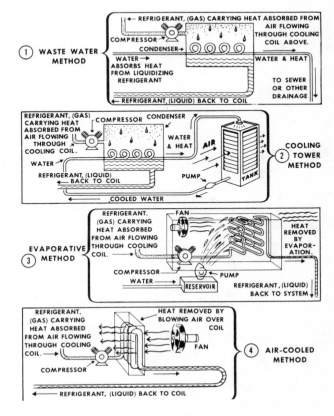

Fig. 47-1. Four basic methods of cooling structures.

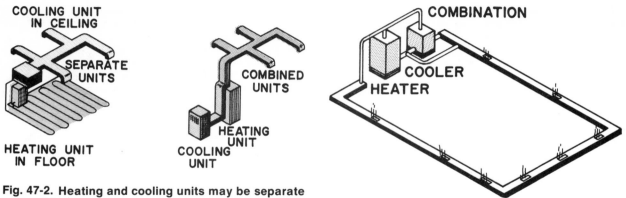

Fig. 47-2. Heating and cooling units may be separate or combined.

Fig. 47-3. Cooling and heating unit combined.

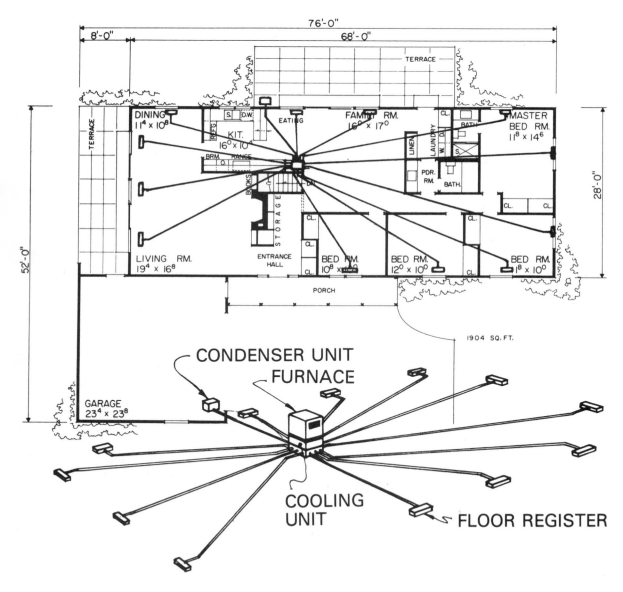

Fig. 47-4. Heating and cooling system using radial distribution system. (Home Planners, Inc.)

CHAPTER 14

Reading Sets of Plans

Previous chapters have covered the principles and practices involved in reading each type of construction drawing. In this unit the basic guidelines for reading and interpreting a complete set of plans are developed with special emphasis on the relationship and consistency among plan features and dimensions.

Unit 48
Relationship of Plans

Drawings used for construction vary from simple floor plans to comprehensive sets of plans complete with details, schedules, and specifications. The number of plans needed to construct a building depends on the complexity of the structure and on the degree to which the designer needs or wants to control the various methods and details of construction. For example, if only a floor plan is prepared, the builder must create the elevation designs. If only the floor plan and elevation drawings are prepared, but no details or specifications are provided, the builder assumes responsibility for many construction details, including selection of framing type, materials used, and many aspects of the interior design. Therefore, the more plans, details, and specifications developed for a structure, the closer the finished building will be to that conceived by the designer.

Set Size

Figure 48-1 shows the types of construction drawings included in a maximum, average, and minimum set of plans. The preparation of only floor plans, elevations, surveys, and a section, as indicated in the minimum set of plans, provides the builder with great latitude in the selection of materials and processes. The maximum set of plans, as suggested in Fig. 48-1, will assure, to the greatest degree possible, agreement between the wishes of the designer and the final constructed building.

Interrelationship of Plans

In previous units, samples of architectural (construction) drawings were shown to illustrate

DRAWINGS	SIZE OF SET OF PLANS		
	MIN.	AVER.	MAX.
FLOOR PLANS	x	x	x
FRONT ELEVATION	x	x	x
REAR ELEVATION		x	x
RIGHT ELEVATION	x	x	x
LEFT ELEVATION		x	x
AUXILIARY ELEVATIONS			x
INTERIOR ELEVATIONS		x	x
EXTERIOR PICTORIAL RENDERINGS		x	x
INTERIOR RENDERINGS			x
PLOT PLAN		x	x
LANDSCAPE PLAN			x
SURVEY PLAN	x	x	x
FULL SECTION	x	x	x
DETAIL SECTIONS		x	x
FLOOR-FRAMING PLANS			x
EXTERIOR-WALL FRAMING PLANS			x
INTERIOR-WALL FRAMING PLANS			x
STUD LAYOUTS			x
ROOF-FRAMING PLAN			x
ELECTRICAL PLAN		x	x
AIR CONDITIONING PLAN			x
PLUMBING DIAGRAM			x
SCHEDULES			x
SPECIFICATIONS			x
COST ANALYSIS			x
SCALE MODEL			x

Fig. 48-1. Drawings necessary for sets of plans.

various principles and practices related to the reading and interpretation of each specific type of drawing. However, these drawings did not necessarily relate to the same structure. Thus, there was no interrelationship between an electrical plan, plumbing plan, and a survey plan. Drawings in this unit, however, are of the same building; therefore, the interpretation and agreement among plans can be studied more easily.

Reading Sequence

It is important in reading and studying a complete set of plans to study them in the correct

1. FLOOR PLANS
2. FOUNDATION PLANS
3. FRONT ELEVATION
4. REAR ELEVATION
5. RIGHT ELEVATION
6. LEFT ELEVATION
7. AUXILIARY ELEVATION
8. EXTERIOR PICTORIAL RENDERING
9. INTERIOR PICTORIAL RENDERING
10. LANDSCAPE ELEVATION
11. PLOT PLANS
12. LANDSCAPE PLANS
13. SURVEY PLANS
14. FULL SECTIONS
15. DETAIL SECTIONS
16. FLOOR-FRAMING PLANS
17. EXTERIOR-WALL FRAMING PLANS
18. INTERIOR-WALL FRAMING PLANS
19. STUD LAYOUTS
20. ROOF-FRAMING PLANS
21. FIREPLACE DETAILS
22. FOUNDATION DETAILS
23. ELECTRICAL PLANS
24. AIR-CONDITIONING PLANS
25. PLUMBING DIAGRAMS
26. DOOR AND WINDOW SCHEDULES
27. FINISH SCHEDULES
28. SPECIFICATIONS
29. BUILDING COST AND ESTIMATES

Fig. 48-2. Sequence of studying sets of construction drawings.

sequence. Since the floor plan reveals more of the arrangements of the design than any other plan, it should be studied first. The floor plan should be studied in the sequence shown in Chap. 5. Be particularly careful to locate the position of bearing partitions, plumbing walls, stairwells, fireplaces, chimney openings, and other components that align vertically. For this reason, study each floor plan in succession before proceeding on to other plans. Then study the foundation plan, which is a floor plan of the basement or support system. Next study the elevations, which provide a good impression of the total appearance of the design.

After this, expand your studies to location plans, including landscape, surveys, and plot plans. This helps orient the structure to its surroundings. After this is done, study construction details and sections; then specialized plans, such as the electrical plans, climate control plans, and plumbing diagrams. Schedules, specifications, and cost estimates should be studied last since they are better understood after a firm grasp of the construction requirements is acquired. Figure 48-2 shows the normal sequence in studying a set of architectural plans.

Selected areas of each plan in this unit have been marked with geometric figures such as circles, rectangles, hexagons, crosses, squares, triangles, and diamonds. Each separate symbol identifies an identical area as shown on each plan. By studying the position of these symbols on each drawing, you can observe how a specific area appears on each drawing in the set. For example, the position of the circle on the main-level floor plan represents the area covered with the same circle on the right elevation, lower-level plan, rear view, pictorial drawing, or any other plan so marked with a circle. Likewise, the part of each drawing covered by a hexagon represents that same area on each drawing where a hexagon appears. The relationship between sectional views and basic drawings can be followed by tracing these geometric symbols and by locating the position of the appropriate cutting-plane line on the basic plans. Figures 48-3 through 48-10 show a related set of residential plans marked with geometric figures. Study the plans in the sequence listed above and trace the location of each of the geometric figures throughout the set of plans. Be sure you understand the positioning and relationship of each area marked with a geometric symbol.

Commercial buildings normally have more floors, cover wider areas, use heavier construction members, and may be more interrelated with other buildings. Nevertheless, the sequence of studying and understanding the relationship between drawings is identical.

Fig. 48-3. Floor plan. (Karren and Seals Architects, Inc.)

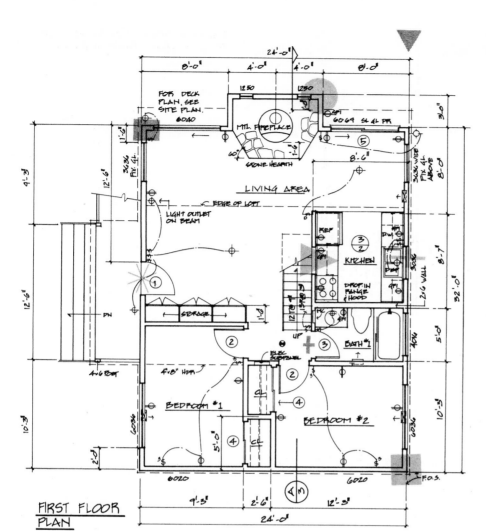

FIRST FLOOR
PLAN

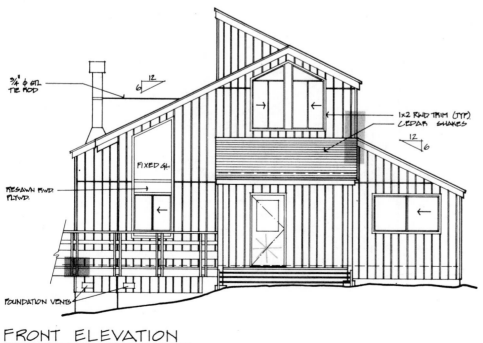

FRONT ELEVATION

Fig. 48-5. Second-floor plan. (Karren and Seals, Architects, Inc.)

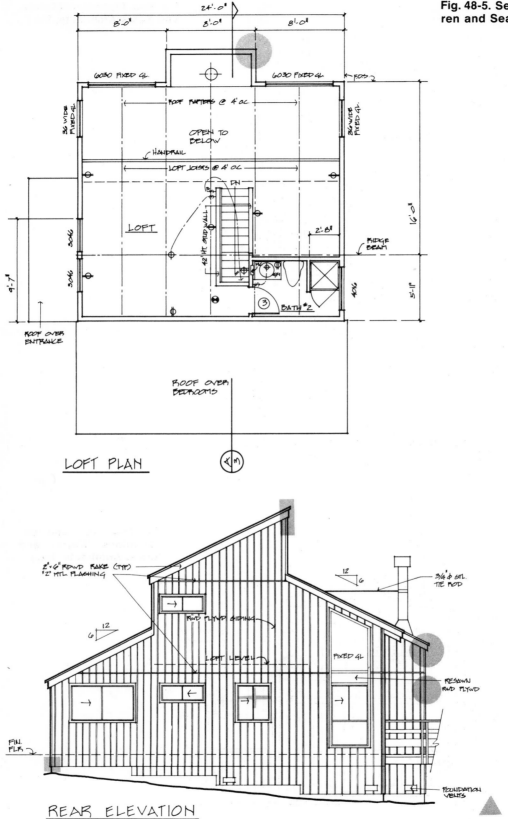

24'-0"

8'-0" 8'-0" 8'-0"

6030 FIXED GL. 6030 FIXED GL. FOS.2

36 WIDE FIXED GL.

ROOF RAFTERS @ 4' OC.

OPEN TO BELOW

HANDRAIL

36 WIDE FIXED GL.

LOFT JOISTS @ 4' OC.

DN

3060

3060

LOFT

42" HT. STUD WALL

16'-0"

2'-8"

RIDGE BEAM

4060

5'-11"

9'-7"

3060

3

BATH #2

ROOF OVER ENTRANCE

LOFT PLAN

A 3

ROOF OVER BEDROOMS

2'×6" RDWD. RAKE (TYP)
"Z" MTL. FLASHING

12
6

3/4 Ø STL. TIE ROD

RDWD. PLYWD. SIDING

RESAWN RDWD. PLYWD.

LOFT LEVEL

FIXED GL.

6
12

FIN. FLR.

FOUNDATION VENTS

REAR ELEVATION

Fig. 48-6. Right and left elevations. (Karren and Seals, Architects, Inc.)

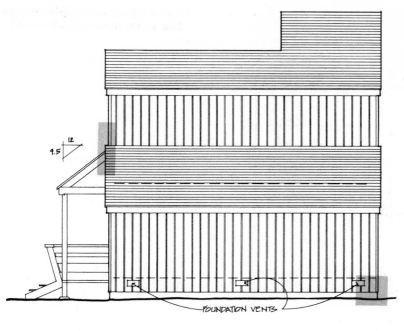

SIDE ELEVATION

Fig. 48-7. Basement plan. (Karren and Seals, Architects, Inc.)

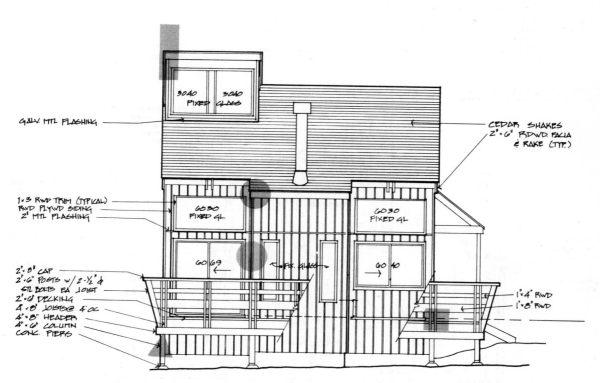

GALV. MTL FLASHING

3040
FIXED GLASS

3040
FIXED GLASS

CEDAR SHAKES
2"×6" RDWD. FACIA
& RAKE (TYP.)

1×3 RWD TRIM (TYPICAL)
RWD PLYWD SIDING
2' MTL FLASHING

6030
FIXED GL.

6030
FIXED GL

2"×8" CAP
2"×6" POSTS w/2-½" ø
STL. BOLTS EA. JOIST
2"×8' DECKING
4"×8' JOISTS @ 4' OC
4"×8" HEADER
4"×8' COLUMN
CONC. PIERS

6069

FIX. GLASS

6040

1"×4" RWD

1"×8" RWD

DECK ELEVATION
SCALE: ¼" = 1'-0"

Fig. 48-8. Full section. (Karren and Seals, Architects, Inc.)

Fig. 48-9. Details. (Karren and
Seals, Architects, Inc.)

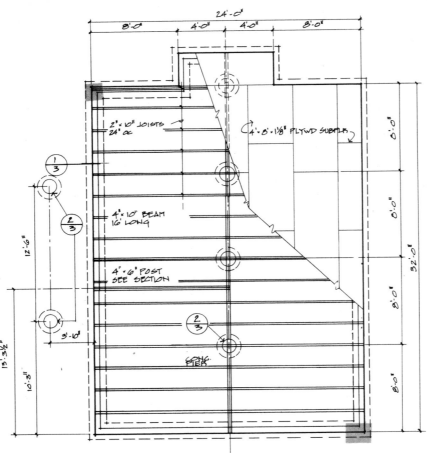

FOUNDATION & FLOOR FRAMING PLAN

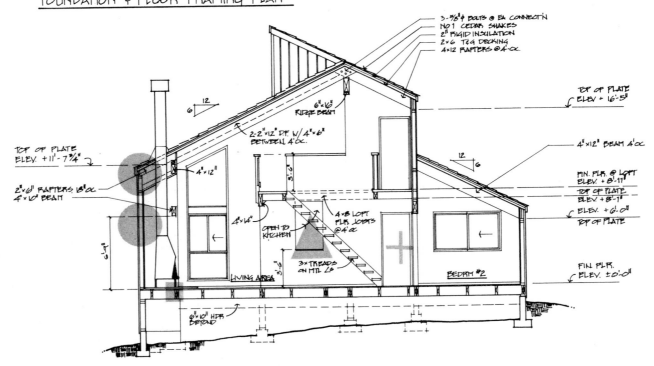

Fig. 48-10. Pictorial rendering. (Karren and Seals Architects, Inc.)

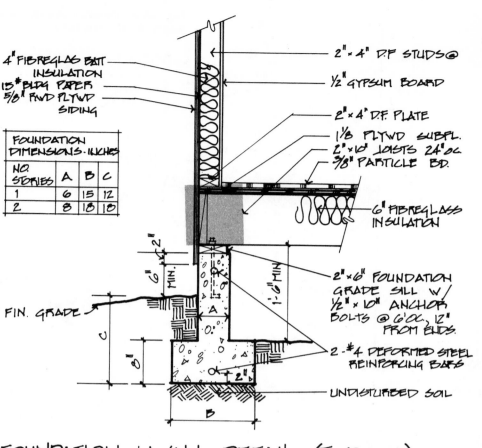

4" FIBREGLAS BATT INSULATION
15# BLDG PAPER
5/8" RWD PLYWD SIDING

2" × 4" D.F. STUDS @

1/2" GYPSUM BOARD

2" × 4" D.F. PLATE

1 1/8" PLYWD SUBFL.

2" × 10" JOISTS 24" OC

3/8" PARTICLE BD.

6" FIBREGLASS INSULATION

2" × 6" FOUNDATION GRADE SILL W/ 1/2" × 10" ANCHOR BOLTS @ 6' OC, 12" FROM ENDS.

2 - #4 DEFORMED STEEL REINFORCING BARS

UNDISTURBED SOIL

FIN. GRADE

FOUNDATION DIMENSIONS · INCHES			
NO. STORIES	A	B	C
1	6	15	12
2	8	18	18

FOUNDATION & WALL DETAIL (TYPICAL)

1" = 1'-0"

1/3

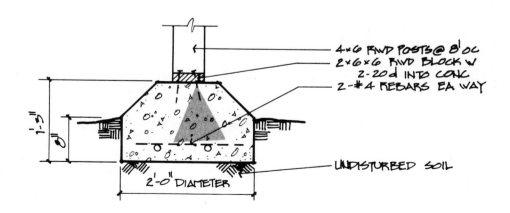

4×6 RWD POSTS @ 8' OC

2×6×6 RWD BLOCK W 2-20d INTO CONC

2 - #4 REBARS EA WAY

UNDISTURBED SOIL

1'-5"
8"

2'-0" DIAMETER

ISOLATED PIER

1" = 1'-0"

2/3

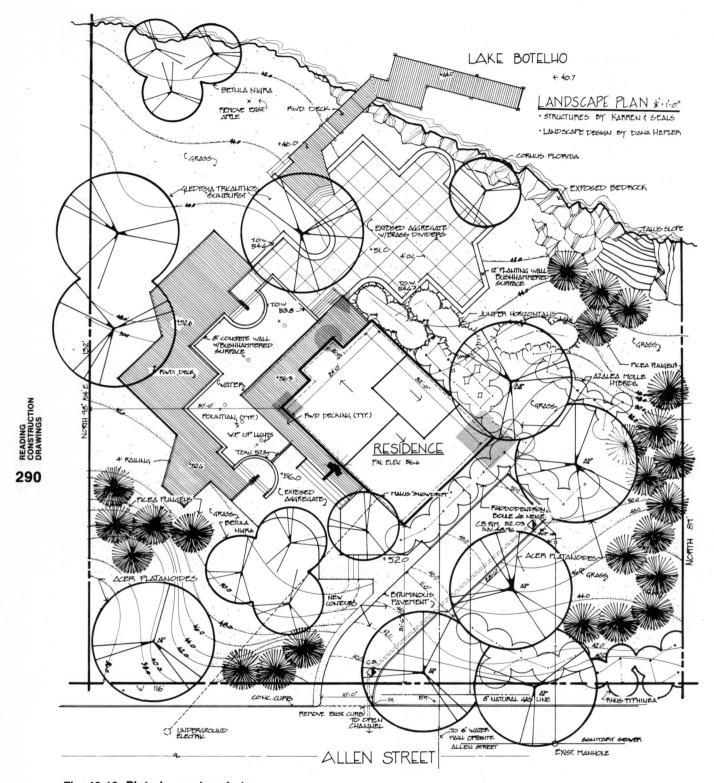

LAKE BOTELHO

+ 40.7

LANDSCAPE PLAN ⅛"·1'-0"
• STRUCTURES BY KARREN & SEALS
• LANDSCAPE DESIGN BY DANA HEFLER

BETULA NIGRA
× REMOVE EXIST ANLE
RWD. DECK
(GRASS)
GLEDITSIA TRIANTHOS "SUNBURST"
CORNUS FLORIDA
EXPOSED BEDROCK
EXPOSED AGGREGATE W/BRASS DIVIDERS
TALUS SLOPE
T.O.W. 54.67
12" PLANTING WALL BUSHHAMMERED SURFACE
+51.0
T.O.W. 54.67
JUNIPER HORIZONTALIS
T.O.W. 53.8
8" CONCRETE WALL W/BUSHHAMMERED SURFACE
(GRASS)
PICEA PUNGENS
RWD. DECK
AZALEA MOLLE HYBRIDA
(WATER)
(GRASS)
FOUNTAIN (TYP.)
RWD. DECKING (TYP.)
W.P. UP LIGHTS
RESIDENCE
FIN. ELEV. 56.6
T.O.W. 57.3
4' RAILING
MALUS "SNOWDRIFT"
RHODODENDRON BOULE de NEIGE
C.B. RIM 52.03
INV. 48.96
ACER PLATANOIDES
PICEA PUNGENS
EXPOSED AGGREGATE
(GRASS)
BETULA NIGRA
+52.0
ACER PLATANOIDES
GRASS
ACER PLATANOIDES
NEW CONTOURS
BITUMINOUS PAVEMENT
W. 116
CONC. CURB
UNDERGROUND ELECTRIC
REMOVE EXIST. CURB TO OPEN CHANNEL
6" NATURAL GAS LINE
RHUS TYPHINEA
TO 6" WATER MAIN OPPOSITE ALLEN STREET
SANITARY SEWER
EXIST. MANHOLE

ALLEN STREET

Fig. 48-12. Plot plan and roof plan.

Fig. 48-13. Door and nailing schedules. (Karren and Seals, Architects, Inc.)

DOOR		SCHEDULE			
DOOR NO.	LOCATION	SIZE	QUANTITY	TYPE	REMARKS
1	ENTRY	3'-0" x 6'-8"	1	SC	8 PANEL OPTIONAL
2	BEDROOMS	2'-8" x 6'-8"	2	HC	
3	BATH	2'-0" x 6'-8"	2	HC	
4	BEDROOM CLOSET	4'-0" x 6'-8"	2	HC	SLIDING
5	DECK	6'-0" x 6'-9"	1	SL GL	

NAILING SCHEDULE

JOIST TO SILL OR GIRDER:
    TOE NAIL   2 - 16d
BRIDGING TO JOIST:
    TOE NAIL   2 - 8d
PLATE TO JOIST OR BLK'G: 16d-16" oc
STUD TO PLATE - TOE NAIL: 4 - 8d
STUD TO PLATE - END NAIL: 2 - 16d
TOP PLATE - SPIKED: 16d-24" oc
TOP PLATE - LAPS & INTERSECTIONS:
    2 - 16d

CEIL'G JOISTS TO STUDS: 2 - 16d
RAFTER TO PLATE: 'SIMPSON'

PLYWOOD SUBFLOOR: 10d - 6" oc
PLYWOOD SIDING: 10d - 6" oc
PARTICLE BOARD: 8d - 6" oc

2" PLANKS   2 - 16d @ BEARING

GENERAL NOTES

1 ALL WORK SHALL CONFORM TO LOCAL BUILDING CODES
2 CONCRETE SHALL DEVELOP 2500 PSI MINIMUM COMPRESSIVE STRENGTH IN 28 DAYS.
3 STRUCTURAL STEEL SHALL CONFORM TO ASTM A615 GRADE 40.

Fig. 48-14. Perspective drawing. (Karren and Seals, Architects, Inc.)

HOME OF	MR & MRS JONES	T FOUNDATION PLAN		DATE 12.1.78	CODE # A2.1
LOT #7	OAKTREE SUBDIV	SCALE 1/4"=1'-0", 1"=1'-0"		CHECKED	SHEET # 3
REDWOOD CITY, CALIFORNIA		DRAWN BY archy Tects		APPROVED	OF 7

Fig. 49-1. Title block with simple indexing system.

Unit 49

Indexing Systems
Large sets of plans which include many different details require an indexing system. For large projects involving many buildings with many components, a detailed indexing system is needed to locate quickly specific plans for any specific building. In these cases, a master index is provided which shows the sheet number where each drawing for each building is found. Figure 49-1 shows a title block with a simple indexing system. It includes a design number, a sheet number, and a total number of drawings in the set.

The American Institute of Architects numbering system provides an access system for cataloging sets of drawings. Drawings are numbered

SK	SKETCHES (USED THROUGH ALL PHASES)
PR	PROGRAMMING
MP	MASTER PLANNING
SC	SCHEMATICS
DD	DESIGN DEVELOPMENT

Fig. 49-2. Alphabetical designations used to identify work phases.

A	ARCHITECTURAL
C	CIVIL
D	INTERIOR DESIGN (COLOR SCHEMES, FURNITURE, FURNISHINGS)
E	ELECTRICAL
F	FIRE PROTECTION (SPRINKLER, STANDPIPES, CO_2, ETC.)
G	GRAPHICS
K	DIETARY (FOOD SERVICE)
L	LANDSCAPE
M	MECHANICAL (HEATING, VENTILATING, AIR CONDITIONING)
P	PLUMBING
S	STRUCTURAL
T	TRANSPORTATION/CONVEYING SYSTEMS

Fig. 49-3. Prefixes used to identify specific disciplines of work.

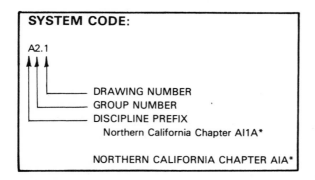

SYSTEM CODE:

A2.1

— DRAWING NUMBER
— GROUP NUMBER
— DISCIPLINE PREFIX
Northern California Chapter AI1A*

NORTHERN CALIFORNIA CHAPTER AIA*

Fig. 49-4. Use of coding system on drawings.

ARCHITECTURAL DRAWINGS

A0.1,2,3	— GENERAL (INDEX, SYMBOLS, ABBREV. NOTES, REFERENCES)
A1.1,2,3	— DEMOLITION, SITE PLAN, TEMPORARY WORK
A2.1,2,3	— PLANS, ROOM MATERIAL SCHEDULE, DOOR SCHEDULE, KEY DRAWINGS
A3.1,2,3	— SECTIONS, EXTERIOR ELEVATIONS
A4.1,2,3	— DETAILED FLOOR PLANS
A5.1,2,3	— INTERIOR ELEVATIONS
A6.1,2,3	— REFLECTED CEILING PLANS
A7.1,2,3	— VERTICAL CIRCULATION, STAIRS (ELEVATORS, ESCALATORS)
A8.1,2,3	— EXTERIOR DETAILS
A9.1,2,3	— INTERIOR DETAILS

STRUCTURAL DRAWINGS

S0.1,2,3	— GENERAL NOTES
S1.1,2,3	— SITE WORK
S2.1,2,3	— FRAMING PLANS
S3.1,2	— ELEVATIONS
S4.1,2	— SCHEDULES
S5.1,2	— CONCRETE
S6.1,2	— MASONRY
S7.1,2	— STRUCTURAL STEEL
S8.1,2	— TIMBER
S9.1,2	— SPECIAL DESIGN

MECHANICAL DRAWINGS

M0.1,2	— GENERAL NOTES
M1.1,2	— SITE/ROOF PLANS
M2.1,2	— FLOOR PLANS
M3.1,2	— RISER DIAGRAMS
M4.1,2	— PIPING FLOW DIAGRAM
M5.1,2	— CONTROL DIAGRAMS
M6.1,2	— DETAILS

PLUMBING DRAWINGS

P0.1,2	— GENERAL NOTES
P1.1,2	— SITE PLAN
P2.1,2	— FLOOR PLANS
P3.1,2	— RISER DIAGRAM
P4.1,2	— PIPING FLOW DIAGRAM
P5.1,2	— DETAILS

ELECTRICAL DRAWINGS

E0.1,2	— GENERAL NOTES
E1.1,2	— SITE PLAN
E2.1,2	— FLOOR PLANS, LIGHTING
E3.1,2	— FLOOR PLANS, POWER
E4.1,2	— ELECTRICAL ROOMS
E5.1,2	— RISER DIAGRAMS
E6.1,2	— FIXTURE/PANEL SCHEDULES
E7.1,2	— DETAILS

Northern California Chapter AIA*

Fig. 49-5. Typical coding system for construction drawings.

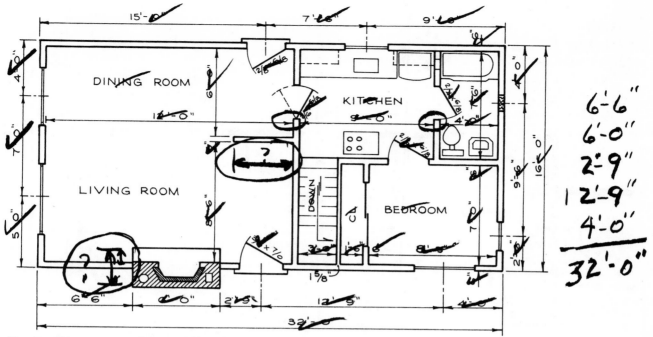

Fig. 49-6. Method of checking dimensions.

for ease of referencing. An alphabetical designation is used to identify the work phases of a project (See Fig. 49-2). A readily identifiable alphabetical prefix is used to denote specific disciplines of work covered by a group of drawings, as shown in Fig. 49-3. In this system construction drawings are divided into ten groups: A0 through A9. The group number always remains the same regardless of the size of the project. Additional drawings are added within groups without interrupting the alphanumerical order. Figure 49-4 shows this coding system as it appears on a drawing. Figure 49-5 shows a listing of these groups and their symbols.

If a building is to be constructed as designed, it is extremely critical that dimensions describing the size of each component agree on each drawing. If the dimensions of a basement plan do not match the related dimensions on a floor plan, prefabricated wall panels will not fit, and the position of stairwell openings, fireplace footings, and so forth will not align. Whether a dimension describes the overall length or width of a structure or only indicates the size of a subdimension of a detail, like dimensions must agree on each drawing in a set. For this reason dimensional accuracy is verified by locating one dimension at a time throughout each drawing in the entire set until each dimension is checked and agreement is verified (Fig. 49-6).

Unit 50
Combination Plans
Sometimes several plans in a complete set of plans are combined. For

example, there may be no specific electrical, plumbing, or air-conditioning plan, but these symbols may be added to the basic floor plan. There may be no specific landscape plan, but the landscape features may be added to a survey plan. Details and sections are also quite often combined into one plan. Combined plans are more difficult to read than separate specialized plans. For this reason, separate plans have been presented throughout this text for instructional purposes. Figure 50-1 shows a combination floor plan which includes not only information normally found on a floor plan, but also includes electrical, air-conditioning, plumbing, some landscape, survey, and plot plan symbols.

In studying this kind of plan the specialized elements must be separated out of the plan through imagination to eliminate confusion. Study only one element at a time. If you are studying the electrical part, refer only to as much of the remainder of the plan as necessary to orient the position of switches, outlets, and so forth. Imagine that the plan is only an electrical plan and that all other features do not exist.

Rooms, doors, windows, and sections are coded and cross-referenced among drawings, as shown in Fig. 50-2. Where sections are shown on the same sheet as the referenced drawing, only alphabetical (A, B, or C) designations are shown. But details that are extensive and fall on several different sheets are located with a system showing the detail, title, and the sheet number where the detail is located. When rooms have the same title, such as bedrooms, classrooms, labs, closets, and so forth, numbers are attached to each room so that schedules, plans, and specifications for that room are consistent and distinctive.

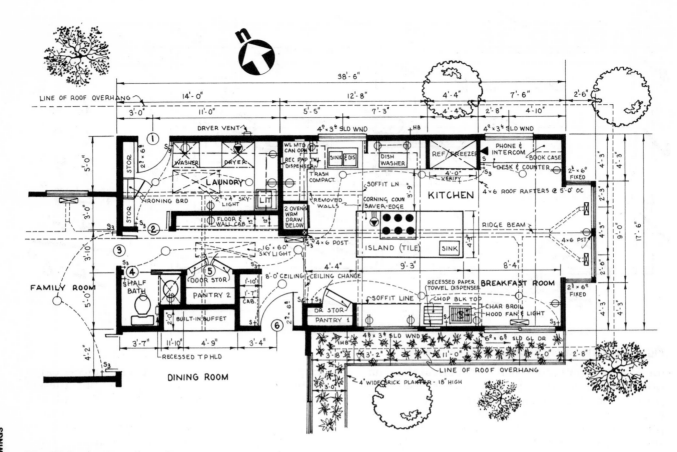

Fig. 50-1. Combination floor plan including electrical, plumbing, climate control, and landscape features.

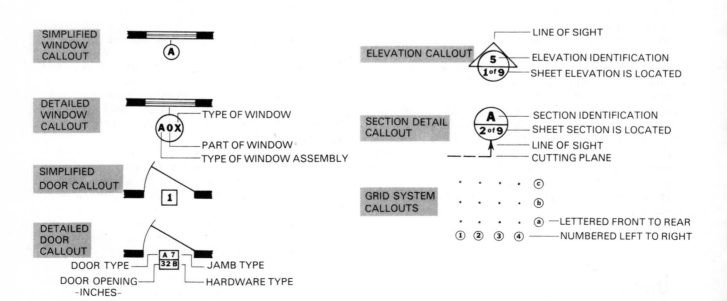

SIMPLIFIED WINDOW CALLOUT	
DETAILED WINDOW CALLOUT	TYPE OF WINDOW — PART OF WINDOW — TYPE OF WINDOW ASSEMBLY
SIMPLIFIED DOOR CALLOUT	
DETAILED DOOR CALLOUT	DOOR TYPE — DOOR OPENING -INCHES- — JAMB TYPE — HARDWARE TYPE

ELEVATION CALLOUT — LINE OF SIGHT — ELEVATION IDENTIFICATION — SHEET ELEVATION IS LOCATED

SECTION DETAIL CALLOUT — SECTION IDENTIFICATION — SHEET SECTION IS LOCATED — LINE OF SIGHT — CUTTING PLANE

GRID SYSTEM CALLOUTS — LETTERED FRONT TO REAR — NUMBERED LEFT TO RIGHT

Fig. 50-2. Room and detail coding.

CHAPTER 15

Understanding Schedules and Specifications

Unit 51

Schedules Schedules conserve time and space on construction drawings by substituting detail information with keys that are related to components on a drawing. When a key number or letter is attached to a component on a drawing, the key is indexed to an entry on a schedule that includes specific sizes, materials, and color of selected components.

Door and Window Schedules

Figure 51-1 shows key numbers indexed from a floor plan to a door and window schedule. The use of these symbols keeps the floor plans relatively clear of many details. Schedules are used because it is virtually impossible to list the width, height, thickness, material, type, screen, quantity, threshold material, manufacture, and other remarks on each door or window shown on a drawing. Figure 51-2 shows a door and window schedule which is indexed to the plan shown in Fig. 51-1.

Window styles and exterior door styles may be shown directly on elevation drawings. However, to conserve time many designers prepare a separate window or door drawing which shows specific details of the design in graphic form. When this is done, only the outline of the door or window is shown on the elevation drawing. A key on the window or door location on the plan relates to a detail drawing of the door (or window), as shown in Fig. 51-3*a*. This procedure eliminates the need for repeating design details of elevation drawings. Figure 51-3*b* shows the dimensions that relate the key to a door.

Floor plans do not show interior door design details. Unless an interior wall elevation is prepared for every partition, it is impossible to determine the design of interior doors without the use of a door schedule which is keyed to a design drawing.

Finish Schedules

To describe the type of finish (enamel, paint, or stain), the degree of gloss, and the specific color hue of each surface in each room requires an exhaustive list with many duplications. A finish

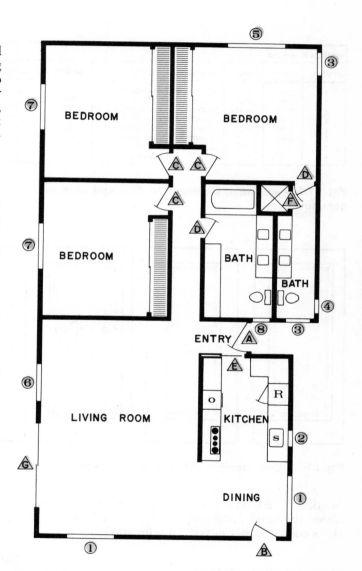

Fig. 51-1. Use of key symbols on floor plans.

schedule in tabular form enables the designer to condense all this information into one chart. The interior finish schedule shown in Fig. 51-4 includes, in the horizontal columns, the parts of each room and the type of finish to be applied. The vertical columns contain a list of rooms. The exact color classification has been noted in the appropriate intersecting blocks. The last column

DOOR SCHEDULE

SYM-BOL	WIDTH	HEIGHT	THICK-NESS	MATERIAL	TYPE	SCREEN	QUAN-TITY	THRESH-OLD	REMARKS	MANUFACTURER
A	3"-0"	7'0"	1-3/4"	WOOD—ASH	SLAB CORE	NO	1	OAK	OUTDOOR VARNISH	A. D. & D. DOOR, INC.
B	2'6"	7'0"	1-3/4"	WOOD—ASH	SLAB CORE	YES	1	OAK	OIL STAIN	A. D. & D. DOOR INC.
C	2'3"	6'8"	1-3/8"	WOOD—OAK	HOLLOW CORE	NO	3	NONE	OIL STAIN	A. D. & D. DOOR, INC.
D	2'0"	6'8"	1-3/8"	WOOD—ASH	HOLLOW CORE	NO	2	NONE	OIL STAIN	A. D. & D. DOOR, INC.
E	2'3"	6'8"	1-1/4"	WOOD—FIR	PLYWOOD	NO	1	NONE	SLIDING DOOR	A. D. & D. DOOR, INC.
F	1'9"	5'6"	1/2"	GLASS & METAL	SHOWER DOOR	NO	1	NONE	FROSTED GLASS	A. D. & D. DOOR, INC.
G	4'6"	6'6"	1/2"	GLASS & METAL	SLIDING	YES	2	METAL	1 SLIDING SCREEN	A. D. & D. DOOR, INC.

WINDOW SCHEDULE

SYMBOL	WIDTH	HEIGHT	MATERIAL	TYPE	SCREEN	QUANTITY	REMARKS	MANUFACTURER	CATALOG NUMBER
1	5'0"	4'0"	ALUMINUM	STATIONARY	NO	2		A & B GLASS CO.	188W
2	2'9"	3'0"	ALUMINUM	LOUVER	YES	1		A & B GLASS CO.	23JW
3	2'6"	3'0"	WOOD	DOUBLE HUNG	YES	2	4 LITES—2 HIGH	A & B GLASS CO.	141PW
4	1'6"	1'6"	ALUMINUM	LOUVER	YES	1		HAMPTON GLASS CO.	972BW
5	6'0"	3'6"	ALUMINUM	LOUVERED SIDES	YES	1		HAMPTON GLASS CO.	417CW
6	4'0"	6'6"	ALUMINUM	STATIONARY	NO	1		H & W WINDOW CO.	57DH
7	5'0"	3'6"	ALUMINUM	SLIDING	YES	2	FROSTED GLASS	H & W WINDOW CO.	22DH
8	1'9"	3'0"	ALUMINUM	AWNING	YES	1		H & W WINDOW CO.	1711JB

Fig. 51-2. Key symbols indexed to door and window schedules.

Fig. 51-3a. Door designs indexed to floor plan keys.

headed "Remarks" is used for making notes about the finish application, sequence of applications cost, or other pertinent information.

Materials Schedule

To ensure that all wall and ceiling coverings blend with the overall decor in each room, an interior finishing materials schedule is prepared listing materials specified for each part of each room. Blocks on the schedule are checked with suitable materials for the ceilings, walls, cove, base, and floors of each room, as shown in Fig. 51-5. This kind of schedule condenses many pages of unrelated materials lists for each room into one chart, thus enabling the builder to see at a glance all materials that are to be ordered for finishing each room.

Special Schedules

To ensure that purchased items, such as appliances and fixtures, will blend with the overall decor of each room, a separate schedule is often prepared as shown in Fig. 51-6. Similar schedules are sometimes prepared for furniture, draperies, and built-in components.

Schedules are not only useful for ensuring that the work is completed as planned, but are also valuable as aids in ordering manufactured items from appropriate manufacturers according to specifications.

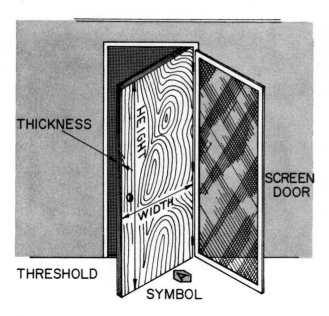

THICKNESS

HEIGHT

WIDTH

SCREEN DOOR

THRESHOLD

SYMBOL

Fig. 51-3b. Dimensions found on door keys.

ROOMS	FLOOR			CEILING					WALL					BASE					TRIM					REMARKS
	FLOOR VARNISH	UNFINISHED	WAXED	ENAMEL GLOSS	ENAMEL SEMIGLOSS	ENAMEL FLAT	FLAT LATEX	STAIN	ENAMEL GLOSS	ENAMEL SEMIGLOSS	ENAMEL FLAT	FLAT LATEX	STAIN	ENAMEL GLOSS	ENAMEL SEMIGLOSS	ENAMEL FLAT	FLAT LATEX	STAIN	ENAMEL GLOSS	ENAMEL SEMIGLOSS	FLAT LATEX	ENAMEL FLAT	STAIN	
ENTRY			✓				OFF WHT					OFF WHT				OFF WHT					OFF WHT				OIL STAIN
HALL			✓					LT BRN			TAN				DRK BRN					DRK BRN					OIL STAIN
BEDROOM 1	✓						OFF WHT					OFF WHT					GREY			GREY					ONE COAT PRIMER & SEALER —PAINTED SURFACE
BEDROOM 2	✓						OFF WHT				LT YEL						YEL					YEL		ONE COAT PRIMER & SEALER —PAINTED SURFACE
BEDROOM 3		✓					OFF WHT						LT BRN			DRK BRN								TAN	ONE COAT PRIMER & SEALER —PAINTED SURFACE
BATH 1				WHT					WHT					LT BLUE					LT BLUE					WATER-RESISTANT FINISHES
BATH 2				WHT					WHT					LT BLUE					LT BLUE					WATER-RESISTANT FINISHES
CLOSETS	✓					BRN					BRN					BRN					BRN			
KITCHEN			✓	WHT				YEL					YEL					YEL						
DINING			✓				TAN		YEL					YEL					YEL					OIL STAIN
LIVING		✓					TAN					LT BRN					LT BRN					LT BRN		OIL STAIN

Fig. 51-4. Finish schedule.

Fig. 51-5. Materials schedule.

ROOMS	FLOOR: ASPHALT TILE	CERAMIC TILE	CORK TILE	LINOLEUM TILE	WOOD STRIP—OAK	WOODS SOS—OAK	PLYWOOD PANEL	CARPETING	SLATE	CEILING: DIATO	PLASTER	WOOD PANEL	ACOUSTICAL TILE	EXPOSED BEAM	WALL: PLASTER	WOOD PANEL	WALL PAPER	WAINSCOT: WOOD	CERAMIC TILE	PAPER	ASPHALT TILE	STONE VENEER	BASE: LINOLEUM	WOOD	RUBBER	TILE—CERAMIC	ASPHALT	REMARKS
ENTRY									✓	✓		✓				✓								✓				TERRAZZO STEP COVERING
HALL			✓								✓				✓					✓				✓				
BEDROOM 1					✓							✓			✓			✓								✓		MAHOGANY WAINSCOT
BEDROOM 2					✓							✓			✓	✓		✓								✓		MAHOGANY WAINSCOT
BEDROOM 3					✓	✓						✓			✓										✓			SEE OWNER FOR GRADE CARPET
BATH 1		✓									✓				✓				✓							✓		WATER-SEAL-TILE EDGES
BATH 2	✓										✓				✓				✓							✓		WATER-SEAL-TILE EDGES
KITCHEN				✓								✓			✓						✓		✓					
DINING				✓									✓	✓	✓	✓					✓		✓					
LIVING							✓	✓					✓		✓							✓		✓				SEE OWNER FOR GRADE CARPET

Fig. 51-6. Appliance and fixture schedules.

APPLIANCE SCHEDULE

ROOM	APPLIANCE	TYPE	SIZE	COLOR	MANUFACTURER	MODEL NO.
KITCHEN	ELECTRIC STOVE	COOK TOP	4 BURNER	YELLOW	IDEALE APPLIANCES	341 MG
KITCHEN	ELECTRIC OVEN	BUILT-IN	30″ × 24″ × 24″	YELLOW	ZEIDLER OVEN MFG.	27 MG
SERVICE	HOT-WATER HEATER	GAS	50 GAL	WHITE	ORATZ WATER HTR.	249 KG

FIXTURE SCHEDULE

ROOM	FIXTURE	TYPE	MATERIAL	MANUFACTURER	MODEL NO.
LIVING	2 ELECTRIC LIGHTS	HANGING	BRASS REFLECTORS	HOT SPARK LTD.	1037 IG
BEDROOM 1	2 SPOT LIGHTS	WALL BRACKET	FLEXIBLE NECK—ALUMINUM	GURIAN & BARRIS INC.	1426 SG
BATHS 1 & 2	2 ELECTRIC LIGHTS	WALL BRACKET	ALUMINUM—WATER RESISTANT	MARKS ELECTRICAL CO.	2432 DG

Unit 52
Specifications

Specifications are written instructions describing the basic requirements for constructing a building. Specifications describe sizes, kinds, and quality of all building materials. The methods of construction, fabrication, or installation are also explicitly described. Specifications are orders to the contractor which show and tell him precisely what materials he must use and exactly how he must use them. Specifications also outline conditions under which he undertakes the job. Specifications guarantee the purchaser that the contractor will finish the building exactly as determined within the time frame specified.

Information that cannot be conveniently included on drawings, such as legal responsibility for materials, methods of purchasing materials, and unit costs of materials, are included in the set of specifications. In order to make an accurate construction estimate, contractors refer to the materials lists that are included in the specifications, as shown in Fig. 52-1.

Specifications ensure that the building will be constructed according to standards that local building laws require. Specifications are also used frequently by banks and federal agencies in appraising the market value of buildings and in making mortgage loan decisions. Since the items included in most sets of specifications are similar, the use of standard forms such as the FHA-VA form shown in Fig. 52-2 is frequently used to conserve time, increase clarity, and promote consistency of interpretation. The following specification outline shows the major divisions and subdivisions of a typical set of specifications. This outline does not include the exact kinds and sizes of materials included under each category because these would vary with each building. The sequence of the outline roughly approximates the sequence of actual construction.

SPECIFICATIONS OUTLINE
Owner's name and address
Contractors's name and address
Location of new structure

Fig. 52-1. Part of a contractor's materials list.

LINE NO.	ITEM COLUMN NO. 2	QUANTITY & UNIT MEAS.	MATERIAL (TYPE and/or SIZE)	UNIT COST	TOTAL COST
		#1575			
1	Interior Partitions				
2		195 Pcs.	2 x 4 x 8'-0" Studs		
3		30 Pcs.	2 x 4 x 10'-0" Studs		
4		34 Pcs.	2 x 4 x 12'-0" Studs		
5		870 Lin. Ft.	2 x 4 Plates		
6		48 Lin. Ft.	2 x 6 Headers		
7		3 Pcs.	2 x 8 x 10'-0" Headers		
8		2 Pcs.	2 x 8 x 14'-0" Headers		
9		2 Pcs.	2 x 12 x 12'-0" Headers		
10		7 Pcs.	2 x 8 x 8'-0" Studs		
11		3 Pcs.	2 x 8 x 6'-0" Plates		
12					
13	Ceiling Framing				
14		26 Pcs.	2 x 6 x 12'-0" Ceiling Joist		
15					
16	Roof Framing				
17		16 Units	24'-0" 3/12 Pitch 2 x 6 Trusses		
18		76 Pcs.	2 x 6 x 16'-0" Rafters		
19		24 Pcs.	2 x 6 x 12'-0" Rake Rafters		
20		2800 Sq. Ft.	Roof Sheathing		
21		5600 Sq. Ft.	15# Felt		
22		9 Pcs.	2 x 6 x 10'-0" Rafters		
23	Balcony Framing				
24		24 Lin. Ft.	2 x 6 Joist Trimmers		
25		230 Lin. Ft.	2 x 6 S 4 S Plank Flooring		
26					
27	Roofing & Sheet Metal				
28		1 Square	Asphalt Ridge Shingles		
29		28 Squares	Asphalt Self Sealing Shingles		
30		284 Lin. Ft.	Metal Drip Edging		
31		90 Pcs.	5" x 7" Metal Drip Edging		
32		16 Lin. Ft.	Window Head Flashing		
33		24 Lin. Ft.	Roof to Wall Flashing		
34		13 Lin. Ft.	Chimney Counter Flashing		
35		140 Lbs.	Roofing Nails		
36					
37	Windows				
38		15 Single	4'-0"x3'-0" Aluminum Gliding Windows-Loose Casing		
39		4 Single	44x80" Fixed Plate Glass - Loose Casing		
40		1 Each Right			
41		& Left	40x24"-36" Fixed Plate Glass 3/12 Pitch		
42			Sloping Head - Loose Casing		
43		1 Each Right			
44		& Left	26"x104"-110½" Fixed Plate Glass 3/12 Pitch		
45			Sloping Head - Loose Casing		
46		2 Singles	26"x92" Fixed Plate Glass - Loose Casing		
47		1 Single	8"x80" Fixed Plate Glass - Loose Casing		
48		4 Single	4'-78" Fixed Plate Glass - Loose Casing		
49	Door Frames	1 Front	7'-0" Rabbeted 1-3/4" Loose Casing		

1. General information

List of all drawings, specifications, legal documents

Allowances of money for special orders, such as wallpaper, carpeting, fixtures

Completion date

Contractor's bid

List of manufactured items bought for the job

Guarantees for all manufactured items

2. Legal responsibilities—contractor

Good workmanship

Adherence to plans and specifications

Fulfillment of building laws

Purchase of materials

Hiring and paying all workers

Obtaining and paying for all permits

Providing owner certificate of passed inspection

Responsibility for correction of errors

Responsibility for complete cleanup

Furnishing all tools and equipment

Providing personal supervision

Having a foreman on the job at all times

Providing a written guarantee of work

3. Legal responsibilities—homeowner

Carrying fire insurance during construction

Paying utilities during construction

Specifying method of payment

4. Earthwork

Excavation, backfills, gradings

Irregularities in soil

Location of house on lot

Clearing of lot

Grading for water drainage

Preparation of ground for foundation

5. Concrete and cement work

Foundation sizes

Concrete and mortar mix

Cement, sand, and gravel

Curing the concrete

Finishing-off concrete flatwork

Vapor seals and locations

Type and size of reinforcing steel and locations

Outside concrete work, sizes and locations

Cleaning of masonry work

Porches, patios, terraces, walks, driveways: sizes and locations

For accurate register of carbon copies, form may be separated along above fold. Staple completed sheets together in original order.

Form approved.
Budget Bureau No. 63-R055.11.

☐ Proposed Construction

☐ Under Construction

DESCRIPTION OF MATERIALS

No. _____
(To be inserted by FHA or VA)

Property address _____ City _____ State _____

Mortgagor or Sponsor _____ _____
                          (Name)   (Address)

Contractor or Builder _____ _____
                          (Name)   (Address)

INSTRUCTIONS

1. For additional information on how this form is to be submitted, number of copies, etc., see the instructions applicable to the FHA Application for Mortgage Insurance or VA Request for Determination of Reasonable Value, as the case may be.

2. Describe all materials and equipment to be used, whether or not shown on the drawings, by marking an X in each appropriate check-box and entering the information called for in each space. If space is inadequate, enter "See misc." and describe under item 27 or on an attached sheet.

3. Work not specifically described or shown will not be considered unless

required, then the minimum acceptable will be assumed. Work exceeding minimum requirements cannot be considered unless specifically described.

4. Include no alternates, "or equal" phrases, or contradictory items. (Consideration of a request for acceptance of substitute materials or equipment is not thereby precluded.)

5. Include signatures required at the end of this form.

6. The construction shall be completed in compliance with the related drawings and specifications, as amended during processing. The specifications include this Description of Materials and the applicable Minimum Construction Requirements.

1. EXCAVATION:

Bearing soil, type _____

2. FOUNDATIONS:

Footings: concrete mix _____; strength psi _____ Reinforcing _____

Foundation wall: material _____ Reinforcing _____

Interior foundation wall: material _____ Party foundation wall _____

Columns: material and sizes _____ Piers: material and reinforcing _____

Girders: material and sizes _____ Sills: material _____

Basement entrance areaway _____ Window areaways _____

Waterproofing _____ Footing drains _____

Termite protection _____

Basementless space: ground cover _____; insulation _____; foundation vents _____

Special foundations _____

Additional information: _____

3. CHIMNEYS:

Material _____ Prefabricated (make and size) _____

Flue lining: material _____ Heater flue size _____ Fireplace flue size _____

Vents (material and size): gas or oil heater _____; water heater _____

Additional information: _____

4. FIREPLACES:

Type: ☐ solid fuel; ☐ gas-burning; ☐ circulator (make and size) _____ Ash dump and clean-out _____

Fireplace: facing _____; lining _____; hearth _____; mantel _____

Additional information: _____

5. EXTERIOR WALLS:

Wood frame: wood grade, and species _____ ☐ Corner bracing. Building paper or felt _____

Sheathing _____; thickness _____; width _____; ☐ solid; ☐ spaced _____" o. c.; ☐ diagonal; _____

Siding _____; grade _____; type _____; size _____; exposure _____"; fastening _____

Shingles _____; grade _____; type _____; size _____; exposure _____"; fastening _____

Stucco _____; thickness _____"; Lath _____; weight _____ lb.

Masonry veneer _____ Sills _____ Lintels _____ Base flashing _____

Masonry: ☐ solid ☐ faced ☐ stuccoed; total wall thickness _____"; facing thickness _____" facing material _____

Backup material _____; thickness _____"; bonding _____

Door sills _____ Window sills _____ Lintels _____ Base flashing _____

Interior surfaces: dampproofing, _____ coats of _____; furring _____

Additional information: _____

Exterior painting: material _____; number of coats _____

Gable wall construction: ☐ same as main walls; ☐ other construction _____

6. FLOOR FRAMING:

Joists: wood, grade. and species _____; other _____; bridging _____; anchors _____

Concrete slab: ☐ basement floor; ☐ first floor; ☐ ground supported; ☐ self-supporting; mix _____; thickness _____

reinforcing _____; insulation _____; membrane _____

Fill under slab: material _____; thickness _____". Additional information: _____

7. SUBFLOORING: (Describe underflooring for special floors under item 21.)

Material: grade and species _____; size _____; type _____

Laid: ☐ first floor; ☐ second floor; ☐ attic _____ sq. ft.; ☐ diagonal; ☐ right angles. Additional information: _____

8. FINISH FLOORING: (Wood only. Describe other finish flooring under item 21.)

LOCATION	ROOMS	GRADE	SPECIES	THICKNESS	WIDTH	BLDG. PAPER	FINISH
First floor							
Second floor							
Attic floor	sq. ft.						

Additional information: _____

Fig. 52-2. FHA and VA description of materials.

9. **PARTITION FRAMING:**
Studs: wood, grade, and species _____ size and spacing _____ Other _____
Additional information: _____

10. **CEILING FRAMING:**
Joists: wood, grade, and species _____ Other _____ Bridging _____
Additional information: _____

11. **ROOF FRAMING:**
Rafters: wood, grade, and species _____ Roof trusses (see detail): grade and species _____
Additional information: _____

12. **ROOFING:**
Sheathing: wood, grade, and species _____ ; ☐ solid; ☐ spaced _____ " o.c.
Roofing _____ ; grade _____ ; size _____ ; type _____
Underlay _____ ; weight or thickness _____ ; size _____ ; fastening _____
Built-up roofing _____ ; number of plies _____ ; surfacing material _____
Flashing: material _____ ; gage or weight _____ ; ☐ gravel stops; ☐ snow guards
Additional information: _____

13. **GUTTERS AND DOWNSPOUTS:**
Gutters: material _____ ; gage or weight _____ ; size _____ ; shape _____
Downspouts: material _____ ; gage or weight _____ ; size _____ ; shape _____ ; number _____
Downspouts connected to: ☐ Storm sewer; ☐ sanitary sewer; ☐ dry-well. ☐ Splash blocks: material and size _____
Additional information: _____

14. **LATH AND PLASTER**
Lath ☐ walls, ☐ ceilings: material _____ ; weight or thickness _____ Plaster: coats _____ ; finish _____
Dry-wall ☐ walls, ☐ ceilings: material _____ ; thickness _____ ; finish _____
Joint treatment _____

15. **DECORATING:** *(Paint, wallpaper, etc.)*

Rooms	Wall Finish Material and Application	Ceiling Finish Material and Application
Kitchen		
Bath		
Other		

Additional information: _____

16. **INTERIOR DOORS AND TRIM:**
Doors: type _____ ; material _____ ; thickness _____
Door trim: type _____ ; material _____ Base: type _____ ; material _____ ; size _____
Finish: doors _____ ; trim _____
Other trim *(item, type and location)* _____
Additional information: _____

17. **WINDOWS:**
Windows: type _____ ; make _____ ; material _____ ; sash thickness _____
Glass: grade _____ ; ☐ sash weights; ☐ balances, type _____ ; head flashing _____
Trim: type _____ ; material _____ Paint _____ ; number coats _____
Weatherstripping: type _____ ; material _____ Storm sash, number _____
Screens: ☐ full; ☐ half; type _____ ; number _____ ; screen cloth material _____
Basement windows: type _____ ; material _____ ; screens, number _____ ; Storm sash, number _____
Special windows _____
Additional information: _____

18. **ENTRANCES AND EXTERIOR DETAIL:**
Main entrance door: material _____ ; width _____ ; thickness _____ ". Frame: material _____ , thickness _____ "
Other entrance doors: material _____ ; width _____ ; thickness _____ ". Frame: material _____ , thickness _____ "
Head flashing _____ Weatherstripping: type _____ ; saddles _____
Screen doors: thickness _____ "; number _____ ; screen cloth material _____ Storm doors: thickness _____ "; number _____
Combination storm and screen doors: thickness _____ "; number _____ ; screen cloth material _____
Shutters: ☐ hinged; ☐ fixed. Railings _____ , Attic louvers _____
Exterior millwork: grade and species _____ Paint _____ ; number coats _____
Additional information: _____

19. **CABINETS AND INTERIOR DETAIL:**
Kitchen cabinets, wall units: material _____ ; lineal feet of shelves _____ ; shelf width _____
Base units: material _____ ; counter top _____ ; edging _____
Back and end splash _____ Finish of cabinets _____ ; number coats _____
Medicine cabinets: make _____ ; model _____
Other cabinets and built-in furniture _____
Additional information: _____

20. **STAIRS:**

Stair	Treads		Risers		Strings		Handrail		Balusters	
	Material	Thickness	Material	Thickness	Material	Size	Material	Size	Material	Size
Basement										
Main										
Attic										

Disappearing: make and model number _____
Additional information: _____

2

Fig. 52-2. (Continued.)

21. SPECIAL FLOORS AND WAINSCOT:

	Location	Material, Color, Border, Sizes, Gage, Etc.	Threshold Material	Wall Base Material	Underfloor Material
Floors	Kitchen ____				
	Bath ____				

	Location	Material, Color, Border, Cap. Sizes, Gage, Etc.	Height	Height Over Tub	Height in Showers (From Floor)
Wainscot	Bath ____				

Bathroom accessories: ☐ Recessed; material _____ ; number _____ ; ☐ Attached; material _____ ; number _____

Additional information: _____

22. PLUMBING:

Fixture	Number	Location	Make	Mfr's Fixture Identification No.	Size	Color
Sink ____						
Lavatory ____						
Water closet ____						
Bathtub ____						
Shower over tub△ ____						
Stall shower△ ____						
Laundry trays ____						

△☐ Curtain rod   △☐ Door   ☐ Shower pan: material _____

Water supply: ☐ public; ☐ community system; ☐ individual (private) system.★

Sewage disposal: ☐ public; ☐ community system; ☐ individual (private) system.★

★Show and describe individual system in complete detail in separate drawings and specifications according to requirements.

House drain (inside): ☐ cast iron; ☐ tile; ☐ other _____   House sewer (outside): ☐ cast iron; ☐ tile; ☐ other _____

Water piping: ☐ galvanized steel; ☐ copper tubing; ☐ other _____   Sill cocks, number _____

Domestic water heater: type _____ ; make and model _____ ; heating capacity _____

_____ gph. 100° rise. Storage tank: material _____ ; capacity _____ gallons.

Gas service: ☐ utility company; ☐ liq. pet. gas; ☐ other _____   Gas piping: ☐ cooking; ☐ house heating.

Footing drains connected to: ☐ storm sewer; ☐ sanitary sewer; ☐ dry well.   Sump pump; make and model _____

_____ ; capacity _____ ; discharges into _____

23. HEATING:

☐ Hot water. ☐ Steam. ☐ Vapor. ☐ One-pipe system. ☐ Two-pipe system.

   ☐ Radiators. ☐ Convectors. ☐ Baseboard radiation. Make and model _____

   Radiant panel: ☐ floor; ☐ wall; ☐ ceiling. Panel coil: material _____

   ☐ Circulator. ☐ Return pump. Make and model _____ ; capacity _____ gpm.

   Boiler: make and model _____   Output _____ Btuh.; net rating _____ Btuh.

Additional information: _____

Warm air: ☐ Gravity. ☐ Forced. Type of system _____

   Duct material: supply _____ ; return _____   Insulation _____ , thickness _____   ☐ Outside air intake.

   Furnace: make and model _____   Input _____ Btuh.; output _____ Btuh.

   ☐ Additional information: _____

☐ Space heater; ☐ floor furnace; ☐ wall heater. Input _____ Btuh.; output _____ Btuh.; number units _____

   Make, model _____   Additional information: _____

Controls: make and types _____

Additional information: _____

Fuel: ☐ Coal; ☐ oil; ☐ gas; ☐ liq. pet. gas; ☐ electric; ☐ other _____ ; storage capacity _____

   Additional information: _____

Firing equipment furnished separately: ☐ Gas burner, conversion type. ☐ Stoker: hopper feed ☐; bin feed ☐

   Oil burner: ☐ pressure atomizing; ☐ vaporizing _____

   Make and model _____   Control _____

   Additional information: _____

Electric heating system: type _____   Input _____ watts; @ _____ volts; output _____ Btuh.

   Additional information: _____

Ventilating equipment: attic fan, make and model _____ ; capacity _____ cfm.

   kitchen exhaust fan, make and model _____

Other heating, ventilating, or cooling equipment _____

24. ELECTRIC WIRING:

Service: ☐ overhead. ☐ underground. Panel: ☐ fuse box; ☐ circuit-breaker: make _____ AMP's _____ No. circuits _____

Wiring: ☐ conduit, ☐ armored cable; ☐ nonmetallic cable; ☐ knob and tube; ☐ other _____

Special outlets: ☐ range; ☐ water heater; ☐ other _____

☐ Doorbell. ☐ Chimes. Push-button locations _____   Additional information: _____

25. LIGHTING FIXTURES:

Total number of fixtures _____   Total allowance for fixtures, typical installation, $ _____

Nontypical installation _____

Additional information: _____

3   DESCRIPTION OF MATERIALS

Fig. 52-2. (Continued.)

26. INSULATION:

LOCATION	THICKNESS	MATERIAL, TYPE, AND METHOD OF INSTALLATION	VAPOR BARRIER
Roof			
Ceiling			
Wall			
Floor			

HARDWARE: (make, material, and finish.) _____

SPECIAL EQUIPMENT: (State material or make, model and quantity. Include only equipment and appliances which are acceptable by local law, custom and applicable FHA standards. Do not include items which, by established custom, are supplied by occupant and removed when he vacates premises or chattels prohibited by law from becoming realty.)_____

27. MISCELLANEOUS: (Describe any main dwelling materials, equipment, or construction items not shown elsewhere; or use to provide additional information where the space provided was inadequate. Always reference by item number to correspond to numbering used on this form.) _____

PORCHES:

TERRACES:

GARAGES:

WALKS AND DRIVEWAYS:

Driveway: width _____ ; base material _____ ; thickness _____"; surfacing material _____ ; thickness _____ "
Front walk: width _____ ; material _____ ; thickness _____". Service walk: width _____ ; material _____ ; thickness _____ "
Steps: material _____ ; treads _____"; risers _____". Cheek walls _____

OTHER ONSITE IMPROVEMENTS:

(Specify all exterior onsite improvements not described elsewhere, including items such as unusual grading, drainage structures, retaining walls, fence, railings, and accessory structures.)

LANDSCAPING, PLANTING, AND FINISH GRADING:

Topsoil _____" thick: ☐ front yard; ☐ side yards; ☐ rear yard to _____ feet behind main building.
Lawns (seeded, sodded, or sprigged): ☐ front yard _____ ; ☐ side yards _____ ; ☐ rear yard_____
Planting: ☐ as specified and shown on drawings; ☐ as follows:

_____ Shade trees, deciduous, _____" caliper.	_____ Evergreen trees. _____ ' to _____ ', B & B.	
_____ Low flowering trees, deciduous, _____ ' to _____'	_____ Evergreen shrubs. _____ ' to _____ ', B & B.	
_____ High-growing shrubs, deciduous, _____ ' to _____'	_____ Vines, 2-year _____	
_____ Medium-growing shrubs, deciduous, _____ ' to _____'		
_____ Low-growing shrubs, deciduous, _____ ' to _____'		

IDENTIFICATION.—This exhibit shall be identified by the signature of the builder, or sponsor, and/or the proposed mortgagor if the latter is known at the time of application.

Date_____ Signature _____

Signature _____

FHA Form 2005
VA Form 26-1852

4

GPO 1968 o48—16—80081-1 296-152

Fig. 52-2. (Continued.)

6. Carpentry, rough

Required types of wood grades
Maximum amount of moisture in wood
List of construction members, sizes, and
  amount of wood needed
Special woods, mill work
Nail sizes for each job

7. Floors

Type, size, and finish of floor
Floor coverings

8. Roofing

Type of coverings
Amount of coverings
Methods to bond coverings
Color of final layer

9. Sheet metal

List of flashing and sizes
List of galvanized iron and sizes
Protective metal paint and where to use
Size and amount of screens for vents, doors,
  and windows

10. Doors and windows

Sizes
Material
Type
Quantity
Manufacturer and model number
Window and door trims
Frames for screens
Amount of window space per room
Amount of openable window space per room
Types of glass and mirrors
Types of sashes
Window and door frames
Weather stripping and caulking

11. Lath and plaster

Type, size, and amount of lath needed
Type, size, and amount of wire mesh, felt
  paper, and nails
Types of interior and exterior plaster
Instructions of manufacturer for mixing and
  applying
Number of coats
Finishing between coats
Drying time

12. Dry walls

Wall covering—types, sizes, manufacturer's
  model number

13. Insulation

List of types, makes, sizes, model number
Instructions for applying

14. Electrical needs

Electrical outlets and their locations

Electrical switches and their locations
Wall brackets and their locations
Ceiling outlets and their locations
Signed certificate that electrical work has
  passed the building inspection
Guarantee for all parts
List of all electrical parts with name, type, size,
  color, model number, and lamp wattage
Locations for television outlet and aerial, tele-
  phone outlet, main switches, panel board, cir-
  cuits, and meter
Size of wire used for wiring
Number of circuits

15. Plumbing

List of fixtures with make, color, style, manu-
  facturer, and catalog number
List of type and size of plumbing lines—gas,
  water, and waste
Vent pipes and sizes
Inspection slips on plumbing
Guarantees for plumbing
Instructions for installing and connecting pipe-
  lines

16. Heating and air conditioning

List of all equipment with make, style, color,
  manufacturer's name, catalog number
Guarantee for all equipment
Signed equipment inspection certificate
List and location of all sheet metal work for
  heat ducts
List of fuels, outlets, exhausts, and registers
Types of insulation
Location for heating and air-conditioning units

17. Stone and brickwork

List and location of all stone and brickwork
  (fireplace, chimney, retaining walls)
Concrete and mortar mix
Reinforcing steel
Kind, size, and name of manufacturer of any
  synthetic stone

18. Built-ins

List of all built-ins to be constructed on the job
Dimensions
Kinds of materials
List of all manufactured objects to be built in
Model number and make
Color
Catalog number
Manufacturer

19. Ceramic tile

List of types, sizes, colors, manufacturers, catalog number
Mortar mix

20. Painting

List of paints to be used—type, color, manufacturer's name, catalog number

Preparation of painted surface
Number of coats and preparation of each
Instructions for stained surfaces or special finishes (type of finish, color, manufacturer, and catalog number)

21. Finish hardware

List of hardware—type, make, material, color, manufacturer's name, catalog number

22. Exterior

List of types of finishes for each exterior wall
Instructions for each type
Color, manufacturer, and catalog number

23. Miscellaneous

List and location of all blacktop areas

CHAPTER 16

Understanding Legal Documents

Unit 53
Building Codes A building code is a collection of laws that outlines the restrictions necessary to maintain minimum standards established by building and health departments of communities. These codes (laws) help control design, construction, materials, maintenance, location of structures, use of structures, number of occupants, quality of materials, and use of materials. To stay within the law, designers and builders must observe the local building codes.

Before any structure is built, altered, or repaired, a building permit must be obtained from the municipality building department. This permit ensures the appearance of an inspector to inspect the work. Inspections are made of plans, grading of land, excavations, foundations, forms, carpentry, plumbing, heating, ventilating, and electrical work.

To be effective codes must be updated continually to keep pace with new types of construction materials; otherwise, the use of antiquated materials and methods is perpetuated by outdated building codes.

Building codes vary in different communities because of geographical differences. Each municipality formulates its own building-code requirements. Building codes are necessary not only to ensure that substandard or unsafe buildings are not erected but that unattractive or architecturally inconsistent buildings are not built in an area. Building codes also help regulate the kinds of structures that can be built in specific areas by zoning.

Zones are usually classified as residential, commercial, or industrial. Building codes also contain regulations pertaining to building permits, fees, inspection requirements, drawings required, property location, and general legal implications connected with the building.

Size restrictions are a vital part of every code. Some of the most common items included in building codes include room sizes, ceiling heights, window areas, foundations, retaining walls, concrete mix, and girders.

The type and composition of foundations are outlined specifically in building codes, as are the sizes and spacing of girders, posts, and joists. The size of door and window openings related to the size of the lintel and the weight of support required of that lintel are also spelled out in building codes. Roof types, pitches, size of materials, and spacing of rafters, especially in cold climates where snow loads are prevalent, are rigidly controlled through building codes.

The maximum amount of load permissible for each kind of structure is always listed in the building codes. Live loads, which include the weight of any movable object on floors, roofs, or ceilings, and dead loads, the weight of the building itself that must be supported, are rigidly controlled through specifying the size and type of materials that are used in foundations to support these loads. The size and spacing of materials in walls that support roof loads are also specified. Maximum loads in tons per square inch as related to the pitch of the roof are specified in building codes.

Unit 54
Contracts and Bids Legal documents define and affix responsibility for various aspects of construction to the designer or architect, builder or contractor, and the owner or purchaser of the building to be erected. Legal documents define agreements reached between the architect, builder, and owner. This agreement indicates the fees to be paid the architect and builder and the general conditions under which the project is undertaken. The following subsection of AIA document A201 describes the areas covered in general agreements between architects and builders.

- **A.** Definitions.
- **B.** Architect's supervision.
- **C.** Architect's decision.
- **D.** Notice.
- **E.** Separate contracts.
- **F.** Intent of plans and specifications.
- **G.** Errors and discrepancies.
- **H.** Drawings and specifications furnished to contractors.
- **I.** Approved drawings.
- **J.** Patents.
- **K.** Permits, licenses, and certificates.

L. Supervision and labor.

M. Public safety and watchmen.

N. Order of completion.

O. Substitution of materials for those called for by specifications.

P. Materials, equipment, and labor.

Q. Inspection.

R. Defective work and materials.

S. Failure to comply with orders of architect.

T. Use of completed parts.

U. Rights of various interests.

V. Suspension of work due to unfavorable conditions.

W. Suspension of work due to fault of contractor.

X. Suspension of work due to unforeseen causes.

Y. Request for extension.

Z. Stoppage of work by architect.

AA. Default on part of contractor.

BB. Removal of equipment.

CC. Monthly estimates and payments.

DD. Acceptance and final payment.

EE. Deviations from contract requirements.

FF. Estoppel and waiver of legal rights.

GG. Approval of subcontractors and sources of material.

HH. Approval of material samples requiring laboratory tests.

II. Arbitration.

JJ. Bonds.

KK. Additional or substitute bonds.

LL. Public liability and property damage insurance.

MM. Workmen's Compensation Act.

NN. Fire insurance and damage due to other hazards.

OO. Explosives and blasting.

PP. Damages to property.

QQ. Mutual responsibility of contractors.

RR. Contractor's liability.

SS. Familiarity with contract documents.

TT. Shop drawings.

UU. Guarantee of work.

VV. Clean up.

WW. Competent workmen (state law).

XX. Prevailing wage act (state law).

YY. Residence of employees.

ZZ. Nondiscrimination in hiring employees (state law).

AAA. Preference to employment of war veterans (state law).

BBB. Hiring and conditions of employment (state law).

Contracts include fees and fees schedules, performance bond, labor and materials bonds, payments, time schedules, estimates, general conditions, and supplementary conditions. Schedules, specifications, and working drawings are also indexed to the contracts so that they become legal adjuncts to the contract.

Contracts describe responsibilities for relevant financial changes which may be affected by time schedules or unavoidable delays, such as acts of God and strikes. Performance bond is offered by the contractor and guarantees that the contractor's responsibilities as builder will be performed according to the conditions of the contract.

Labor and materials bonds posted by the contractor guarantee that invoices for materials, supplies, and services of subcontractors will be paid by the major contractors (prime contractor) according to the terms of the contract.

Payment schedules are an important part of any contract. Payments are directly related to completion of various phases of the work, such as acceptance of the bid, beginning of work, completion of working phases of construction, and final approval by building inspector.

Just as most sets of plans must contain an architect's seal, licensed subcontractors are specified and required on most construction jobs. Indications of licensing requirements are usually included in construction contracts. Supervision of the labor force by journeymen in each specific area and the use of licensed electricians and licensed plumbers are also specified in contracts.

Contractors receive invitations to bid by mail, through newspaper advertisements, or through private resources such as McGraw-Hill Dodge reports. Construction bid forms are very specific in indicating the availability of documents, when the documents can be examined, provision for the resolution of questions, approval for submission of materials, specific dates for bids submission, and the form for preparing bids. The bid form includes specific instructions to the bidders, price of the bid, substitution, restrictions, and involvement of subcontractors.

The bid form is a letter which is sent from the bidder to the party responsible for issuing the construction bid. This may be the architect, general contractor, or owner. The letter covers the following points: verification of receipt of all drawings and documents, specific length of time bid will be held open, price quotation for the entire project or stipulation of the part being bid, and a listing of substitute materials or components if any item varies from specified requirements. Upon signing this bid form, the bidder agrees to abide by all of the conditions of the bid, including the price, time, quality of work, materials as specified in the contract documents, and drawings.

Sample instructions to bidder outlined by the Construction Specification Institute are as follows:

DOCUMENTS Bonafide prime bidders may obtain _____ sets of Drawings and Specifications from the Architect upon deposit of $_____ per set. Those who submit prime bids may obtain refund of deposits by returning sets in good condition no more than _____ days after Bids have been opened. Those who do not submit prime bids will forfeit deposits unless sets are returned in good condition at least _____ days before Bids are opened. No partial sets will be issued; no sets will be issued to sub-bidders by the Architect. Prime bidders may obtain additional copies upon deposit of $_____ per set.

EXAMINATION Bidders shall carefully examine the documents and the construction site to obtain first-hand knowledge of existing conditions. Contractors will not be given extra payments for conditions which can be determined by examining the site and documents.

QUESTIONS Submit all questions about the Drawings and Specifications to the Architect, in writing. Replies will be issued to all prime bidders of record as Addenda to the Drawings and Specifications and will become part of the Contract. The Architect and Owner will not be responsible for oral clarification. Questions received less than _____ hours before the bid opening cannot be answered.

SUBSTITUTIONS To obtain approval to use unspecified products, bidders shall submit written requests at least ten days before the bid date and hour. Requests received after this time will not be considered. Requests shall clearly describe the product for which approval is asked, including all data necessary to demonstrate acceptability. If the product is acceptable, the Architect will approve it in an Addendum issued to all prime bidders on record.

BASIS OF BID The bidder must include all unit cost items and all alternatives shown on the Bid Forms; failure to comply may be cause for rejection. No segregated Bids or assignments will be considered.

PREPARATION OF BIDS Bids shall be made on unaltered Bid Forms furnished by the Architect. Fill in all blank spaces and submit two copies. Bids shall be signed with name typed below signature. Where bidder is a corporation, Bids must be signed with the legal name of the corporation followed by the name of the State of incorporation and legal signatures of an officer authorized to bind the corporation to a contract.

BID SECURITY Bid Security shall be made payable to the _____ _____ _____, in the amount of _____ percent of the Bid sum. Security shall be either certified check or bid bond issued by surety licensed to conduct business in the State of New York. The successful bidder's security will be retained until he has signed the Contract and furnished the required payment and performance bonds. The Owner reserves the right to retain the security of the next _____ bidders until the lowest bidder enters into contract or until _____ days after bid opening, whichever is the shorter. All other bid security will be returned as soon as practicable. If any bidder refuses to enter into a Contract, the Owner will retain his Bid Security as liquidated damages, but not as a penalty. The Bid Security is to be submitted _____ day(s) prior to the Submission of Bids.

PERFORMANCE BOND AND LABOR AND MATERIAL PAYMENT BOND Furnish and pay for bonds covering faithful performance of the Contract and payment of all obligations arising thereunder. Furnish bonds in such form as the Owner may prescribe and with a surety company acceptable to the Owner. The bidder shall deliver said bonds to the Owner not later than the date of execution of the Contract. Failure or neglecting to deliver said bonds, as specified, shall be considered as having abandoned the Contract and the Bid Security will be retained as liquidated damages.

SUBCONTRACTORS Names of principal subcontractors must be listed and attached to the Bid. There shall be only one subcontractor named for each classification listed.

SUBMITTAL Submit Bid and Subcontractor Listing in an opaque, sealed envelope. Identify the envelope with: (1) project name, (2) name of bidder. Submit Bids in accord with the Invitation to Bid.

MODIFICATION AND WITHDRAWAL Bids may not be modified after submittal. Bidders may withdraw Bids at any time before bid opening, but may not resubmit them. No Bid may be withdrawn or modified after the bid opening except where the award of Contract has been delayed for _____ days.

DISQUALIFICATION The Owner reserves the right to disqualify Bids, before or after opening, upon evidence of collusion with intent to defraud or other illegal practices upon the part of the bidder.

GOVERNING LAWS AND REGULATIONS: NON DISCRIMINATORY PRACTICES Contracts for work under the bid will obligate the contractor and subcontractors not to discriminate in employment practices. Bidders must submit a compliance report in conformity with the President's Executive Order No. 11246.

U.S. GOVERNMENT REQUIREMENTS This contract is Federally assisted. The Contractor must comply with the Davis Bacon Act, the Anti-Kickback Act, and the Contract Work Hours Standards.

STATE EXCISE TAX Bidders should be aware of any state laws as it relates to tax assessments on construction equipment.

OPENING Bids will be opened as announced in the Invitation to Bid.

AWARD The Contract will be awarded on the basis of low bid, including full consideration of unit prices and alternatives.

EXECUTION OF CONTRACT The Owner reserves the right to accept any Bid, and to reject any and all Bids, or to negotiate Contract Terms with the various Bidders, when such is deemed by the Owner to be in his best interest.

Each Bidder shall be prepared, if so requested by the Owner, to present evidence of his experience, qualifications, and financial ability to carry out the terms of the Contract.

Notwithstanding any delay in the preparation and execution of the formal Contract Agreement, each Bidder shall be prepared, upon written notice of bid acceptance, to commence work within _____ days following receipt of official written order of the Owner to proceed, or on date stipulated in such order.

The accepted bidder shall assist and cooperate with the Owner in preparing the formal Contract Agreement, and within _____ days following its presentation shall execute same and return it to the Owner.

INDEX

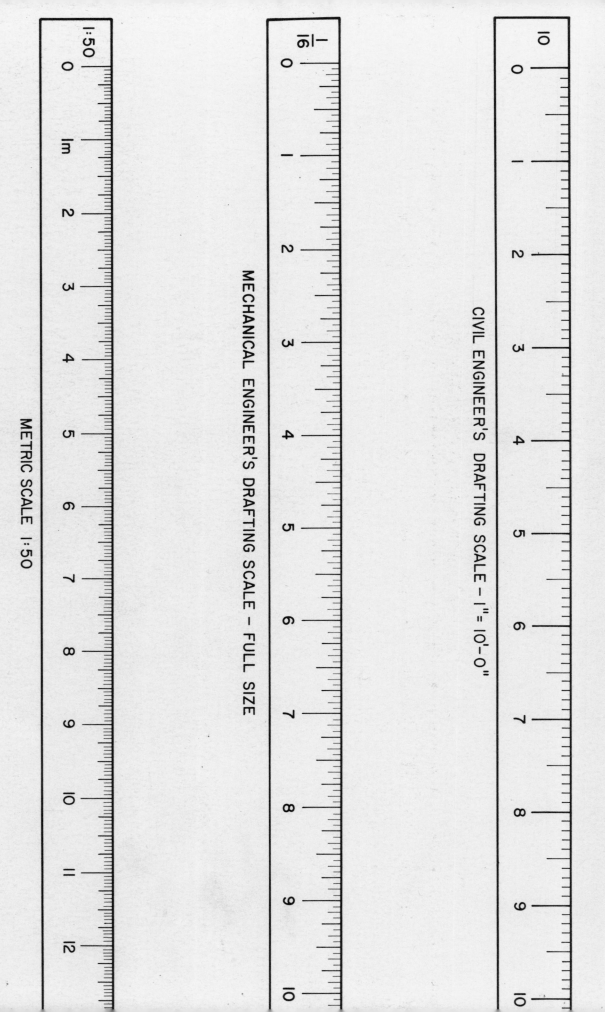

CIVIL ENGINEER'S DRAFTING SCALE – 1" = 10'–0"

MECHANICAL ENGINEER'S DRAFTING SCALE – FULL SIZE

METRIC SCALE 1:50